History of Science in Latin America: The Construction of an Intellectual Field (20th century)

María de la Paz Ramos-Lara
Luis Carlos Arboleda
Editors

History of Science in Latin America: The Construction of an Intellectual Field (20th century)

Essays in Honor of Juan José Saldaña González

Springer

Editors
María de la Paz Ramos-Lara
Centro de Investigaciones Interdisciplinarias en Ciencias y Humanidades
Universidad Nacional Autónoma de México
Mexico City, Mexico

Luis Carlos Arboleda
Universidad del Valle
Cali, Colombia

ISBN 978-3-031-75535-4 ISBN 978-3-031-75536-1 (eBook)
https://doi.org/10.1007/978-3-031-75536-1

This Springer imprint is published by the registered company Springer Nature Switzerland AG
The registered company address is: Gewerbestrasse 11, 6330 Cham, Switzerland

Dr. Juan José Saldaña González (1944–2022)

Foreword

In Memoriam of Juan José Saldaña

I met Juan José Saldaña on his arrival in Paris when he was preparing his doctoral thesis at the Institut d'Histoire des Sciences (Université Paris I). During his stay, he joined our research team in the history and philosophy of science, headed by Georges Canguilhem, and got to know most of the French researchers in this field. In 1980, he defended his thesis on *La notion de révolution scientifique: le cas de l'introduction des conceptions ondulatoires dans l'optique du début du XIXe siècle*. I later learned that he was the first Mexican to earn a doctorate in the history of science. I couldn't foresee at the time that this was the first act of a very long and beautiful story. That of a great friendship with an outstanding scholar; that of constant cooperation between the two of us; and, finally, that of an ongoing exchange between Parisian research teams and their Latin American counterparts, a veritable dialectic of mutual enrichment.

Between Juan José Saldaña and myself, then, it's a story of friendship and shared interests, punctuated by successive meetings at the various international congresses we've attended.

At the time, I was a researcher at the CNRS (Centre National de la Recherche Scientifique), working on the creation of the REHSEIS research team (Recherche en Épistémologie, Histoire des Sciences et des Institutions Scientifiques, CNRS), founded in 1984 and co-directed by Christian Houzel, Michel Paty, and myself.[1] It didn't take long for this team to become the basis of a shared history between France and Latin America.[2] REHSEIS quickly launched a research program entitled

[1] (The notes have been incorporated into the Foreword by the book's editors in agreement with the author.) Since 2009, REHSEIS has been transformed into Laboratoire SPHere, Centre d'Histoire des Sciences et de la Technologie, CNRS, and Université Paris-Cité. See the website: http://www.sphere.univ-paris-diderot.fr/spip.php?rubrique1&lang=fr

[2] Several historians from Brazil, Colombia, and other countries in the region have received doctoral training under the direction of REHSEIS researchers. Agreements with cooperative research projects, as well as reciprocal internships, symposia, and joint publishing, have been drawn up.

"Sciences and Empires," to forge new ways of thinking about the transfer of science and technology, and their appropriation by the recipients. The aim was to carry out a historical and comparative study of science during European expansion in Asia, Africa, and America in the nineteenth and early years of the following century. Indeed, knowing this history is not only essential to understanding the past; it is also the means to shed light on the present situation, and to grasp the challenges that had to be overcome in the appropriation of science and technology. This is a subject that has been a constant concern of economists, sociologists, anthropologists, politicians, and others ever since the Second World War.

This concern was echoed in a movement that began in Latin America in the 1980s, following the 16th International Congress on the History of Science, held in Bucharest in 1981. At that time, a group of Latin American researchers, finding themselves isolated and lacking a professional framework, had begun an effort to gain recognition for the history of science as an academic discipline. This is where Juan José Saldaña comes in: he is one of the main instigators of this effort to professionalize the history of science in Latin America, particularly in Mexico. I'm not in the best position to recall everything that came into being in the 1980s: Suffice it to point out the establishment of a Chair in the History of Science at UNAM, the *Research and Thesis Seminar in the History of Science and Technology*, the birth of the Latin American Society for the History of Science and Technology (SLHCT) in 1982, the journal *Quipu* in 1984, *Cuadernos de Quipu* in 1986, the holding of congresses and creation of societies and journals in Mexico, Brazil, Colombia, Argentina, etc.

On the initiative of REHSEIS, a colloquium on this theme, *Science and Empires*, was held at UNESCO in Paris in April 1990. As the initiator of this meeting, I was keen to encourage the participation of a large SLHCT delegation, and I recall with great pleasure the enthusiastic activity of Juan José Saldaña, Ubiratan D'Ambrosio, and José Leite Lópes on the International Scientific Committee, and Juan José's outstanding lecture on "Science and power in the 19th century: France and Mexico in perspective." I also recall his contribution to the final round table: "Science and development: can science policy learn from the history of science?," and his exchanges with, among others, José Leite Lopes, Nakayama Shigeru, and Abdur Rahman. Our Latin American friends, Patricia Aceves, Diana Obregón, Luis Carlos Arboleda, ... were able to exchange views with Michel Paty, Lewis Pyenson, ... The proceedings of this symposium were published in the United States, and my lecture was translated into Portuguese and published in Brazil.[3]

[3] P. Petitjean, C. Jami et A. M. Moulin (éds), *Science and Empires*, Boston Studies in the Philosophy of Science, Kluwer Academic Publishers, 1992. Rashed's lecture at the congress ("Science classique et science moderne à l'époque de l'expansion de la science européenne", p. 19–30) was translated into Portuguese and published in São Paulo. (R. Rashed, "As condições de transferência e difusão da ciência moderna", A. Garibaldi (ed.), *Principios*, n°. 27, São Paulo, p. 39–47). It's worth recalling that one theme among others particularly struck the Latin Americans present at this conference: his radical critique of the idea of the "transfer" of European science to historically and culturally different contexts. Indeed, a significant number of articles published in the journal *Quipu* since its creation in 1984 have been precisely aimed at criticizing the idea of "scientific diffusionism" from the center to the periphery.

In the meantime, I met Juan José again in São Paulo in 1988, on the occasion of the Second Latin American Congress on the History of Science and Technology, where I had the chance to engage in a lengthy discussion with him on a question that was beginning to preoccupy me and on which I subsequently worked at length, that of the periodization of classical mathematics.[4]

Then there was the series of international colloquia on the history of science, in Zaragoza in 1993 and Liège in 1997, where, if I remember correctly, Juan José spoke about *Newton in 18th-century Mexico*.[5] It was precisely in Liège that Mexico City was chosen to host the XXI International Congress on the History of Science. I remember how hard the Mexican delegation argued for the next Congress to be hosted by a Latin American country. It was a great success for Juan José and the Mexican delegation. For all of us, it was a priceless opportunity.

The XXI International Congress on the History of Science, held in Mexico City in 2001, was a major event, on the theme of *Science and Cultural Diversity*. During the organization of the congress and at the congress itself, Juan José Saldaña displayed the organizational talents and ability to coordinate large teams that have accompanied his entire career. The congress was a resounding success, providing a fertile forum for academic exchanges on science and cultural diversity.

Allow me to recall the formula with which I began the inaugural lecture I was honored to deliver[6]:

> It is the first congress to be held in a country with an ancient culture that is neither Mediterranean nor Asian. It's also the first congress in the history of science not hosted by an industrial country in the North. Our presence here is by no means purely circumstantial; it's the effect of an evolution in the discipline, and also in prevailing ideas.

For me, the 2001 meeting is a landmark in several respects. It embodies the ever-closer links between historians of science and their Latin American colleagues. The number of encounters has multiplied: I recall, for the record and with pleasure, my relationships with Ubiratan D'Ambrosio, Luis Carlos Arboleda, Martha Cecilia Bustamante, Ana Maria Alfonso-Goldfarb, etc., whether in France or Brazil; my various missions to Brazil; and so on. Researchers in the history and philosophy of science have found in this openness to the world a ferment that has enriched their

---

[4] R. Rashed, "Periodization in Classical Mathematics", in U. D'Ambrosio (ed), Anais do segundo congresso Latino-Americano de História da Ciência e da Tecnologia (30 de junho a 4 de Julho, 1988). São Paulo: Nova Stella, 1989; p. 491–499.

[5] Saldaña presented two communications at the XXth International Congress on the History of Science (Liège, July 20-26, 1997): The plenary lecture "Recent Historical Writing on Provincial Science: Latin America", and the lectures about Newton in New Spain and "The Backwardness of Physics in 19th Century Mexico", given jointly with Mary de la Paz Ramos-Lara in the symposium *The Spread of the Scientific Revolution in the European Periphery, Latin America and East Asia* organized as part of the congress by Celina Lértora, Efthymios Nicolaïdis, and Jan Vandersmissen.

[6] "History of Science and Diversity at the Beginning of the 21st Century" by R. Rashed in: Juan José Saldaña (ed.). *Science and Cultural Diversity, Proceedings of the XXIst International Congress of History Science*, Plenary Lectures, Vol. I, México, UNAM-SMHCT, 2003.

conception of their field, endowing it with a new, infinitely richer dimension. But the 2001 congress was also the date of my last meeting with Juan José Saldaña, which is why I can only evoke it with great emotion.

Roshdi Rashed
He is the author of several books and scientific articles in History of Science. He is currently Emeritus Director of Research (special class) at CNRS (France). He was Director of the Centre for History of Arab and Medieval Science and Philosophies (until 2001), Paris, and the doctoral formation in epistemology and history of science at Paris Director University (until 2001). He is an Emeritus Professor at Tokyo University, Mansoura University, and also at Paris Diderot University. He was a founder (1984) and Director (until May 1993) of the REHSEIS (Research Epistemology and History of Science and Scientific Institutions) research team, at CNRS, Paris.

Paris
Spring 2024

Roshdi Rashed

Introduction

This book is dedicated to the memory of Juan José Saldaña González (1944–2022),[1] whose contributions to the History of Science and Technology have been significant in Mexico, Latin America, and internationally. His legacy is linked to the changes this field of study has undergone worldwide from the 1980s to the beginning of the twenty-first century. During this time, Latin America transitioned from being practically excluded from the world map of the History of Science to occupying a leading position. The editors invited Saldaña's mentors, colleagues, and closest disciples to contribute to this book, guiding the reader through the paths, roads, and arid terrains that he explored in his tireless efforts to institutionalize this field in Latin America and to gain visibility and recognition at the international level.

The texts in this publication highlight the scientific research, academic contributions, and dedicated teaching of the first Mexican philosopher and professional historian of science. They also underscore his significant leadership and organizational skills, crucial in creating academic and intellectual spaces for developing this discipline in Latin America and Europe. Reference is made to several of his initiatives to seek local and international resources to establish favorable conditions for the creation of critical masses of researchers and scholars of the history of science in the countries of the region, as well as a minimum infrastructure to ensure communication and scientific exchange among historians, at a time when face-to-face meetings were predominant. These meetings were instrumental in shaping theoretical and methodological discussions on the most effective models for studying the history of scientific practice in Latin America.

In various forums, Saldaña stated that studying the region from a Eurocentric perspective led to the *historicization of European science in Latin America rather than the historicization of Latin American scientific practice*. He questioned the "center-periphery" model and the idea that the history of universal scientific

[1] Saldaña passed away in Mexico City on October 31, 2022, at the age of 77. See a biographical sketch of him at: Ramos-Lara, María de la Paz and Arboleda, Luis Carlos. 2024. News of the profession. Eloges. Juan José Saldaña González, 1944-2022. *ISIS*, Volume 115, Number 2, May 2024. https://www.journals.uchicago.edu/doi/10.1086/730306

practice originated in the West and developed linearly and cumulatively over the centuries. He used *Quipu* to share the findings of his research and of historians of science in the region, showing that scientific practice in Latin America was centuries old, that it had its dynamics, and that, therefore, it deserved to be legitimized and recognized on the world stage. Several chapters of this book emphasize the critical role of *Quipu*, the *Cuadernos de Quipu*, and other publications created and directed by Saldaña in promoting and disseminating systematic studies on scientific practice in local contexts using diverse methodological approaches. This effort has helped to establish the right place for these studies on a global scale.[2]

The pages of this book showcase Saldaña's remarkable work as an organizer, which led to him becoming the first Latin American to hold the position of Secretary General of the International Union of History and Philosophy of Science and Technology/Division of History of Science and Technology (IUHPST/DHST). He played a key role in founding the International Association for Scientific Cultural Diversity (IASCUD), which remains active and continues to make significant contributions in the fields of history of science and cultural history. Readers will appreciate several important episodes of his academic endeavors in the Americas. He was the first president of the Latin American Society for the History of Science and Technology (LASHST),[3] the editor of the journal *Quipu*, and the aforementioned *Cuadernos de Quipu*. He also participated in many commissions and regional and international editorial projects. Additionally, he directed the Mexican Society of History of Science and Technology (MSHST) for 20 years. Notably, he succeeded in holding, for the first time in a Latin American country (Mexico 2001), the international congress of the IUHPST/DHST, making history with this achievement.

In the fall of 1981, during the XVI International Congress of History of Science (ICHS), seven participants of Latin American origin met at the event. They realized that Latin American scientific practice did not have the presence it deserved in the international community. As a result, they developed a program titled the "Bucharest Declaration" to promote this field in the region. Various authors in the book review different aspects related to this document. They comment on the main components of the program, which include the establishment of working groups and commissions focused on strategic areas of action, the creation of a newsletter and an editorial policy centered on *Quipu* as a specialized journal, the holding of frequent meetings to exchange development strategies and research advances, the organization of thematic colloquia for theoretical and methodological discussion, the creation of seminars and chairs specializing in postgraduate studies for the training of professionals, and the search for resources to promote research centers.

[2] One of the most elaborate results of these dialogues among Latin American colleagues was the publication of the book *Science in Latin America. A History* (J. J. Saldaña, editor). 2006. Austin: University of Texas Press.

[3] Sociedad Latinoamericana de Historia de la Ciencia y la Tecnología (SLHCT). In this book, some authors use the acronyms LASHST, SLAHCT, and LSHCT in English, while others SLHCT in Spanish.

Saldaña proposed holding the first event at the University of Puebla (State of Puebla in Mexico) in August 1982, where LASHST was established. He assumed the presidency for two terms. His impactful six-year leadership was instrumental in solidifying the institutionalization of the History of Science and Technology in Latin America. Before this initiative, although there were some local societies for the history of medicine and other disciplines, national societies were scarce in the region, operating in isolation with limited resources and minimal social recognition. Notable examples include societies in Argentina and Mexico. Saldaña took on the task of invigorating these societies and promoting the creation of new ones, as we will see later. This labor was done to integrate the majority of Latin American historians of science into LASHST, thus laying the foundations for a thriving community of Latin American historians with close ties to various organizations and academic peers in other parts of the world. The book's authors provide detailed accounts of the internal workings of the societies in different countries of the region and their relationships over the years with the LASHST. They also relate the activities of prominent members of the regional community of historians who, through their positions in committees and different bodies of the organization, played an important role in the development and consolidation of this academic project.

Saldaña consistently supported these colleagues in implementing these initiatives, particularly in establishing scientific societies in their respective countries and affiliating them with existing ones to LASHST. By 1984, LASHST had incorporated four societies: the Sociedade Brasileira de História da Ciéncia, the Asociación Costarricense de Historia y Filosofía de la Ciencia, the Grupo Argentino de Historia de la Ciencia, and the Sociedad Ecuatoriana de Historia de las Ciencias y la Tecnología. The following year, four more were added: the Sociedad Cubana de Historia de la Ciencia y la Tecnología, the Sociedad Colombiana de Historia de la Ciencia y la Tecnología, the Grupo Chileno de Historia de la Ciencia, and the Sociedad Mexicana de Historia de la Ciencia y la Tecnología (SMHCT). A year later, the Sociedad Peruana de Historia de la Ciencia y la Tecnología and the Grupo Venezolano de Historia y Sociología de la Ciencia also became affiliated. In only four years, Latin America already had the history of science societies that would accompany LASHST during its trajectory and in its projects.

These national societies in the history of science played a crucial role in developing the Latin American association through their delegates, who served as presidents. After Saldaña, the presidents of LASHST were Ubiratan D'Ambrosio from Brazil, Luis Carlos Arboleda from Colombia, Silvia Fernanda de Mendonça Figueirôa from Brazil, and Carlos D. Galles from Argentina. One of its functions was to organize the Latin American Congresses on the History of Science and Technology (LACHST) every three years. The first congress was held in 1985 in Havana, Cuba; the second in 1988 in São Paulo, Brazil; the third in 1992 in Mexico City, Mexico; the fourth in 1995 in Santiago de Cali, Colombia; the fifth in 1998 in Rio de Janeiro, Brazil; and the sixth in 2004 in Buenos Aires, Argentina. The book contains numerous testimonies demonstrating how these congresses, along with the specialized symposia and articles published in *Quipu*, successfully fostered an enthusiastic and participatory community engaged in global debates on common interest topics.

Since its beginning, LASHST aimed to engage with the international community. Its members were determined to conquer this space of legitimization that they considered Latin American history of science deserved. Accordingly, in line with the "Declaración de Bucharest" (Bucharest Declaration), the participation of Latin Americans in international events outside the region became increasingly common. For example, the *Ibero-American Colloquium on the History of Science* in Spain in 1984, the International Congress *Ciencia, descubrimiento y mundo colonial* held in Madrid in 1991, as part of the of the V Centenary of the Encounter between two Worlds, at the XIX ICHS held in Zaragoza in 1993, among others. A special event to which a chapter is dedicated in this book was the International Colloquium *Sciences and Empires* held in Paris in 1990, where Latin Americans played a leading role. Some chapters mention the region's presence at the XVII ICHS in Berkeley (USA) in 1985. At this congress, Saldaña arranged a symposium on the *Cross-Cultural Diffusion of Science in Latin America*, which gained international prominence with the publication of the proceedings in the *Cuadernos de Quipu*.

At the XVIII ICHS held in Hamburg, Germany, in 1989, Saldaña successfully incorporated the SMHCT as a member of the IUHPST/DHST. He also made an unsuccessful proposal for Mexico to host the next international congress, ultimately awarded to Spain (Zaragoza). At the 20th ICHS held in Liège, Belgium, Mexico City won the bid, beating New York, USA, and Beijing, China, for the venue. Securing the hosting rights for the 21st Congress marked a significant achievement for Mexico, signifying global recognition of Latin American scientific contributions to the history of science. Likewise, from that moment on, it increasingly expanded its influence in the global arena. From July 8 to 14, 2001, Saldaña and his team successfully hosted the XXI ICHS in Mexico City. The international community had the opportunity to visit a city with a millennia-old tradition in the ancient knowledge of the pre-Hispanic peoples, significant scientific development during the colonial period, and rich cultural heritage following Mexico's independence. The contributions of the XXI ICHS 2001 Mexico City were diverse and of great regional and international relevance. Here is a list of some of them highlighted by the authors in this book:

- Rashed mentioned in the preface that the congress was a place for valuable academic discussions on science and cultural diversity. It was the first congress to be held in a country with an ancient culture that is neither Mediterranean nor Asian. This congress marked the first time in the history of science that a non-industrial country in the North hosted such an event. "Our presence here is by no means purely circumstantial; it's the effect of an evolution in the discipline, and also in prevailing ideas."
- Other authors have mentioned that it was the first international congress of the new millennium and the first to be held outside the Western countries, where it had been held for 68 years. Paris was the inaugural city to host the ICHS in 1929. At the beginning of the twenty-first century, Latin America had shifted the

balance of the ICHS in its favor, and for the first time the congress was held in one of its countries, Mexico.[4] The presence of the Latin American delegation was numerous, surpassing even the European community.

- The overall theme of the congress *Science and Cultural Diversity* was the first in the history of congresses in a country with a non-Western scientific and cultural tradition. Various local practices in the history of science and technology were considered, and the complexity of the problem of the conditions of transfer and reception of European science to other regions of the world became even more evident to the academic community.
- Symposia were explicitly organized to discuss the treatment of documentary collections and sources of information related to scientific and cultural practices in regions with unique traditions, such as Latin America. These discussions have provided valuable research opportunities in our field of studies.
- The Mexican State and its governmental, scientific, and educational institutions recognized the importance of the history of science and technology in the nation's development by sponsoring this international congress.[5] Likewise, the participation of the Mexican academic community in both the sciences and the humanities was stimulated in such a way that the Mexican delegation, together with the Latin American delegation, were the most numerous, as discussed in Chap. 4 of the book. The congress supported existing dynamics in the country's discipline and encouraged new research topics, chairs, societies, publications, and collaborations.
- In Latin America, it had a favorable impact both on the professional and institutional consolidation of groups, societies, and institutional spaces of the same researchers in the region and on the recognition of the field by the States, specifically in terms of the use of the history of science in the design of scientific policies.

[4] The IUHPST/DHST fully endorsed the event at its 1997 assembly.

[5] The congress program lists 20 institutions sponsoring the congress, with 14 from Mexico and 6 from abroad: the Latin American Academy of Sciences, UNESCO, ICSU, Jorge Juan Foundation, Juanelo Turriano Foundation, and Wellcome Trust.

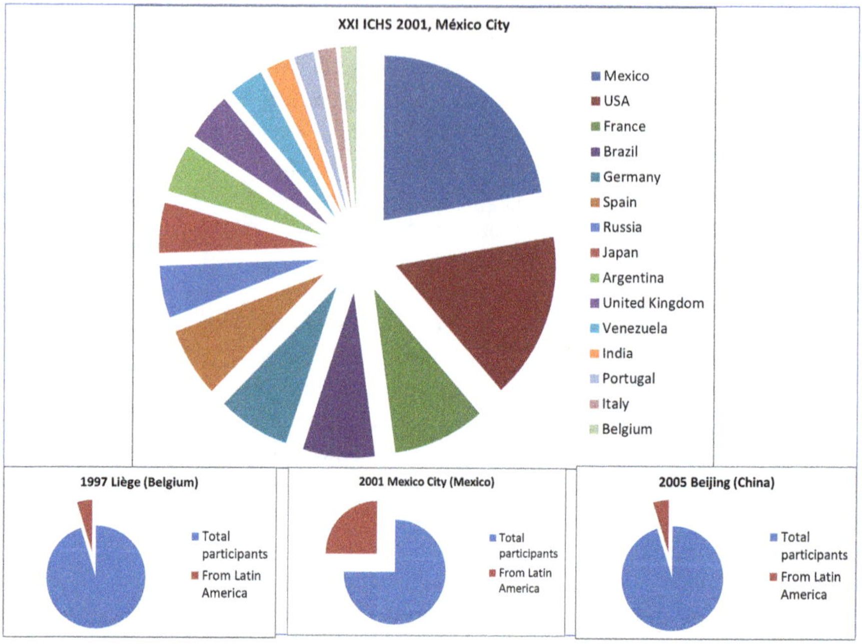

Total number of participants in the XXI ICHS held in Mexico City in 2001 (above). Comparison of the Latin American delegation concerning the previous and subsequent congresses in Liège and Beijing, respectively (below)

After successfully hosting the *XXI ICHS 2001 in Mexico City*, Saldaña was appointed Secretary General of the IUHPST/DHST, making him the first Latin American to hold this position. The way was open for a country of the region to once again host an international congress. Indeed, the XXV ICHS took place in Rio de Janeiro, Brazil, in 2017. Currently, Marcos Cueto, a distinguished Peruvian historian of science, holds the presidency of the IUHPST and is president of the DHST. These facts are extensively discussed in multiple works within this book. In various ways, these essays provide an account of Saldaña's role in the history of science and technology in Latin America and its impact on the international stage.

The book consists of three sections. The first section, titled "France and Spain Relations in the Institutionalization of the History of Science in Latin America," showcases the works of Martha Cecilia Bustamante De La Ossa (Colombia/France) and Leoncio López-Ocón Cabrera (Spain). This section follows the preface written by Roshdi Rashed, an Egyptian mathematician, philosopher, and historian of science. He organized the *Sciences and Empires* Congress in France, the first of its kind in the world. Bustamante provides additional information about it and exposes how several members of LASHST attended this congress. For some, this meeting was so significant that they initiated collaborative projects between European and Latin American countries. For example, some French researchers spent extended periods in Brazil and Colombia, while young Latin American students pursued postgraduate studies in France. Some even joined the teaching and research staff in Paris, as she did. For his part, López-Ocón focuses on three congresses held in Spain in different years and cities, during which collaboration with Latin American historians of science was strengthened. Some Spanish researchers from LASHST collaborated on projects of common interest, such as the colonial period, and published their findings in *Quipu*. López-Ocón shares the same view as other authors in this book, emphasizing that before the establishment of LASHST, only a small number of intellectuals, such as Germán Somolinos in Mexico, worked independently on the history of science. They primarily published books and articles that were circulated predominantly within the country.

In the second section entitled "The Latin American Society for the History of Science and Technology (LASHST): Foundation, Challenges, and Perspectives," papers are presented by Luis Carlos Arboleda (Colombia), María de la Paz Ramos-Lara (Mexico), María Amélia Mascarenhas Dantes (Brazil), Carlos D. Galles (Argentina), Norma Isabel Sánchez (Argentina), and Silvia Figueirôa (Brazil, interviewed). The authors discuss LASHST's objectives, formal constitution, forums organized in its first years in several Latin American countries, participation in international events, and influence on the local population. This influence included encouraging new generations to work in this field and stimulating the creation of local societies dedicated to the history of science, as seen above. The participation of three former presidents of LASHST (Figueirôa, Galles, and Arboleda) in this section provides firsthand information on the challenges and adversities faced in organizing the Latin American congresses. It also sheds light on the benefits obtained to promote the development of our field of study. Dantes, Ramos, and Arboleda recognize Saldaña's leadership in establishing LASHST, creating *Quipu* and *Cuadernos de Quipu*, and transitioning to digital media. The primary goal was to keep the academic community connected, facilitate communication, and promote its activities and projects. They also emphasize the significance of the discussion forums that guided the LASHST during its presidency, especially regarding theoretical and methodological issues. They also provide information about other efforts to enhance the professionalization of the history of science and solidify its institutionalization in different countries of the region.

Lewis R. Pyenson (USA), Regis Cabral (Brazil/Sweden), Olival Freire Jr. (Brazil), Michel Paty (France), Ana María Carrillo (Mexico), and Maria Rachel Fróes da Fonseca (Brazil) present their papers in the third section titled "Latin American Science Policy and Participation in the International Community: Incorporating Latin American History of Science into a Worldwide Agenda." This section aims to provide an overview of the extent and influence of Saldaña's work, whether through LASHST or independently, in various contexts. Pyenson presents a scholarly study of the opposition to cubism and relativity shown by specific counter-disciplinary movements in Buenos Aires during the early twentieth century. These movements were exemplified by the architect and geometrician Claro C. Dassen and the artist and geometrician Emilio Coutaret. In a brief conclusion to his work, Pyenson compares the stances of these two individuals with Saldaña's role as an organizer who successfully brought together academics from various backgrounds and practices into a unified community of interests.

For the first time, Cabral, Freire, and Paty address a fundamental theme in Saldaña's life and work. This theme is treated with care and dedication in his academic works: the theoretical and practical characterization of the identity of the Latin American region and, by extension, of other non-Western cultures. Cabral associates it with the concept of "Scientific Diplomacy" in the sense that historians of science, by recovering their past, reconstruct their identity. If they use this knowledge to help in decision-making, they would, in turn, participate in the construction of the future. He claims to be inspired by Saldaña's ideals and his professional practice as a historian of science. Freire and Paty, for their part, study a significant episode of this same theme, Saldaña's participation in the creation of the International Association for Scientific and Cultural Diversity (IASCUD) at the beginning of the twenty-first century in Mexico City. The organizational vitality of IASCUD and its international projection allow them to value Saldaña's efforts to integrate in the same field the studies of the history of science and cultural diversity at a global level.

Carrillo remembers the influence of the activities organized by LASHST, especially its congresses, the IASCUD meeting in Mexico in 2000, and the hosting of the XXI ICSH 2001. These events significantly established *Hispalc-L*, the List of History of Public Health in Latin America and the Caribbean. While this list was active, it facilitated the exchange of recent publications, experiences, and work techniques that led to comparative studies. Fróes da Fonseca, in line with Saldaña's "Thinking Our Science," presents theoretical and methodological considerations on the unique historical development of scientific practice in colonial Latin America. This approach is based on studying the intellectual journeys of two Enlightenment scholars: the Mexican José Antonio Alzate and the Brazilian Manuel Arruda da Câmara.

Finally, the fourth section was dedicated to "The emergence of local History of Science projects, associations, journals, educational and research programs" in countries such as Argentina, Brazil, Mexico, Colombia, Venezuela, Chile, Peru, and Cuba. The participants are Pablo Lorenzano (Argentina), Carlos Galles (Argentina), Heloisa M. Bertol Domingues (Brazil), María de la Paz Ramos-Lara (Mexico), Emilio Quevedo V. (Colombia), Julieth Nathalia Barrera-Díaz (Colombia), Hebe Vessuri (Argentina/Venezuela), Isabelle Sánchez-Rose (Venezuela), Zenobio Saldivia M. (Chile), Pablo Rojas C. (Chile), Leticia Quiñones (Peru), Jorge Lossio (Peru), Lizardo Seiner (Peru), Diego de Jesús Alamino Ortega (Cuba), and Yenile Aguilar Rodríguez (Cuba). This section's diversity covers a wide range of topics, offering information on the various approaches taken by different countries to institutionalize the history of science. Some of these approaches share common points across different social and cultural contexts. In the cases of Argentina and Cuba, a historically educational and scientific account is provided, spanning from the colonial period to demonstrate that societies for the history of medicine (and other disciplines) were established before the creation of national societies. For instance, the Argentine Group for the History of Science was founded in 1933, while the Cuban Society for the History of Science and Technology was established in 1985. The case of Peru is similar in the field of health sciences, and it demonstrates how other relevant societies were involved in the development of engineering and architecture.

In the 1980s, the Peruvian Society for the History of Science and Technology was founded, and in 2018, the Peruvian Association for the History of Science, Technology, and Health was established.

The papers from Mexico and Brazil both discuss the influence of LASHST on the reactivation of the SMHCT (1964) and the establishment of the Sociedade Brasileira de História da Ciência (1985), both fundamental in the process of consolidation of the institutionalization of the history of science in both countries, particularly in Brazil, in addition to the inter-institutional support with France. The papers also highlight the availability of institutional resources for training and research. In the case of Colombia, the interaction between the Seminar on Philosophy and History of Science and Medicine (1979) and LASHST is evident. This interaction and other initiatives in the early 1980s led to the establishment of the Colombian Society for the History of Science and Technology in 1984. This development resulted in the implementation of various projects that significantly impacted the discipline, including the publication of works such as the *Social History of Science in Colombia* (ten volumes) and the *History of Medicine in Colombia* (four volumes). Concerning Venezuela, the authors discuss the significance of LASHST in advancing the history of science studies, which led to the establishment of the Venezuelan Group of History and Sociology of Science in the 1980s. Particular emphasis is placed on the importance of *Quipu*, which enables the exchange of knowledge and facilitates local and international connections. The articles authored by Venezuelan historians of science in this journal are being analyzed. For Chile, the authors present an account of the Chilean historians of science who have contributed to the process of institutionalizing the history of science in that country.

We express our sincere gratitude to all the authors who contributed to this book and extend our warm appreciation to those authors who participated in the posthumous tribute we organized for Juan José Saldaña in March 2023.[6] In most of the contributions in this work, the reader will notice the authors' ongoing commitment to LASHST and *Quipu* journal. Their efforts to promote the visibility of Latin American history of science are evident through their roles in national and international organizations. Finally, these works testify to the relationship of friendship and scientific collaboration with Saldaña for many years. Special mention should be made of Roshdi Rashed, who, with immense kindness, willingness, and affection, generously agreed to write the preface to this book under delicate health conditions.

[6] A group of researchers and professors from Latin America, Canada, Sweden, and Spain gathered virtually on March 23 and 24, 2023, through the Zoom platform of both the FFyL and the CEIICH of UNAM to carry out the *First International Conference on the History of Science and Technology in Latin America in tribute to Juan José Saldaña González*. María de la Paz Ramos-Lara and Luis Carlos Arboleda took charge of organizing the event.

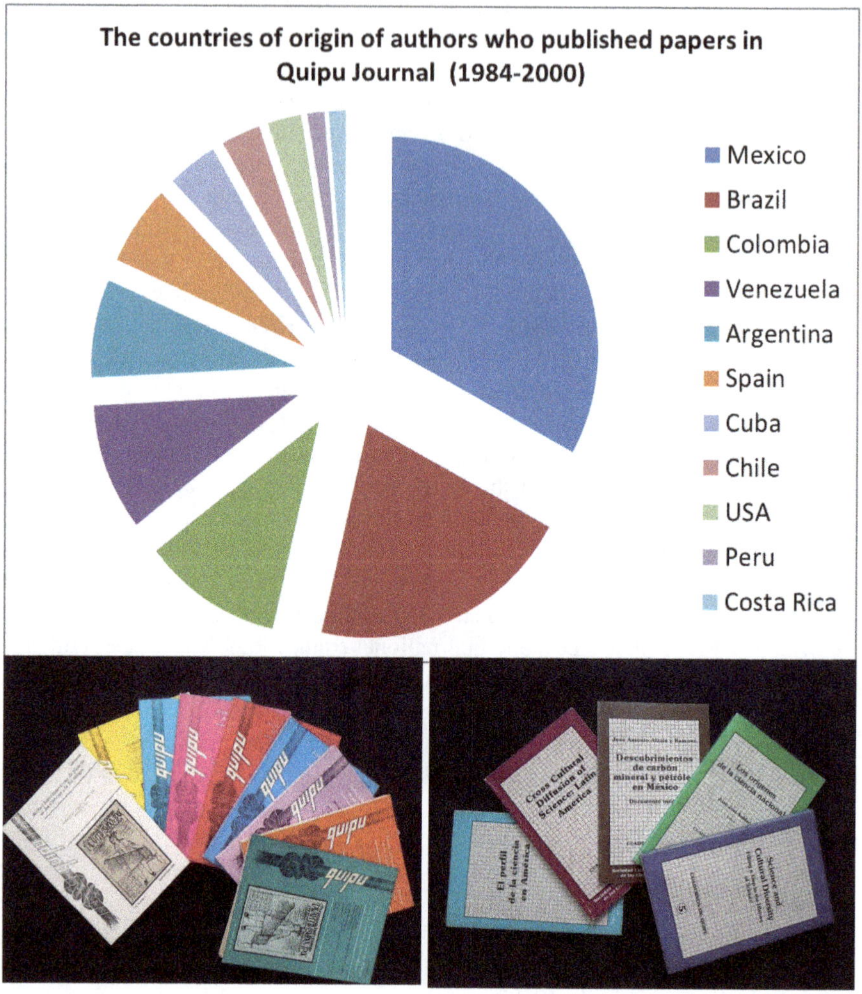

Number of papers published in *Quipu* by country from 1984 to 2000 (in the middle). Covers of *Quipu* and *Cuadernos de Quipu* journals

We are grateful to Springer for agreeing to publish this book, which provides an overview of the community of historians of science in Latin America. The book includes contributions from scholars who have been involved in developing and strengthening this field, both within the region and internationally.

We are pleased to announce that Dra. Clara Isabel Carpi Navarro, widow of Dr. Saldaña, has decided, on behalf of the family, to donate the *Juan José Saldaña González Library* and his archive to the Universidad Nacional Autónoma Metropolitana Iztapalapa in Mexico City. This university and the Universidad Nacional Autónoma de México have expressed interest in digitizing the journal *Quipu Primera Época* once the copyright procedures are completed. We consider

Quipu's content to be a historical and scientific heritage of the historians of the science community. Therefore, it would be highly beneficial to circulate it freely through digital networks. We hope that another digital publishing project will soon emerge, uniting the Latin American community in the history of science and technology, as *Quipu* did.

Spring 2024

María de la Paz Ramos-Lara
Luis Carlos Arboleda

Contents

About the Editors

María de la Paz Ramos-Lara is a researcher at CEIICH-UNAM, where she founded the History of Science Department and two book collections: *Bibliotheca Mexicana Historiae Scientiarum* and *Science and Technology in the History of Mexico*. Head of the Laboratory for Interdisciplinary Research and Complexity Sciences, and the Science and Technology Department. She is also a Philosophy of Science Postgraduate Program professor at UNAM and a member of the National System of Researchers (SNI-Conahcyt, Mexico). She has authored three books, published 16 more, and written a hundred chapters and articles on the History of Science in Mexico. She was awarded the Dr. Enrique Beltrán Prize, the Sor Juana Inés de la Cruz, and José Joaquín Izquierdo Medals. She uses the complex systems to study the history of science.

Luis Carlos Arboleda is Emeritus Professor at the Universidad del Valle, Cali, Colombia. For a little more than half a century he has taught the history of mathematics and science at this university, where he was also Vice Rector of Research. From 1991 to 1994 he was subdirector of Colciencias (today Ministry of Science), an entity that designated him as Emeritus Researcher in 2018. Since 2017 he has been a full member of the Colombian Academy of Exact, Physical and Natural Sciences. He was awarded the *Sábato Medal* of the Inter-American Commission on Mathematical Education (CIAEM for its acronym in Spanish) in 2023 in recognition of his services in history and mathematics education.

Part I
France and Spain Relations in the Institutionalization of the History of Science in Latin America

Chapter 1
Science and Empires (Paris–April 1990): A Significant Moment in the History of Latin American Science

Martha Cecilia Bustamante De La Ossa

Abstract In this chapter, we shall consider the international colloquium *Sciences and Empires* held in Paris from April 3rd to 6th, 1990. This was a significant moment in the history of science, technology and medicine in the twentieth century. It was organized by UNESCO and had a strong theoretical and epistemological symbolism. Researchers specializing in the sciences of the "periphery" and historical memory attended the *Science and Empires Colloquium*, Latin American historians of science were working to professionalize their discipline and to disseminate their studies widely. The revival of Latin American science history has been taking shape since the Bucharest Declaration, born in the context of *The XVIth International Congress of the History of Science*. We also take a look back at this meeting attended by young and promising Latin American academics. This congress, held in Bucharest (26 August–3 September 1981), was organized by "The International Union of History and Philosophy of Science/History of Science Division (IUHPS/DHS)", under the auspices of UNESCO.

Keywords Science and empires · Bucharest declaration · France · Romania · Latin America · Professionalization · *Quipu* · Local-knowledge

1.1 Presentation

The publication of a book dedicated to Juan José Saldaña provides an invaluable opportunity to reflect on a significant moment in the history of science, technology and medicine in the twentieth century. Thirty-four years ago the international colloquium *Sciences and Empires—A comparative History of scientific Exchanges: European Expansion and Scientific Development in Asian, African, American and*

M. C. Bustamante De La Ossa (✉)
Laboratoire (SPHERE)-UMR 7219, CNRS et Université Paris-Cité, Paris, France
e-mail: mcbusta@univ-paris-diderot.fr

M. de la P. Ramos-Lara, L. C. Arboleda (eds.), *History of Science in Latin America: The Construction of an Intellectual Field (20th century)*,
https://doi.org/10.1007/978-3-031-75536-1_1

Oceanic Countries took place in France. This event, the first of its kind in the world, held particular significance for Latin American scholars.

A total of one hundred and twenty scientists from diverse geographic locations worldwide meet at UNESCO in Paris to discuss the dissemination of modern science in the wake of European expansion. The diffusionist thesis proposed by George Basalla was over 20 years old and a number of studies had already begun to demystify Basalla's paradigm, which posited that Western Europe was the center of creation and dissemination of modern science. Published works explored the construction and exchange of knowledge beyond what had been denominated the centers and their peripheries. The idea was to focus on a scientific phenomenon in any relevant setting. Simultaneously, Latin American historians of science were working to professionalize their discipline and to disseminate their studies widely. The revival of Latin American science history has been taking shape since the Bucharest Declaration (26 August–3 September 1981).

Researchers specializing in the sciences of the "periphery" and historical memory attended the *Science and Empires Colloquium*, held from 3 to 6 April 1990. Scholars came from all over the world, including India, China, Japan, Latin America, France, Spain, and Canada. More than 20 nationalities were present, thus the cultural diversity the colloquium sought to investigate was represented. Juan José Saldaña, Maria Acevedes, Luis Carlos Arboleda, Diana Obregon, Lewis Pyenson and Xavier Polanco, among others, were part of the Mexican, Colombian, Canadian and French delegations. J. J. Saldaña was a key contributor to discussions. *Science and Empires Colloquium,* with its international scope, attracted a global audience. The event was of interest to the scientific press and achieved notable success (Bonneuil, 1994). We have two subsequent publications that may be regarded as an extension of the concerns expressed by the event.

A first illustrative example is Michel Paty's work published 20 years later. French historian and philosopher of science M. Paty produced a comprehensive paper on the history of science in Latin America. History of science is considered in its double sense: firstly, the description of sciences in these countries, as they exist and have existed in history until the present day; and secondly, the history of sciences as a discipline in itself (Paty, 1992). His work focuses on the institutions, relationships, and procedures involved in generating scientific historical knowledge. M. Paty emphasizes on the significance of scholars and intellectuals from Mexico and other Latin American countries in promoting a dynamic and proactive approach to historical studies of sciences in peripheral regions. He asserts that the center is not necessarily the most intellectually advanced, and that a more balanced approach to the matter is required. In Paty's view, the history of science, medicine, and technology in Latin America has expanded significantly in recent years, becoming a more professionalized field. Furthermore, the historiography of science has become increasingly complex by a slightly different dichotomy, namely the distinction between indigenous science and the focus on interactions with Europeans (Paty, 1992). The second example is a collective book published in Paris in 1996 (ORSTOM) and edited by P. Petitjean. This publication presents the proceedings of a meeting held in Paris. A review of the book notes that this collective work presents itself as an exploration of a field that has not been adequately addressed by historians of science, as well as by historians of colonization. This field concerns the colonial sciences, which asserted themselves and developed between the beginning of the twentieth century and the decolonization of the colonies (Petitjean, 1992). A short paper signed by L. C. Arboleda, *Cauchy en Colombie*, elucidates the intricacies of the reception and dissemination of mathematical theories originating in France in the peripheries (Arboleda, 1992).

The *Sciences and Empires Colloquium* was made possible by the collective efforts of a team of researchers. As reported by P. Petitjean, C. Jami and A. M. Moulin in their edited and issued account in 1992, the meeting was organized by: P. Petitjean, Roshdi Rashed and the REHSEIS group (Research on Epistemology and History of Exact Sciences and Scientific Institutions) of the CNRS (National Center for Scientific Research). The colloquium was “the initiative of Professor Roshdi Rashed”, who had pioneered this field of study in France. He had proposed the research programme “Sciences and Empires” as one of the main research projects of the REHSEIS group. M. Paty, R. Rashed and Christian Houzel had founded REHSEIS in 1984. Towards the end of the decade, the most favorable conditions existed at REHSEIS for research on Science and Empires. Its various protagonists, including M. Paty and P. Petitjean, had already begun to forge strong ties of friendship with Brazil and other Latin American countries (Freire, 2010).

During this time, students from Brazil, as well as from Colombia and other Latin American countries, had already begun spending short or long research periods within the REHSEIS team, many of them preparing their doctoral thesis under the supervision of M. Paty.[1] Myself, having arrived in Paris in 1982 and now living permanently in France, became a member of REHSEIS when *Sciences and Empires* took place. I first met Juan José Saldaña, Luis Carlos Arboleda and Lewis Pyenson when they were preparing to give their respective interventions. We had been walking together to UNESCO. I had just completed my doctoral thesis in epistemology and the history of science within the team and was starting post-doctoral research. Many conversations or formal meetings took place in the premises of REHSEIS, which was then housed at the Université Paris 7-Jussieu. It was then that I began my intellectual journey that will always be associated with the names of M. Paty, P. Petitjean, R. Rashed, C. Houzel and a number of foreigners who became associate members and regulars of the team.

It was at this time that I met Olival Freire from Brazil, and we have never stopped collaborating. Finally, M. Paty constantly guided my research. We organised a summer school in Colombia in 1992, at the Universidad Pedagógica Nacional. Other French and Italian researchers participated in this summer school.[2] Moreover, a number of French researchers regularly visited Colombia and Brazil. M. Paty,

[1]Among others, from Brazil: Antonio Augusto Passos-Videira, Mauricio Pietrocola Pinto de Olivera and from Colombia: Olga Luz Dary Rodriguez and Angel Enrique Romero.

[2]Among them: Olivier Darrigol and Giovanni Battimelli.

among others, collaborated systematically with Brazilian, Colombian, and other Latin American colleagues.[3]

As we currently have evidence to suggest, the history of science in Latin America is inextricably linked to the interconnections between a diverse array of actors. M. Paty's specific interest in Southern countries, however, began well before 1983. During his tenure at Strasbourg University, M. Paty forged connections with Brazilians, who had relocated to France in response to the military junta in Brazil (Freire, 2010). R. Rashed, for his part, was renewing the terminology when he thought of the encounter between Europe and America: in his view, the term "discovery" was justified, strictly speaking, only in terms of the new knowledge acquired by Europeans (Paty, 1992).[4]

This Paris colloquium, convened under the auspices of UNESCO, had a strong theoretical and epistemological symbolism. The organizers wanted to mark their desire for change and openness. P. Petitjean reports on two meetings that had already taken place on themes related to science and empires; these were more limited in scope. First of all, a conference was held at the University of Melbourne in May 1981. A selection of papers from the conference was published, titled *Scientific Colonialism: a Cross-Cultural Comparison* (Reingold & Rothenberg, 1987). As Ian Inskter says, the book's title is misleading: the history of science, technology and medicine in under-developed colonies in Asia, Africa and Latin America is patently not included (Inskter ,1988).[5] The editors were led to define colonialism as embracing "cultural hegemony" but not "political sovereignty" (Inkster, 1988). A later

---

[3] It's worth noting that Colombian historian and physicist Regino Martinez-Chavanz and Michel Paty began their work in the late 1990s, using archival documents to describe the transfer, implementation and reception of relativity theory and, secondarily, quantum theory, in the social, political, economic and academic environment of late nineteenth-century Colombia (Paty & Martínez-Chavanz, 2004). This is one of the results of the cooperative project "Formation of scientific culture in Colombia", carried out between REHSEIS and the History of Mathematics Group of the Universidad del Valle, Cali, Colombia (ECOS-Nord, 97PCF08). See (Arboleda & Paty, 2004). Regino Martinez-Chavanz delved deeper into the case of Colombia's first physicist, Julio Garavito Armero (1865–1920), his role as a university professor and his many works in physics, mathematics and astronomy. See Chemla, 2007. For the notion of transfer used by Martinez and Paty, see note 5 below.

[4] The issue of transferring European science beyond the borders of Europe is at the heart of the theme of *Sciences et Empires,* which concerns R. Rashed during this period. He articulates his position on this matter in his conference in Paris at the colloquium. Amongst anthropologists and orientalists, he specifies, the transfer was conceptualized as the valued displacement of a Western product, foreign by its nature as well as by its history, to its recipients. The second assertion [...] is that this transfer occurred on virgin territory, in a scientific void, thanks to various vectors of Empires, such as missions, merchants, armies, etc. In the remainder of his lecture, Rashed seeks to demonstrate "the inaccuracy of the void postulate." In order to do this, he addresses two situations represented respectively by Iran and Egypt (Rashed, 1992). It is therefore in line with this central idea that the work of Juan José Saldaña and more generally of Latin American historians of science is situated. Moreover, a void postulate is also inaccurate when applied to the problem of quantum physics in France at the beginning of the twentieth century. See for instance: Bustamante De La Ossa 2019.

[5] See also: Buchanan (1988).

conference was held in January 1985 in New Delhi under the same title, *Science and Empire, and similarly*, it focused on phases in British imperial science. As a specialist on the subject would put it, it examined how British colonizers had to dethrone and delegitimize several pre-colonial symbols and totems (Kumar, 1991).

In 1990, the *Science and Empires Colloquium* encompassed more than simply the opening-up to other continents. It also entailed acknowledging the world's cultural diversity. The participation of Juan José Saldaña and other personalities from Latin America specializing in the role of science in their societies as an element of politics and culture, allows us to look back at the intellectual and scientific context of the time and put the realization of the congress into perspective. This also gives us the opportunity to explore the relationship that Saldaña and other actors from the continent had established with the theoretical current that had inspired them. Around 1990, the subject reflected in the title of the colloquium, *Sciences and Empires*, was not an accepted sub-discipline of the history of science; "rather, it refers to a set of questions which found autonomy only recently." The very idea of an international scientific congress on this subject, so commonplace in recent years, was highly exceptional. "The project to organize such a colloquium was a gamble" (Petitjean, 1992).

The following section will highlight two main topics of the colloquium: firstly, discussions on models of the spread of modern science, and secondly, studies focused on "national science". I will set aside the contributions on science in Brazil, under Portuguese domination and later, as well as the contributions from the members of the Brazilian delegation. I will confine myself to regions under Spanish domination. Let's note that M. Paty pointed out that it is important to bear in mind certain constants and differences between Latin American countries of Hispanic or Portuguese origin. In both cases, colonization was organized with the intention of exploiting wealth for the benefit of the metropolis, rather than for the local population or for the development of the region. Brazil became independent in 1822 thanks to very different circumstances (Paty, 1992).

1.2 Three Main Epistemological Trends in the Agenda

A review of the analytical frameworks for understanding science discussed during the colloquium may be useful here, although it will not be exhaustive. Regardless of the *a priori* mode in which they are presented, this will facilitate a more profound comprehension of the prevailing ideas about the spread of science circa 1990.

The chart illustrates three major epistemological trends. The first one is exemplified by Basalla's analysis of the dissemination of Western science from its center to peripheral regions (Basalla, 1967). According to the proceedings and as previously noted, many studies of science and imperialism commence with the "diffusionist" thesis. The "Basalla model", referenced in P. Petitjean's inaugural address, is presented as a general schema that presupposes the diffusion of science on a virgin substratum, in a scientific vacuum.[6] Alternatively, as had been explained by Ian Inkster 2 years prior, in the initial phase, the periphery serves as a database for the further development of European-based scientific theory. According to Petitjean, Basalla's comprehensive theory of the spread of science had been regarded as a universal and timeless concept (Petitjean, 1992). In the context of the colloquium, other scholars have drawn attention to this subject.

It is clear that Basalla's model represents the fundamental tenet of V. V. Krishna's talk. Krishna puts it to empirical testing, focusing on the emergence of national science in India. He focused his investigation on a period spanning from 1876 to 1920, a timeframe that Basalla categorizes as colonial science. His analysis focuses on some individual scientists, scientific institutions and the practice of science. A look at the author's concluding remark is highly instructive: Krishna argues persuasively enough that the concept of colonial science as employed by Basalla lacks relevance in the Indian context. As will be subsequently demonstrated, the metaphor of diffusion and its associated stages, fails to adequately account for the cultural and intellectual processes that underpin the development of new spaces of knowledge (Kleiche, 2018). It would appear that a remark from Krihsna is required. It is essential for historians of science to adopt a social approach to science (Krishna, 1991).

The second major trend evoked during the congress was represented by Lewis Pyenson's work on the development of exact sciences in several colonial and imperial contexts. This view has been held for a few years. Let's quickly overview Pyenson's most outstanding points, starting with his trilogy about cultural imperialism. In the 1980s, the Canadian philosopher and historian showed that the exact sciences—astronomy, physics and geophysics, which appeared in the first division of the Paris Academy of Sciences—were an imperialist practice for Germany, the Netherlands and France in the nineteenth and twentieth centuries. His extensive work comparing these countries led to the typology of expansion strategies he presented at a conference in Utrecht in 1986 (Pyenson, 1993a, b).

Pyenson discussed the resolution of "imperialist patterns of scientific activity along three orthogonal axes," which included a research axis, a functionary axis, and a mercantilist axis. He then positioned these axes as vectors on the resulting diagram. The German pattern was a magnitude along the research axis, the French pattern was a magnitude along the functionary axis, and the Dutch pattern was a vector with positive components along all three axes. On the basis of preliminary surveys, it appeared that the British pattern was a vector with strong research and mercantilist components. There was little direct control of British colonial physics and astronomy from England. Physicists and astronomers in Canada and Australasia, while responding to commercial demands, followed their karma in choosing research topics; their laboratories and observatories arose from local funding (Pyenson, 1989).

In his inaugural address, P. Petitjean presents Pyenson's approach as less general and more limited in scope than other models, prompting a different set of questions.

[6] See note 5.

This perspective anticipates the position held by historians of science and colonization in the 2000s.[7] Pyenson was not interested in the construction of national scientific traditions nor in the integration aspect, or even in the problem of the reception of modern science, but in the use of science in the development of empires. Pyenson's model allows us to study the transplantation of European scientists to the periphery, their practices and their results. "The functionary" is portrayed as an agent of closely related academic, military and religious interests, with a weak or non-existent commitment to original research (Pyenson, 1993a, b).[8] The scientist is portrayed as serving the interest of science, with a real commitment to original work. The mercantilist responds to commercial demands. Petitjean posits that this classification reflects generally accepted ideas or prejudices about the specificities of each of these countries as colonizing powers. Rien ne différence l'expansion scientifique (l'impérialisme culturel selon les termes de Lewis Pyenson) de l'impérialisme dans son ensemble (Petitjean, 1992).

Pyenson's approach is of significant importance in the context of the Paris colloquium. Pyenson himself chaired a round table discussion entitled *Typologie des stratégies d'expansion en sciences exactes*. It is not necessary to revisit the content of the discussions. Pyenson was able to elaborate on his model. But let's remember the precise definition of the term "exact sciences" as elucidated by Pyenson: "any scholarly statement that uses mathematical formalization on the one hand, or that, on the other hand, produces as a provisional result a series of precise figures". He sees research in the exact sciences as a particularly appropriate field for studying the interaction between scientific knowledge and the scope of political power. Such a study will provide information above all about the ideological orientations behind political-economic expansion. It seemed to him to be the right terrain for identifying the ideologies that are linked in the first instance to imperialist expansion proper (Pyenson, 1992).[9]

In Paris, a third epistemological trend emerged and gained prominence, one which was represented by Xavier Polanco's theory. In a talk entitled *World-Science: How is the History of World-Science to be Written?*, Polanco provided a detailed and comprehensive account of his work. The congress presents an opportunity to gain insight into emerging concepts. At that time "world-science" represented a newly-invented concept coined on the model of Fernand Braudel's world-economy. This last point was consistently emphasized throughout the discussions (Polanco, 1992).

As Polanco wrote, he had exposed, for the first time, the world-science notion in a collective book that appeared the same year in which the congress took place (Polanco, 1990). Let's remember that in Polanco's conception, the modern scientific world system originated from a world-science that emerged in Europe during the

[7] See for instance: Dray (2018).

[8] "Pyenson will concur that French colonial scientists served as agents of French imperial interests". Pyenson cited by Christopher Carter (Carter, 2009).

[9] For recent comments on Pyenson's typology: Dray (2018).

sixteenth and seventeenth centuries. Polanco wanted to emphasize the fact that modern science is practiced in networks, within a framework of global integration that generates dependencies. He said in Paris: "world-science [...] is marked by a hierarchy, and it has a network structure." In a more general sense, Polanco's primary concerns were twofold: firstly, the globalization of European science and, subsequently, the construction of national scientific traditions. In his published inaugural address, Petitjean devotes attention to this model. He argues that Polanco doesn't want to reduce one phenomenon to the other. Furthermore, the meaning of "world science" is not the same as international science, world science is not the international scientific community: "c'est une réunion d'espaces et de sciences". Polanco's idea is to study the rules of evolution and transformation of the world sciences, just as Braudel studied those of the world economy. "World-science is a sociological and historical concept". As Polanco notes, it is not a new label to stick onto old methods of analysis. According to Petitjean "c'est une incitation à repenser différentes études d'histoire nationale des sciences dans ce nouveau cadre, et il s'agit moins d'un nouveau modèle que d'une nouvelle méthodologie". It is not possible here to extend this topic; I can only note that Latin American scholars were active in discussions on models, none of these were the main subject of the presentations, but the authors commented on them.

1.3 From Bucharest to Paris

From the point of view that interests us here, an overview of the *Sciences and Empires Colloquium* cannot overlook the presented works that focused on science in Latin America. Most of them were based on the idea of "national science". Contributors were willing to deal with their local scientific cultures seriously, the centre does not necessarily represent the pinnacle of intellectual advancement (Paty, 1992). Their studies follow on from the Bucharest Declaration, the platform of commitments signed by Saldaña, Arboleda and a small community of scholars. Feeling isolated and lacking a professional framework within which to operate, they sought to "professionalize their field" and disseminate the studies they had carried out as widely as possible. As M. Paty shows in the paper we have mentioned, and as we will see later, in the late 1980s and early 1990s, there was a notable increase in the history of sciences in Latin America. The Bucharest declaration, as it is known, was born in the context of *The XVIth International Congress of the History of Science*. Let's take a look back at this significant meeting attended by young and promising Latin American academics.

This congress, held in Bucharest from August 26th to September third 1981[10] was organized by "The International Union of History and Philosophy of Science/

[10]The 16th International Congress of the History of Science, under the auspices of UNESCO, IUHPS/DHS, and the Academy of the Socialist Republic of Romania: Bucharest, Romania. Pub. House of the Academy of the Socialist Republic of Romania, [Bucharest], 1981–1981: proceedings

History of Science Division (IUHPS/DHS)", under the auspices of UNESCO. In Romania, an organizing committee headed by the President of the Romanian Academy was set up, with the active participation of the Romanian Committee for the History and Philosophy of Science. In addition, "Substantial support has been given by several national educational and cultural institutions in Romania."[11] The theme of the congress was: Science and technology, humanism and progress.[12] Of the 1200 participants, half were Romanians. There were around 40 French participants and a few from Latin America. The Local Organizing Committee had made preparations on a grand scale, exceptional success was certain.[13]

The main purpose of the congress, defined by the executive Council of the Historical Division of IUHPS, was to examine the social and cultural conditions, past and present, enabling science and technology to take root and develop; to promote a greater awareness of the social cultural and ethical implications of scientific and technological advance and to promote a broader public understanding of science. From my point of view, the objectives of the Congress, established over 40 years ago, are pioneering in many ways. They foreshadow new trends and studies. Five of the congress symposia would be concerned with issues affecting science and society, science and technology education, intellectual creativity, and the role of women in science.[14] Actually, Academic Work was divided into four main categories; (14) scientific sections, (12) symposia, (7) thematic sessions and (4) commemorations.

In the official speeches, the Congress was held "under the patronage of Mrs. Elena Ceausescu, First Deputy Prime Minister and wife of the Head of State" (Bruylants, 1981). It is presented as a historical introduction to salutary reflections on the progress of science and the scientist's responsibility for maintaining peace. Today we know that this alliance between science and politics resulted in "a vibrating call to all ministers of science and culture, to politicians, governments and parliaments of people to make everything possible to safeguard peace—mankind's most precious asset, to secure man's and mankind's fundamental right—the right to life."[15] The congress published four volumes for participants, containing summaries of papers, reports and discussions, as well as a special issue of *Revue Roumaine* (1981, 6–7) devoted to the history of science in Romania. These proceedings do not include the Bucharest Declaration. The discussions of Latin America scholars took place during the conference but the meeting was impromptu. It doesn't appear in the CAPHES archives in Paris. The declaration is part of the collective memory, it is,

[11] « Un concours essentiel ». Original documents are in French. CAPHES archives: Dossier IUHPS, S7–1. 16°-ICHS. Bucharest,1981

[12] CAPHES archives: Dossier IUHPS, S7–1. 16°-ICHS. Bucharest, 1981.

[13] CAPHES archives: Dossier IUHPS, S7–1. 16°-ICHS. Bucharest, 1981.

[14] CAPHES archives: Dossier IUHPS, S7–1. 16°-ICHS. Bucharest, 1981.

[15] CAPHES archives: Dossier IUHPS, S7–1. 16°-ICHS. Bucharest, 1981.

however, known that a written source exists.[16] For those lucky enough to have been there, it was a successful call.

Two decades later much research has been dedicated to the study of the scientific and technological national science in Latin America. History of science was growing institutionally. The *Sociedad Latinoamericana de historia de la ciencia y la tecnología* (SLAHCT) was created in 1982, with Saldaña as founding Member and First President (1982–1985) (Aguileros, 2000).[17] In 1986, a chair was created in the Faculty of Philosophy and Letters of the National Autonomous University of Mexico (UNAM) to promote historical research on science and technology. The seminar brought together a group of Mexican historians and scientists who were interested in analyzing the facts, the protagonists and the circumstances in which this knowledge was produced, and they began to research this subject. In 1991, from 24 to 25 October, the *II Coloquio Regional de Historia de la Ciencia y de la Tecnología* was held (Universidad Autónoma de Puebla y *Sociedad Mexicana de Historia de la Ciencia y de la Tecnología*, which had emerged in the 1960s). Once again, the meeting brought together historians and scientists.

In the meantime, as early as 1983, the scientific and international journal *Quipu* was created, edited by Juan José Saldaña as the official publication of the society, with Spanish as the primary language of communication, while English summaries were provided as well as a few articles in English and Portuguese. A key driving force was the aim to foster ongoing communication among Latin American historians of science and technology, thereby advancing the professionalization of their field (Aguileros, 2000). Let's look at a paper published by D. Chambers in *Isis*. It talks about *Quipu* during its early life.

Published three times a year in Mexico City, the first issue in 1984, *Quipu* was giving voice, in particular, to the professional community of Latin America. The journal is depicted as contributing to the wider recognition of the "remarkable" scientific and technological heritage of the countries of Latin America (Chambers, 1991). It has no chronological focus, although around half of the papers deal with the nineteenth and twentieth centuries. According to Chambers, the time frames of nearly half the papers cover more than a single century. A significant number of them look at the interesting events and institutions of eighteenth century Latin America. The vast majority of articles limit their analysis to a single country. "Only about 10 percent are comparative studies of more than one country, though there is some indication that interest in comparative history and in the colonial context as such is on the increase." In 1994, *Quipu* was recognized as a scientific journal of excellence by the *Consejo Nacional de Humanidades, Ciencias y Tecnologías* (CONACYT), and included in the index of Mexican scientific journals.

This paper by Chambers allows us to understand the special aura that cloaked *Quipu*. It was published in 1990, 8 years after the creation of the journal. At the

[16] See for instance, https://issuu.com/cihcytal/docs/declaracion_de_bucarestb

[17] Among other founders were: Elena Pennini of Argentina, Ruy Gama and Ubiratan D'Ambrosio of Brazil, Luis Carlos Arboleda of Colombia. See: Aguileros Rios (2000).

time, Brazilian Urbiratán d'Ambrosio (University of Campinas) was president of the society with which *Quipu* is officially associated, *Sociedad Latinoamericana de historia de la ciencia y la tecnología,* U. d'Ambrosio gave an opening address at the *Sciences and Empires Colloquium* that French historians of science had organized in Paris. What occurred at this meeting?

1.4 Talking in Paris on Latin American Science

In Paris, scientific work was divided into various categories: nine sessions, two round tables and a final round table, which aimed at coordinating conclusions and organizing the future. The scientific programme reflects a promising trend in the history of science in a colonial context, as previously highlighted by the author of the report on the *Quipu* journal published in *Isis.* In order to promote cultural comparisons, only two out of the nine sessions were limited to distinct cultural areas: Brazil and Far west. As far as the notion of empires is concerned, lecturer B. Schroeder-Gudehus says: the term refers first and foremost to colonial empires—metropolises and dependent territories (Schroeder-Gudehus, 1992). The presentations discussed territories under the British, Portuguese and Spanish empires. We will only look at papers dealing with New Spanish societies, territories under Spanish control but in the process of being liberated. These include papers presented by Colombian scholars Arboleda, Obregon, and Mexican scholars Aceves and Saldaña.

L. C. Arboleda's paper is entitled *Science and nationalism in New Granada on the eve of the revolution of independence* (Arboleda, 1992). For those versed in Colombian history and in scientific practices and theories in the late-eighteenth-century Hispanic world, this work is a much-welcomed contribution. Arboleda dealt with "El sabio Caldas", a prominent intellectual and prime example of Creole science. As quoted by A. Lafuente, Francisco José de Caldas (1768–1816), brought up in the Colombian city of Popayan, experienced feelings of isolation and frustration (Lafuente, 2000). He asked J. C. Mutis to include him in the expedition he was leading through the New Kingdom of Granada. Caldas discovered that the boiling point of water changed with altitude, a discovery that Humboldt acknowledged in 1801 during his stay in the province of Quito. As a member of the intellectual elite of the colony, he is of significant importance in the context of the new form of patriotism that began to take shape, giving priority to the useful and public dimensions of knowledge. Lafuente talks on priorities that shifted in the scientific disciplines: from astronomy to geodesy, from botanical taxonomy to the geography of plants, and from exploration of territories to regional econometrics. In short, having identified New Grenada with a new technical standard, the Creole elite fought to substitute more integrated views for classificatory practices; to replace observation with measurement, and the university chair with the laboratory (Lafuente, 2000).

Arboleda's analysis is based on the writings of Caldas, with particular emphasis on the letters he wrote to his friend Santiago Arroyo, which are a primary source of first-order importance. The *a priori* objective is to elucidate the process that led to the formation of Caldas's philosophical vision of nature. Arboleda observes that certain historians who have dedicated years to the study of Caldas' life and work have served as sources of inspiration for him. However, one recent work by Colombian historians of the mathematical sciences have proven particularly useful, as they offer insight into Caldas's geophysical experiments and his understanding of the state of the question in Europe.[18]

The problem underlying these studies on Caldas is the emergence, among intellectuals with nationalist views, of a realistic programme of specialized training and research in their countries. It is key to understanding the dynamics of scientific activity in peripheral areas towards the end of the colonial regime. Arboleda examines Caldas's nationalist, pragmatic and political objectives. "The desire to be useful in science and with science to contribute to the social projects of the *criollo élite*, is what gives to Caldas way of knowing his special style (Arboleda, 1992). Let me sum up Arboleda's point of view in a few words: the historical case of Caldas bears witness to the emergence, in the peripheral areas towards the end of the eighteenth century, of a realistic trend of scientific endeavor among the intellectuals with nationalistic tendencies.

D. Obregon shared Arboleda's concerns about nineteenth-century Colombia. His talk, *Trade and the natural science in the United States of Colombia*, did not address local knowledge produced by indigenous people, focusing instead on two key points: the Colombian economy at the end of the century and the scientific movement, including the passion for science by local political leaders. Great importance was given to studies applicable to agriculture, mining and industry. The presentation also addressed the institutionalization of science. In this instance, it has sought to emphasize the failure of the process because it was unable to consolidate itself or form a scientific community (Obregon, 1992).

One more paper in Paris dealt with a "single local knowledge".' This talk addressed modern chemistry, as a native Mexican form of scientific culture. As M. Paty argues, the description of science in Latin America, as it exists and has existed historically up to the present day, involves themes linked to the exploitation of local wealth and resources as part of organized colonization. Aceves focused on the features of the chair of chemistry in the Royal School of Mines at the turn of the 18th to the nineteenth century. It does not tend to take the West in general or the world's most advanced level of research (that of France in the past) as the model. It does not want to compare the state of affairs of the local scientific community with that of France to bewail local institutional and cultural defects. It talks about a process of diffusion-reception of European science, which it assumes is complex. During the period under study, the chair of chemistry showed a marked French influence and, internally, adopted and diffused the new theories. In the mining sector, the necessity arose for the reorganization of mining as a whole, at an earlier date than the founding of the Royal School of Mining (Aceves, 1992). As Arboleda's

[18] Arboleda talks about mathematician Victor Albis and physicist Regino Martinez.

study shows (Arboleda, 1992), Aceves' is a successful example in which primary written sources, hitherto unexploited by twentieth-century historians, are consulted.

As for Saldaña and his talk, he has endeavored to align his approach with that of the congress organizers, who have expressed a preference for a comparative studies methodology. The point is apparent in the title of his talk: *Science et pouvoir au XIXe siècle. La France et le Mexique en perspective.* His work discusses the important role of science in the conception of the *Etat*, both in France and Mexico. We have identified key elements that define Saldaña's approach to Mexican case studies. In the second half of the nineteenth century, science became an ideological weapon against obscurantism, a political instrument for shaping the new nation, "a technique of government" (Saldaña, 1992). The scientific effort of the nineteenth century had an effect on the integration of the country, which was an important political goal at a time when the territory was being amputated by the United States or coveted by France. He speaks of a structuring function of scientific activity on the part of the *Etat*, based on the political philosophy of the Republic. Additionally, talking about Mexican science, which was to be compared with the French model, he argues: nineteenth-century Mexican science was quite far removed from the French model in terms of actions and results, even if their motivations were quite similar according to the texts. The question of context appears to be an essential issue for the comparative history of science (Saldaña, 1992). Over the last years, Saldaña has dedicated much research to the study of these questions. He has worked hard to spread these ideas throughout Latin America, and the consequences are already there: recent scholarship emphasizes local contexts, independent of their place within the structure of an empire.

1.5 Final Remarks

In conclusion, studying the international colloquium *Sciences and Empires* (Paris April 1990) and looking back at the preceding *International Congress of the History of Science* (Bucharest 26 August–3 September 1981) organized under the auspices of UNESCO enabled us to address two outstanding points: the part played by Latin American scholars in the institutionalization of history of science, technology and medicine in the second half of twentieth century; more generally the consequences of the Bucharest Declaration (26 August–3 September 1981): the creation of the scientific and international journal *Quipu* edited by Saldaña and of the *Sociedad Latinoamericana de historia de la ciencia y la technología* with Saldaña as founding Member and First President (1982–1985), and secondly, the works presented in Paris in April 1990, focused on knowledge produced in territories under Spanish control but in the process of being liberated. We offered a detailed treatment of these questions, illustrating how Latin American's scholars emphasized on the significance of History of science as a discipline at how they described science of their countries as it existed in history. This brings us back to the questions addressed by M. Paty in 1992 in his article published in the French journal *La Pensée*. Finally, as a Franco-Colombian historian of science, and therefore “in the middle”, I stressed the constant relations that M. Paty and French researchers have with Latin American scholars.

Acknowledgements I warmly thank Luis Carlos Arboleda and Marypaz Ramos who invited me to participate to this book and gave me the benefit of their comments. It is impossible not to have a fond thought for Professor Robert Halleux who for many years worked with J. J. Saldaña at the International Academy of History of Science (París, France). Our regular exchanges at the Salon des Académiciens of the Institut de France have generously contributed to nourishing my reflection. He did not live long enough to see this book completed. This final version of my text owes a lot to Daniela Matiz and Julian Medrano, who gave me valuable suggestions. I thank them for that.

Primary Sources

CAPHES archives: Dossier IUHPS, S7-1.16°-ICHS. Bucharest (1981)

References

Aceves, P. (1992). The first chair of chemistry in Mexico (1796–1810). In P. Petitjean, C. Jami, & A. M. Moulin (Eds.), *Science and Empires: Historical studies about scientific development and European expansion* (pp. 137–146). Springer Netherlands.

Arboleda, L. C. (1992). Science and nationalism in New Granada on the eve of the revolution of independence. In P. Petitjean, C. Jami, & A. M. Moulin (Eds.), *Science and Empires: Historical studies about scientific development and European expansion* (pp. 247–258). Springer.

Arboleda, L. C., & Paty, M. (Eds.). (2004). *Formación de Cultura Científica en Colombia. Ensayos sobre Matemáticas y Física*. Ed. Universidad del Valle.

Aguileros Rios, S. (2000). *Biblio 3W. Revista Bibliográfica de Geografía y Ciencias Sociales* Universidad de Barcelona [ISSN 1138–9796] N° 212, 28 de febrero de 2000.

Basalla, G. (1967). The spread of Western science: A three-stage model describes the introduction of modern science into any non-European nation. *Science, 156*, 611–622.

Bonneuil, C. (1994). *Science and Empires: Historical studies about scientific development and European expansion*, ed. Patrick Petitjean, Catherine Jami and Anne-Marie Moulin. Kluwer Academic Publishers. In: *Revue d'histoire des sciences*, tome 47, n°3–4 : 522–523.

Bruylants, A. (1981). XVIe Congrès international d'Histoire de la Science (Bucarest, 26 août-3 septembre 1981). Rapport du délégué de l'Académie. *Bulletin de la Classe des sciences, 67*, 623–625.

Buchanan, R. A. (1988). Nathan Reingold and Marc Rothenberg, eds., scientific colonialism: A cross-cultural comparison (book review). *Technology and Culture, 29*(3), 685.

Bustamante De la Ossa, M. C., & Borel, É. (2019). *À l'aube de la théorie des quanta. Notes inédites d'Emile Borel sur un cours de Paul Langevin au Collège de France (1912–1913)*. Brepols.

Carter, C. (2009). Magnetic fever: Global imperialism and empiricism in the nineteenth century. *Transactions of the American Philosophical Society, 99*(4), III.

Chambers, D. W. (1991). Quipu: Revista Latinoamericana de Historia de las Ciencias y la Tecnologia. *Isis, 82*(2), 323–324.

Chemla, K. (2007). REHSEIS. Rapport Quadriennal 2004–2007.

Dray, M. K. (2018). Les ancrages nationaux de la science mondiale, XVIIIe-XXIe siècles. EAC; IRD.

Freire, O. (2010). Michel Paty et l'Amérique Latine: une histoire d'amitié et d'échanges. Exposé au Colloque d'hommage à Michel Paty, Paris, le 7 décembre 2010. Unpublished.
Inkster, I. (1985). Scientific Enterprise and the colonial 'model': Observations on Australian experience in historical context. *Social Studies of Science, 15*(4), 677–704.
Inkster, I. (1988). Nathan Reingold & Marc Rothenberg (eds). Scientific colonialism: A cross-cultural comparison. Washington: Smithsonian Institution press, 1987. *The British Journal for the History of Science, 21*(3), 382–384.
Kumar, D. (Ed.). (1991). *Science and Empires: Essays in Indian context, 1700–1947*. Anamika Pub & Distributors.
Lafuente, A. (2000). Enlightenment in an imperial context: Local science in the late-eighteenth-century Hispanic world. *Osiris, 15*, 155–173.
Obregon, D. (1992). Trade and the natural sciences in the United States of Colombia. In P. Petitjean, C. Jami, & A. M. Moulin (Eds.), *Science and Empires: Historical studies about scientific development and European expansion* (pp. 147–152). Springer.
Paty, M. (1992). L'histoire des sciences en Amérique latine. *La Pensée, 288-289*, 21–45.
Paty, M., & Martínez-Chavanz, R. (2004). Formación y desarrollo de la cultura científica en Colombia: la física de 1880 a 1940. In L. C. Arboleda & M. Paty (Eds.), *Formación de cultura científica en Colombia. Ensayos sobre Matemáticas y Física* (p. 111). Ed. Universidad del Valle.
Petitjean, P. (1992). Sciences et Empires: un thème prometteur, des enjeux cruciaux. In P. Petitjean, C. Jami, & A. M. Moulin (Eds.), *Science and empires: Historical studies about scientific development and European expansion* (pp. 3–12). Springer.
Polanco, X. (1990). *Naissance et développement de la science-monde*. Éditions la découverte/Conseil de l'Europe/UNESCO.
Polanco, X. (1992). World-science: How is the history of world-science to be written? In P. Petitjean, C. Jami, & A. M. Moulin (Eds.), *Science and empires: Historical studies about scientific development and European expansion* (pp. 225–242). Springer.
Pyenson, L. R. (1989). Pure learning and political economy: Science and European expansion in the age of imperialism.
Pyenson, L. R. (1992). Typologie des Stratégies d'Expansion en Sciences Exactes. In P. Petitjean, C. Jami, & A. M. Moulin (Eds.), *Science and empires: Historical studies about scientific development and European expansion* (pp. 211–217). Springer.
Pyenson, L. R. (1993a). *Civilizing mission. Exact sciences and French overseas expansion 1830–1940* (Vol. 47, p. 74). The Johns Hopkins University Press.
Pyenson, L. R. (1993b). Cultural imperialism and exact sciences revisited. *Isis, 84*(1), 103–108.
Rashed, R. (1992). Science classique et science moderne à l'époque de l'expansion de la science européenne. In P. Petitjean, C. Jami, & A. M. Moulin (Eds.), *Science and empires: Historical studies about scientific development and European expansion* (pp. 19–30). Springer.
Saldaña, J.-J. (1992). Science et pouvoir au XIXe siècle. La France et le Mexique en perspective. In P. Petitjean, C. Jami, & A. M. Moulin (Eds.), *Science and empires: Historical studies about scientific development and European expansion* (pp. 153–164). Springer.
Schroeder-Gudehus, B. (1992). Sciences exactes et politique extérieure. In P. Petitjean, C. Jami, & A. M. Moulin (Eds.), *Science and Empires: Historical studies about scientific development and European expansion* (pp. 219–223). Springer.

Martha Cecilia Bustamante De La Ossa, Physicist, Universidad Pedagogica Nacional de Bogota, Colombia. Master in Astrophysics, Université Paris7, Denis Diderot, Paris, France. PhD in Epistemoloy and History of Sciences and Scientific Institutions, Université Paris7 Denis Diderot, Paris, France. Post-Doctorate, Laboratoire de l'accélérateur Linaire, IN2P3, Orsay, France.

Chapter 2
Three Moments of Spanish Contributions to the History of Science in Latin America During the Second Half of the Twentieth Century

Leoncio López-Ocón Cabrera

Abstract Scientific congresses fulfill multiple functions. They allow the exchange of ideas and projects among members of a scientific community. They favor the sociability of their participants. Behind their celebration lies an organizational effort that requires resources and negotiations with funding institutions. If it has an international projection, it becomes another manifestation of scientific diplomacy. Based on these considerations, this contribution analyzes the international congress "Science, Discovery and the Colonial World" held in Madrid in June 1991 from a threefold perspective. Firstly, the context of its celebration will be shown, placing it in the framework of the fifth centenary of the encounter of two worlds that was commemorated in 1992. Secondly, it will be explained how this congress was the culminating manifestation of an interaction carried out in the 1980s between members of the Latin American Society for the History of Science and Technology, chaired by Juan José Saldaña, the Spanish Society for the History of Science and Technology and the History of Science Society of the United States. Finally, the presentation of the issues discussed at the congress highlights how and why throughout its development an interesting antecedent of the trialogue that took place in Madrid in 1991 was not considered: I am referring to the historiographic work carried out by a group of Spanish scientists living in America between 1940 and 1975, mostly linked to the journal *Ciencia*, the exponent of the republican scientific diaspora.

Keywords Scientific congresses · Scientific societies · Latin American society for the history of science and technology · Historiography of science · 1992

L. López-Ocón Cabrera (✉)
Institute of History, Center for Human and Social Sciences, Spanish National Research Council (CSIC), Madrid, Spain
e-mail: leoncio.lopez-ocon@cchs.csic.es

M. de la P. Ramos-Lara, L. C. Arboleda (eds.), *History of Science in Latin America: The Construction of an Intellectual Field (20th century)*,
https://doi.org/10.1007/978-3-031-75536-1_2

2.1 Introduction

Scientific practice requires a communicative action oriented to multiple audiences. One of these audiences is the colleagues themselves, who are reached in various ways. Some are indirect, through, for example, publications of various kinds. Others are more direct. Among these forms of communication, congresses stand out, which favor face-to-face relationships among participants, and the forging of sociable relationships, elements that can be relevant for the construction of scientific communities that are formed around new disciplinary fields sustained in shared and collaborative projects.

Given their importance in the communicative dimension of science and their role in the construction of scientific communities and disciplinary fields, it is not surprising that historiography has paid attention to international scientific congresses, considering them as an object of study in themselves, and not only as a theater of scientific practice.[1] Thus, they have begun to be considered as relevant instruments of the internationalization of science.[2] Attention has also been paid to the fact that their organization and development require the implementation of diplomatic actions, and their staging has become a manifestation of scientific diplomacy.[3]

Taking these considerations into account, an approach will be made to the International Congress "Science, Discovery and the Colonial World", held in Madrid for 4 days -between Tuesday 25 and Friday 28 June 1991-, whose proceedings were published in the book "Mundialización de la ciencia y cultura nacional" (Globalization of Science and National Culture).[4] The analysis of this congress aims to highlight three issues.

First, it is worth highlighting how the holding of this congress was the culmination of a series of preceding initiatives connected to two cultural projects, born almost simultaneously on both sides of the Atlantic. On the one hand, the mobilizing program launched by the Consejo Superior de Investigaciones Científicas (CSIC) since 1985 to promote the study of cultural and scientific relations between Spain and Latin America in the framework of the commemoration of the V Centenary of the Encounter between Two Worlds that took place in 1992.[5] This program would be directed by José Luis Peset, a member of the Department of History of Science of the Institute of History of the CSIC. On the other hand, members of this mobilizing program, such as the Spanish historians of science Antonio Lafuente and José

---

[1] Bigg et al. (2023).

[2] See, among other contributions, *Les congrès scientifiques internationaux,* special issue of *Relations internationales* (1990) 62 (2); Pascale Rabault-Feuerhahn, P. & Feuerhahn, W. (eds) (2010), *La fabrique internationale de la science: Les congrès scientifiques de 1865 à 1945,* special issue or *Revue germanique internationale* 12; Geyer and Paulmann (2001); Reinisch et al. (2022).

[3] An approach to this issue in Turchetti, S., Adamson, M., Rispoli G., Olsáková. D. & Robinson, S. (eds) (2020). *Science Diplomacy,* special issue of *Historical Studies in the Natural Sciences* 50 (4).

[4] See Lafuente et al. (1993).

[5] Historiography has begun to take stock of what this commemorative program meant for Spanish society at that time. See for example Canal (2021) and Sanz López (2023).

Sala Catalá, also researchers of the aforementioned department, were founding members and active promoters of the Latin American Society for the History of Science and Technology, founded in the Mexican city of Puebla in 1982, under the leadership of Juan José Saldaña.

Secondly, to show the participation of Spanish historians, mainly attached to the CSIC, in the multilateral reflection or three-way dialogue that was being carried out in 1992 by European, Latin American, and North American historiography on the role played by Latin America in a global history of knowledge and its role in the creation and development of a science-world. This will be demonstrated through the analysis of the design of the congress, in whose organization the Spanish Society for the History of Science and Technology, the Latin American Society for the History of Science and Technology and the History of Science Society participated, and through the exposition of the contents of its proceedings.

Thirdly, to show how this interest in establishing a multiparty dialogue was also manifested in the central decades of the twentieth century in various places, such as in the pages of the journal *Ciencia*, created in Mexico in 1940 by Spanish Republican scientists exiled in that country.

2.2 The Immediate Antecedents of a Congress

"A three-way dialogue". This was the title of the preamble of the book "Globalization of science and culture" by its signatories: Antonio Lafuente, Alberto Elena and María Luisa Ortega. The aim was to highlight how the international congress "Science, Discovery and the Colonial World" held in Madrid in June 1991 was the result of a trilateral conversation that members of the North American History of Science Society had been having for some time with members of the Latin American Society for the History of Science and Technology (LASHST)[6] and the Spanish Society for the History of Science and Technology (SEHCYT) on the role played by the Latin American region in the creation of a science-world.[7] This dialogue was held by representatives of these three scientific societies. Juan José Saldaña played an important role in it. This was recognized by the editors of the proceedings of the congress, who stated that "it is fair to recognize that no one has contributed as much as Juan J. Saldaña, first president of the LASHST, so that the academic exchange of ideas between the three areas could take place. His enthusiasm has been decisive for us to meet in several places in America and Europe, as well as to ensure the continuity in the edition of Quipu, a Latin American journal that has already broken a record on its tenth anniversary".[8]

---

[6] Sociedad Latinoamericana de Historia de la Ciencia y la Tecnología (SLHCT), original name in Spanish.

[7] See in this regard Polanco (1990).

[8] Lafuente et al. (1993). "Un diálogo a tres bandas", 15.

In fact, shortly after the Latin American Society for the History of Science and Technology (LASHST) was constituted in Puebla, in a founding meeting held between August 23 and 26, 1982,[9] the journal Quipu was promoted as the Society's leading voice, and was published continuously from 1984 to 1994,[10] a decade that can be considered the one of greatest institutional strength of the LASHST and the one of greatest intensity in the three-way dialogue promoted by Spanish, Latin American and North American researchers interested in the historical development of knowledge in the Latin American region.

A group of Spanish historians, mainly linked to the Department of History of Science of the Institute of History of the Consejo Superior de Investigaciones Científicas (CSIC), contributed significantly to this period of flourishing and dynamism of the SLCHT. This contribution had a twofold motivation. On the one hand, the network of human relationships forged between Spanish and Latin American colleagues was relevant. Thus, for example, the advice given in Paris by René Taton at various times to Juan José Saldaña, Luis Carlos Arboleda and Antonio Lafuente to promote research on the history of science in Latin America, to which he was unable to pay attention in his *A general history of sciences,*[11] although he became aware of its importance when he directed the ICSU between 1954 and 1971 and presided over the International Union of History and Philosophy of Science between 1975 and 1977, had a certain importance in the constitution and initial development of the Latin American Society for the History of Science and Technology. Simultaneously, the Consejo Superior de Investigaciones Científicas (CSIC), the main public body in Spain dedicated to scientific and technical research, launched during the three-year period 1985–1987 a mobilizing program on cultural and scientific relations between Spain and America, which continued with another research plan in the four-year period 1988–1991 entitled "Spain-America: science and culture between two worlds", coordinated by José Luis Peset, from the Department of History of Science of what was then the CSIC's Centro de Estudios Históricos (Center for Historical Studies). During those years, prominent Latin American historians of science, such as Luis Carlos Arboleda and Eduardo Estrella, worked there as visiting professors. Some of the scientific results achieved by the ten research projects carried out at the CSIC during that six-year period, in which more than a hundred researchers were involved, were presented at the international colloquium "Spanish Science and Ibero-America" in November 1987.[12]

---

[9] This foundational moment was presented by Beltrán, E. (1984). A list of all the 279 articles published by Quipu in its two main stages from 1984 to 1993 and from 1999 to 2000 in https://biblat.unam.mx/es/revista/quipu.

[10] A brief overview of the evolution of this publication can be found at Aguilera Ríos, S. (2000). "*Quipu*: una revista latinoamericana de la historia de las ciencias y la tecnología", in *Biblio 3 W. Revista Bibliográfica de Geografía y Ciencias Sociales*, Universidad de Barcelona, 212. On line. https://www.ub.edu/geocrit/b3w-212.htm.

[11] Taton (1963–1966).

[12] It was the origin of the collective book, coordinated by Peset (1989).

This context favored dialogue and interaction between representatives of the Latin American Society for the History of Science and Technology and Spanish historians of science, linked to the aforementioned mobilizing program, located in various cities, such as Madrid, Valencia, Seville and Barcelona, where the historian of geography Horacio Capel played a very active role.[13] They shared seminars, designed joint publications, collaborated in their journals to exchange knowledge and reflections on the role played by scientific-technical knowledge in the historical development of Latin American societies in the different phases of their evolution and on the situation of historical studies of science in Iberian and Latin American countries as a whole.

The start of this dialogue took place, according to José Luis Peset,[14] around 1983 when historians of science attached at the time to the Arnau de Vilanova Institute of the Consejo Superior de Investigaciones Científicas (CSIC) began conversations with the Latin American Society for the History of Science and Technology to hold an Ibero-American colloquium on the history of science, which would be held at the headquarters of the Institute for Ibero-American Cooperation in September 1984. The results were published in a collective book in which both Latin American and Spanish historians collaborated. They addressed issues related to the practice of science. Sections were organized on the teaching and dissemination of science, scientific expeditions and the institutionalization of science, as well as the relationship between the different levels and components of science, for which proposals were presented in three round tables entitled "Science and ideology", "Science and technology" and "Science and health".[15]

This dialogue was sustained in the following years in various scenarios, generating varied results that facilitated the interaction of Spanish and Latin American historians of science with colleagues from other parts of the world. The foundations were then laid for the great scientific meeting held in Madrid in June 1991, one of whose animators was Juan José Saldaña, as mentioned above. Thus, Antonio Lafuente and Juan José Saldaña coordinated a collective book that was a milestone in the historiography of science in the Spanish language, since in it outstanding historians of science from different parts of the world offered their points of view and reflected on the new trends that were taking shape in the multifaceted field of the history of science. The 17 participants offered their views on the state of the discipline and/or their work programs and advanced proposals for the future. Five

[13] This work was reviewed in López-Ocón (2015).

[14] José Luis Peset (1985). "Introducción", 5.

[15] *Op. cit.* This collective book was written by a Portuguese historian Luiz Alte da Veiga, 13 Spanish historians: Agustín Albarracín, Emiliano Fernández de Pinedo, Mariano Peset, Antonio Lafuente, José Luis Peset, Manuel A. Sellés, Lola Higueras, José Sala Catalá, Raquel Álvarez, Delfín García García, Javier Puerto Sarmiento, Guillermo Olagüe de Ros, Esteban Rodríguez Ocaña and 12 Latin American historians: Mexican women historians Martha Eugenia Rodríguez, Josefina Mac Gregor and Elsa Malvido, the Mexican historians Enrique Beltrán, Juan José Saldaña and Carlos Viesca Treviño, the Colombians Luis Carlos Arboleda, Emilio Quevedo and Jorge Arias De Greiff, the Brazilian Ruy Gama, the Ecuatorian Eduardo Estrella and the Argentinian Celina A. Lértora Mendoza.

of them were North Americans, Roger Hahn, Thomas S. Kuhn, Nathan Reingold, Roy MacLeod—who was then working in Australia—and Lewis Pyenson, then a professor at the Canadian University of Montreal; seven Europeans, five of them Spaniards such as Juan Vernet, Horacio Capel, Antonio Lafuente, José María López Piñero and José Luis Peset, as well as the French René Taton and the Italian Pietro Redondi—linked to French institutions; four Latin Americans, such as the Brazilians Ruy Gama and Shozo Motoyama and the Mexicans Juan José Saldaña and Elías Trabulse, and one Asian: Chinese Qiu Renzog[16] Another example of the international echo of this dialogue between Latin American and Spanish historians of science was the invitation received by a group of them to participate in a colloquium organized by the AMELA association (acronym of Aire méditerranéenne et latino-américaine), created in 1979 with the aim of promoting the analysis of the processes of scientific and technical production and correct appropriation of their results in Mediterranean and Latin American countries. The results of this colloquium were presented in a relevant collective book. [17] But the most stable and relevant space where a permanent dialogue and interaction took place throughout the 1980s among those interested in the historical development of knowledge in the Latin American region was in the journal *Quipu* and its appendix *Cuadernos de Quipu*. During those years, some Spanish historians of science collaborated in its pages, particularly Antonio Lafuente and José Sala Catalá. The former had an active presence in the various *Cuadernos de Quipu*.[18] The latter, who died prematurely at the end of December 1991, in addition to collaborating in these *Cuadernos*,[19] presented in various issues of *Quipu* previews[20] of what would be his posthumous work: his book dedicated to comparing the role of scientific-technical knowledge in three emblematic places of colonial America: the valley of Mexico, the Callao-Huancavelica axis and Dutch Brazil.[21] It was precisely José Sala Catalá and Antonio Lafuente who represented the Spanish Society for the History of Science and Technology on the scientific committee that organized the international congress "Science, Discovery and the Colonial World" held in Madrid between June 25 and 28, 1991. They were joined on that committee by Luis Carlos Arboleda and Hebe Vessuri, for the Latin American Society for the History of Science and Technology, and Michael McVaugh and Seymour Mauskopt for the History of Science Society.

[16] Lafuente and Saldaña (1987).

[17] Xavier Polanco, coord. (1990).

[18] On this subject, see, among others, the following works by Lafuente (1986); Lafuente and Estrella (1987). Lafuente and Saraiva (2001).

[19] Sala Catalá (1986a).

[20] Sala Catalá (1986b, 1988).

[21] Sala Catalá (1994).

2.3 Actors and Themes of an International Scientific Meeting

The International Congress "Science, Discovery and the Colonial World" was held in Madrid from June 25 to 28, 1991. It was organized around 15 symposia, accompanied by four conferences or lectures.

One of them was given by Juan José Saldaña, who spoke on "The discovery of the other as a discovery of the self. The meaning of the polemics of science in America". He was introduced by María Luisa Ortega, professor at the Universidad Autónoma de Madrid.

The inaugural conference was held by Professor Bernard Cohen who spoke on "The New World as a source of science and ideas for Europeans". He was presented by Javier Ordóñez, professor at the Universidad Autónoma de Madrid. The other two lectures were given by Thomas F. Glick, who spoke on "Establishing Scientific Disciplines in Latin America: Genetics in Brazil, 1943-1960" and by José Luis Peset who spoke on "Science and Independence in Spanish America". They were introduced respectively by Spanish historians of science Antonio González Bueno and Leoncio López-Ocón.

As mentioned above, the more than 50 participants in the congress organized about 15 symposia. They dealt with problems and themes connected with the cutting-edge research being carried out at the time by the scientific societies organizing the congress.

Some were related to the role played by science and technology in the creation of an Atlantic history from the beginning of the European colonization of the American continent. Others were interested in the dialectic process generated from the arrival of Europeans in the "New World" between the knowledge of the Native Americans and the global expansion of Western science. In this field there was an interest in the development of colonial science and national science in the American territories before and after their independence processes. There was also a notable interest focused on issues related to the use and management of natural resources and the challenges related to human health care. Another line of research and reflection was oriented towards evaluating the role played by the Latin American region in the world dynamics of science.

Some symposia were oriented towards the modern era, others were confined to the contemporary era, and some intermingled temporalities and spaces, taking place on broad spatial scales and over long periods of time. All of them had their corresponding moderator and commentator, some of whom would make contributions to the book resulting from the congress.

Among those oriented towards the modern era, the following five symposia are worth mentioning:

1. *Cartography and Cosmology in the Age of Discovery* with four contributions: by American Norman J.W. Thrower, "Projections of the maps of the fifteenth and sixteenth-century European discoveries", by British Helen Margaret Wallis, "The measure of the Earth from Ptomely to Columbus" and by Spanish Nicolás García Tapia, "Engineers and cosmographers in 16th and 17th century Spain"

and Mariano Esteban Piñeiro, " The Academy of Mathematics of Philip II. Amendment of the maritime instruments".

2. *The Cultural and Scientific Impact of the Discovery on the Islamic World*, with four contributions by American Thomas D. Goodrich, "The new world in the Ottoman world: some Ottoman thoughts and some new world things", by Brazilian Ana María Alfonso-Goldfarb, "The Salvador-Calicut route: an encounter between two worlds", by Spanish Arabist Nieves Paradela Alonso, "The New World in the chronicles of Arab travelers", and by CSIC researcher Miguel Ángel Bunes, "America and the Islamic world in the natural and political descriptions of the Spaniards (16th and 17th centuries)".
3. *Silver Production in the Modern World: Technical Development and Colonial Context* with four contributions: by Spanish Rafael Dobado's, "Almadén and quicksilver revenues in New Spain", by Mexican Ramón Sánchez Flores, "Evolution of coinage techniques in colonial Mexico (16th-18th centuries)", by Spanish Ignacio González Casasnovas, "Silver and the reasons for colonial power. Changes in the structure of Potosí at the end of the 17th century" and the joint contribution of the Peruvian Carlos Contreras and the Argentinian Guillermo Mira, "Transfer of mining technology from Europe to the Andes".
4. *The Diffusion of Modern Scientific Traditions in Colonial America*, with five contributions: by Argentinian Celina A. Lértora Mendoza, "The introduction of Newtonian theories in the Río de la Plata", by Mexican María de la Paz Ramos Lara in collaboration with Juan José Saldaña, "Diffusion of Newtonian mechanics in New Spain" by Spanish Miguel Ángel Puig-Samper, "Diffusion and institutionalization of the Linnaean system in Spain and America", by Ecuadorian Eduardo Estrella, "Linnaeus in the viceroyalty of Peru" and by Colombian Emilio Quevedo, "The conflict between modern European and American scientific traditions in the field of health in Latin America".
5. *Natural History in the Colonial Politics of the Enlightenment* with three contributions. One was presented jointly by Spanish historians Javier Puerto and Antonio González Bueno, "Scientific policy and botanical expeditions in the Spanish colonial program", another by Mexican Patricia Aceves, "Metropolitan botanical policy in the viceroyalties of Peru and New Spain" and a third by Spanish historian Manuel Lucena Giraldo, "Attempts to promote agriculture in Spanish Guiana (1784-1789)".

The following four symposia are among those oriented towards the contemporary period:

6. *Associationism in Latin America*, with five contributions: by Mexican Leonel Rodriguez, "Geography in independent Mexico, 1824-1835: The National Institute of Geography and Statistics", by Brazilian Silvia Figueiroa, "The Brazilian Historical and Geographical Institute as an institutional space for natural sciences in the 19th century", by Spanish Leoncio López-Ocón, "Half a century of scientific activities of the Geographic Society of Lima", by Colombian Diana Obregón, "The Society of Neogranadinos Naturalists or the invention of a tradition" and by Venezuelan Yajaira Freites, "The Venezuelan Academy of Physical, Mathematical and Natural Sciences (1917-1979)".

7. *The new social history of medicine in Latin America: local, national and international contexts*, with four contributions: by the Argentinians Diego Armús, "The hygienic city: Latin American and European visions" and Eduardo Zimmerman, "Race, medicine and social reform in Argentina, 1900-1920", by Colombian S. Restrepo, "Infectious diseases in Colombia and the diffusion of medical knowledge, 1890-1930", and by the North American Julyan G. Peard, "The Tropicalist School in Brazil in the national context".
8. *The Changing Face of South American Astronomy: Observations and Research in Peru and Chile, 1849 to the Present* with four contributions: by the North Americans Philip Keenan, "Astronomical ideas in the Viceroyalty of Peru", Owen Gingerich, "Contributions to Astronomy from Peru, 1889-1927", and Howard Plotkin, "Harvard's Boyden Station in Peru, 1889-1927" and by Chilean Héctor Álvarez, "Contributions to Astronomy from Chile, 1849-1965".
9. *Toward Modernity: Science in Underdeveloped Scientific Communities*, with three contributions: by Spaniards Antonio Moreno and José Manuel Sánchez Ron, "Spanish science: from regenerationist optimism to patriotic exaltation"; by Peruvian Marcos Cueto, "Between modernity and stereotype: North American visions of Latin American medical sciences" and by Brazilian Joao Carlos Vitor García: "Brazil and scientific modernity"

   Six other symposia mixed temporalities and spatial scales:
10. *Worldwide Diffusion of Science* with three contributions: by Venezuelan Hebe Vessuri, "International exchanges and national styles: aspects of the globalization of science"; by the American with long teaching experience in Australia David Wade Chambers, "Locality and Science: Myths of center and periphery" and by Frenchman Patrick Petitjean, "Scientific relations as a crossing of supplies and demands of science: Franco-Brazilian cases between 1870 and 1940".
11. *European Science in the New World* with three contributions: by American Robert Kargon, "Colonizing the New World and the roots of modern science", by Mexican Virginia González Claverán, "The exploration of the Mexican Pacific in the 18th century" and by American Joy Harvey, "Colonialism and French Anthropology in the Nineteenth Century.
12. *Interdependence and "dependence"* with three contributions by the Dutchman Karel Davids, "Dutch expansion and agricultural technology in South America, 17th-18th centuries", by North American Daniel Headrick, "British technology in India and Latin America: domination and dependence" and by Brazilian Maria Amelia Dantes, "The presence of European scientific traditions in Brazil".
13. *Comparative Perspectives on European Science and New World Societies* with three contributions: by the Swedes Sverker Sörlin and Par Eliasson, "Swedish Science and New World", by American Teresa A. Meade, "Cultural Imperialism in the Old Republic Rio de Janeiro: the urban renewal and public health project" and by Mexican Omar Moncada, "Science in action: engineering and land management in New Spain in the 18th century".
14. *Context in Colonial Medicine*, with four contributions: three by Americans Caroline Hannaway, "Distinctive or derivative? The French colonial experience,

1740-1790"; Ann La Berge, "The conquest of Algeria and the discourse on public health in France, 1830-1850" and Molly Sutphen, "Plague, race, and segregation: British Colonial response to contamination in Cape Town" and another by American researcher John Harley Warner, "Cultural nationalism and tropical fevers: models of colonial medicine in pre-civil war American South".

15. *Literature and Science: The Old World and the New* with three contributions by American researchers Kenneth J. Knoespel, "Deciphering the Barbarians: Ovid's Metamorphoses and the Assimilation of the New World" and George Sebastian Rousseau, "Science, Travel, and the New World: from Bosman to Freud" and by American historian of science Nancy Stepan, "Nature as a form of writing: the construction tropical nature by scientists in the Nineteenth Century".

Finally, the book resulting from the congress was organized in five parts. The contributions presented in the 15 symposia presented during the 4 days of the congress were inserted in these parts. They were denominated by the editors of the book as follows:

New World Perspectives, with 16 contributions. It included eight contributions from symposia 1 and 2, one from symposium 3,[22] one from symposium 11 (by Robert Kargon), two from symposium 15 (by K.J. Knoespel and G. S. Rousseau[23]), a commentary by Victor S. Thomas on Bernard Cohen's inaugural lecture entitled "What Columbus 'Saw' in 1492 by Bernard Cohen," and three contributions by Víctor Navarro Brotons, " Cartography and cosmography in the age of discovery", Alberto Elena, "The configuration of scientific peripheries: Latin America and the Islamic world" and Jaime Vilchis, "Symbolization and natural history in colonial Ibero-America", who were respectively the commentators of symposia 1, 2 and 5.

Colonial elites and Enlightenment, with 13 contributions from various symposia: the five from symposium 4, the three from symposium 5, one from symposium 3 -that of Carlos Contreras and Guillermo Mira, one from symposium 8 -that of Philip C. Keenan-, one from symposium 11, the one signed by Virginia González Claverán who finally titled her contribution "Matutaray: a Hawaiian in New Spain", another from symposium 13, the one by Omar Moncada, and the lecture given by José Luis Peset "Science and independence in Spanish America".

Formation of national scientific communities, with 11 contributions. This part grouped the five contributions of symposium 6 plus the contribution of its organizer and commentator Horacio Capel, entitled " Scientific associationism in Ibero-America. The need for a globalizing approach"; the three from symposium 9; one from symposium 8, by Héctor Ávarez, and the lecture by Thomas F. Glick, "Establishing scientific disciplines in Latin America: genetics in Brazil, 1943-1960".

[22] It was signed by Julio Sánchez Gómez and Guillermo Mira and entitled "American mining and European mining, 1750–1820: a comparative perspective".

[23] The title of his contribution was finally "Magic, science and the "New World" origins of Freud's fetishes".

Nature, healthiness and urban culture, with 11 contributions corresponding to the following symposia: four from symposium 14; three from symposium 7 corresponding to the following authors: Diego Armús, "The hygienic city between Europe and Latin America", Eduardo A. Zimmerman, "Race, Medicine and Social Reform in Argentina, 1890-1920", and Julyan G. Peard, "Tropical medicine, society and race: the case of the Escola Tropicalista Bahiana, Brazil, 1860-1889"; one from symposium 11 -Joy Harvey, entitled "Types and races: the case of the Escola Tropicalista Bahiana, Brazil, 1860-1889"; one from symposium 13, that of Teresa Meade; one from symposium 15—that of Nancy Leys Stepan, who finally entitled her contribution "Tropical nature as a way of writing". Also included was the essay by historian James E. McClellan III, in which he commented on the contributions of symposium 13, Comparative Perspectives on European Science and New World Societies.

Global dynamics of science, with 13 contributions. Here were grouped the three contributions from symposium 10, two from symposium 8—those of Howard Plotkin and Owen Gingerich with the contribution of their commentator Antonio E. Ten entitled "Pure sciences and national prestige. Colonial astronomy and republican astronomy in South America", two from symposium 12—those of Karel Davids and Daniel R. Headrick, one from symposium 3, that of Mexican Ramón Sánchez Flores, one from symposium 9, that of Marcos Cueto, one from symposium 13, that of the Swedes Pär Eliasson and Sverker Sörlin, and the contributions of George Basalla "The spread of western science revisited" and Roy MacLeod, "The worldwide diffusion of science" as commentators of symposium 10.

These 64 contributions were introduced by a broad reflection by the historian of Spanish science José María López Piñero, who entitled the prologue of the book being presented "The tradition of the historiography of science and its current situation: the conditioning factors of a congress".

2.4 The Distant Antecedents: Memories and Oblivion

Given the importance of José María López Piñero in the institutionalization of the history of science in the Spanish academic panorama of the last third of the twentieth century,[24] the preeminent place assigned to him as prologue to the book *Mundialización de la ciencia y cultura nacional* should come as no surprise. In that prologue he set out the three basic conditioning factors that configured the context of the 1991 congress "Science, Discovery and the Colonial World". Namely: its connection with the activities promoted to commemorate the fifth centenary of the so-called encounter between two worlds, the participation in its organization of the Spanish, Latin American and North American associations for the history of

[24] An overview of his work in Capel and Jori (2010).

science, and the situation of the discipline of the history of science in the last decade of the twentieth century.

From this presentation it is worth highlighting a comparison made by López Piñero between the aforementioned 1991 congress and the IX Congress of the Académie Internationale d'Histoire des Sciences held in Barcelona and Madrid in 1959, considering them the most important international scientific congresses held in Spain in the field of the history of science in the second half of the twentieth century. To these congresses should be added the 19th International Congress of History of Science, organized by Mariano Hormigón (1946–2004),[25] which brought together more than a thousand participants in the city of Zaragoza from 22 to 29 August 1993.

The 1959 Barcelona and Madrid congress was chaired by the tandem formed by two great specialists in the history of medieval, Arab, and Jewish science, the Catalans José María Millás Villacrosa (1897–1970) and his disciple the Arabist Juan Vernet (1923–2011), who received a distinction at the Madrid congress in 1991. The vice presidents were Joaquín Carreras Artau (1894–1968), a historian of philosophy very interested in scientific topics, and the sailor and historian Julio F. Guillén Tato (1897–1972) who had organized a seminar on medieval maritime cartography in which participated, among others, Salvador García Franco (1884–1961) and Ernesto García Camarero (1932–2022). Spanish medical historians were also present at the Barcelona congress. Among them were Pedro Laín Entralgo, Juan Antonio Paniagua, Luis S. Granjel and Agustín Albarracín.

López Piñero did not participate in the 1959 congress because he was then completing his studies as a medical historian in Germany. But in his eagerness to establish connections between the 1959 and the 1991 congress, he explained how delegations from many countries came to Barcelona and Madrid, of which he gave timely news. Thus, the French group present at the congress was formed by Alexandre Koyré, Guy Beaujouan, Pierre Costabel, Maurice Daumas, Pierre Huard, François Russo, René Taton and Jean Théodorides. Among the Germans were Willy Hartner, Rudolph Zaunick and Ernst Zinner. The Italian delegation included medical historian Luigi Belloni and astronomy historian Maria Luisa Bonelli. The United Kingdom was represented by Allistair C. Crombie and Joseph Needham, and the Soviet Union was represented by Vassili Zoubov. The American delegates were also numerous as veteran medical historians such as Richard Harrison Shryock attended the 1959 congress along with other scholars such as Marshall Clagett, I. Bernard Cohen, Charles C. Gillispie, Henry Guerlac, Doris Hellmann, Gerald Holton, Thomas F. Kuhn, Everet Mendelsohn, Charles D.O. Malley, Derek J.S. Price and Edward Rosen. Representatives from other European, Asian and Arab countries, as well as Canada and Australia, also attended. But Latin American participation was

[25] An appraisal of his work in Español (2004).

minimal. López Piñero only records the presence of one congressman representing Brazil. He was Ladislao Reti, an Italian-born researcher.[26]

Was this scarce Latin American presence at the Barcelona congress due to a weak historiographic production on the history of science in Latin America at the end of the 1950s, as López Piñero seems to point out when comparing the trajectory of the discipline in the United States, Spain and Latin America between 1959 and 1991? Or was it due to other factors?

There is an element to consider that López Piñero suggests but does not develop in his writing. In fact, in passing, he points out that in the 1959 congress held in Barcelona and Madrid, the Italian Francesco Carta gave news of an ambitious Mexican enterprise that was then beginning: the edition of the complete works of Francisco Hernández headed by Germán Somolinos, an issue that López Piñero noticed since throughout his research career he would devote great attention to the work of this humanist physician.[27] Indeed, in 1959 he began the publication of a work that would become one of the historiographic monuments of the history of science in the American continent. For more than a quarter of a century—since the publishing project ended in 1985—a large interdisciplinary team formed in Mexico took great pains to gather and make known all the writings of the "Protomedico general de todas las Yndias, islas y tierra firme del Mar Océano" during the reign of Philip II of Spain, and the most outstanding naturalist and physician of sixteenth-century Mexico.

The driving force behind this great project was the Republican doctor exiled in Mexico, Germán Somolinos. His identification with the Spain defeated in the civil war and his opposition to Franco's dictatorship prevented Germán Somolinos and other colleagues of his who formed part of the "Pilgrim Spain" contingent and who practiced the history of science in Latin American countries, particularly in Mexico, from attending the IX Congress of the Académie Internationale d'Histoire des Sciences in Spain in 1959. So López Piñero did not explain well in 1991 that the Latin American absence at that 1959 congress was not so much due to the weakness of the history of science in Latin America at that time as to the fact that the main cultivators of that discipline on the American continent were at that time exiled Spanish republicans who opposed Franco's dictatorship.

The start in 1959 of the publication of the complete works of Francisco Hernández, protagonist of the first scientific expedition to the American continent in which he travelled through a large part of the viceroyalty of New Spain during 7 years in the 1570s,[28] was the result of more than a decade of research by Germán Somolinos and the creation by the National Autonomous University of Mexico, on

[26] *Actes du Ixe Congrès International d'Histoire des Sciences, Barcelona-Madrid, 1–7 septembre 1959*, Barcelona, Universidad de Barcelona. París, Hermann, 1960.

[27] This is revealed, among other works, by those he published in collaboration with José Pardo Tomás in (1994) and (1996).

[28] The historiography on this expedition is abundant. See Varey et al. (2000) and Varey (2000). A recent overview in Pimentel (2020), 65–95. *Fantasmas de la ciencia española*, Madrid, Marcial Pons Historia, 2020, pp. 65–95.

10 October 1956, of an editorial commission for the Complete Works of Francisco Hernández. This editorial commission was formed by a significant number of physicians together with biologists, such as botanists and zoologists, dentists, geographers, historians, anthropologists, translators and philologists. At this founding meeting, the secretary general of the UNAM, the notable physician Efrén del Pozo, who valued the prestige and permanence of native medicinal botany, called together a large group of Mexican researchers such as Enrique Beltrán, Angel Mª Garibay, Wigberto Jiménez Moreno, Samuel Fastlicht—who was of Austrian origin -, Roberto Llamas, Roberto Weitlaner, José Rojo Navarro, Alejandro Stolsols, Henrique González Casanova and Miguel León Portilla together with the Spanish Republican exiles, the historian José Miranda, his brother the botanist Faustino Miranda, the biologist Enrique Rioja, the Latinist and palaeographer Agustín Millares Carlo, the anthropologist Juan Comas and the doctor and medical historian Germán Somolinos d'Ardois.[29]

Germán Somolinos had arrived in Mexico at the end of June 1939.[30] He earned his living as a doctor thanks to a clinical analysis laboratory that he set up in his home and tried to resume the cardiological research he had begun in Madrid before the outbreak of the civil war and to initiate a work plan on the HR factor in the Mexican population. [31] But, with a passion for the history of medicine, a hobby that probably arose from his participation in the organization of one of the bibliographical exhibitions held in Madrid in September 1935 at the 10th International Congress on the History of Medicine,[32] he found time after settling in Mexico to research a subject that obsessed and preoccupied him for years: the reconstruction of the life and work of Francisco Hernández, whom he tended to see as an alter ego.[33] Thus his son Juan Somolinos Palencia, when editing volume VI of the Complete Works of Francisco Hernández, recovered a letter written by his father, back on 15 October 1946, to one of his teachers, the cardiologist Luis Calandre, who had remained in Madrid as part of the internal exile. In it, Germán Somolinos told him: "For a long time I have been accumulating data on a very interesting figure in medicine; I am referring to Dr. Francisco Hernández, Felipe II's protomedic who came to these

[29] See the news provided by Somolinos Palencia (1984) and by Germán Somolinos (1965–1966), 287, note 27.

[30] A presentation of his life and scientific career in "Germán Somolinos d'Ardois: memoria de un médico exiliado". Website made by Leoncio López-Ocón, Irati Herrera y Teresa López. Available at http://biblioteca.cchs.csic.es/GermanSomolinos/index.html.

[31] See for example his work (1947) and different reviews of works such as the following: J.M. Hill y W. Dameshek (1948). *El factor Rh en la clínica y el laboratorio,* New York and E.L. Potter (1947), *Rh, su relación con la enfermedad hemolítica congénita y las reacciones transfusionales intragrupo,* Chicago published in *Ciencia,* 9 (4–6), 171 and that of P. Cazal and J. Eliot (1951). Los grupos sanguíneos del sistema Rh (Les Groupes sanguins du système Rh), Ed. de *Expansion scient franç.*, 168 pp., illustr. París, *Ciencia,* 12 (7–8), December 15th 1952, 213.

[32] A review of this congress in López-Ocón (2023), 121–148.

[33] López López, T.C., "Tras la huella de Germán Somolinos: hacia el retrato de Francisco Hernández", thesis end-of-degree presented at Universidad Carlos III, June 2023, under the supervisión of Gonzalo Velasco and Leoncio López-Ocón.

lands sent by the King to study their natural history and obtain what was beneficial for medicine. Very little is known about him and I would like to be able to write his biography, I already have his works and quite a lot of data but it is necessary to look for even more; what I regret most in this case is not being able to consult the Spanish archives, above all the Simancas archive where there must undoubtedly be some interesting data".[34]

This effort culminated in 1960 when UNAM published his Vida y obra de Francisco Hernández, the first volume of the eight that would constitute the corpus of the Obras completas de Francisco Hernández. This first volume of 500 pages in folio—the production of which had taken Germán Somolinos 14 years,[35] coinciding with a period of remarkable development of the production on the history of Mexican science[36]—had been temporally preceded by the publication in 1959—the year of the IX Congress of the Académie Internationale d'Histoire des Sciences in Barcelona and Madrid—of the two volumes of the Historia Natural de Nueva España, a Spanish translation of the Latin edition that the botanist Casimiro Gómez Ortega had ordered to be made of this work by Francisco Hernández at the end of the eighteenth century.

But there were other oversights at the 1991 congress. While the congress participants from the English-speaking world did acknowledge the tradition of historical-scientific studies from which they came, those from the Spanish-speaking world—be they Spanish or Latin American—did not see fit to give consideration to the efforts that their respective historiographies had made throughout the twentieth century to address questions related to the scientific-technical connections between the Old and the New World. And yet, Spanish Republican exiles or Spanish scientists working in Latin America in the middle decades of the twentieth century did pay attention to these problems.

In Argentina, for example, the mathematician Julio Rey Pastor published in 1942 his study on science and technique in the discovery of America, which would go through successive editions[37] and stimulated historical-scientific research among his disciples, whether Spanish[38]or Argentinean, such as José Babini,[39] who would collaborate with Aldo Mieli and Desidero Papp in the elaboration of the first general work on the history of science published in Spanish: Panorama general de historia de la ciencia. Babini would become a member of the International Academy for the History of Science.[40]

[34] Somolinos Palencia. (1985), 11.

[35] Somolinos (1962), quoted by Teresa López (2023), op. cit., 42.

[36] Somolinos (1965–1966).

[37] Rey Pastor (1942).

[38] Rey Pastor and García Camarero (1959).

[39] Rey Pastor and Babin (1951). On the relationship between José Babini and Julio Rey Pastor and an assessment of José Babini's contributions as a historian of science see Ortiz and Lewis (1985).

[40] Cueto and Silva (2020).

In Mexico, where the bulk of the Spanish Republican scientific diaspora disembarked, which as soon as they settled in that country created the journal *Ciencia*, with financial support from the publishing house Atlante,[41] several Spanish scientists became interested in the history of science in Mexico and in the American continent. Those who consult the 294 issues of its 29 volumes published between 1940 and 1975[42] will be able to see the presence of contributions related to the history of science. Germán Somolinos was present in its pages, writing reviews of numerous works in which it could be seen how this author carried out an intense dialogue with the North American, Latin American and Spanish historiographic production in the field of the history of medicine or natural sciences,[43] or making known, from 1949 onwards, advances in his research on Francisco Hernández.[44] His active work in that publication in the early 1950s led to his appointment as a member of the editorial board of that publication in 1956,[45] although from then on, his presence in that publication declined.

Also relevant in the pages of the journal *Ciencia* was the presence of another Republican scientist exiled in Mexico, the chemist, physic-chemical scientist, and historian of technology Modesto Bargalló,[46] as he wrote in an important book published in 1955.[47] Before the publication of this work, at the beginning of the 1950s,

---

[41] I showed these connections in López-Ocón (2014).

[42] They can be accessed at http://www.edaddeplata.org/tierrafirme_jae/revistaciencia/index.html. This publication was analyzed by Carapeto et al. (2002); Argueta Prado (2010).

[43] See, for example, among others, his reviews of books such as those by "Mettler, C.C.: History of Medicine, Filadelfia, 1947", *Ciencia,* 9, (7–10), 1948, 267–268; "Lastres, J.B.: Historia de la medicina peruana, 3 vols., Lima, Universidad Mayor de San Marcos, 1951", *Ciencia,* 12 (5–6), 1952, 162; Fernández del Castillo, F., "La Facultad de Medicina según el Archivo de la Real y Pontificia Universidad de México", México,., Imprenta Universitaria. Ediciones del IV Centenario de la Universidad de México, 1953 and Bantug, J.P., "Bosquejo histórico de la Medicina Hispano-Filipina, Madrid, Ed. Cultura Hispánica, 1952", *Ciencia,* 13, (4–6) n° 4–6, 1953, 119; "Guerra, F.: Historiografía de la Medicina Colonial Hispanoamericana, México. 1953, Ed. del Patronato para el estudio de la cultura colonial hispanoamericana", *Ciencia,* 13, 11–12, 1954, 295. Or Works published in journals such as the articles of Enrique Álvarez López, "Noticias y papeles de la Expedición científica mejicana dirigida por Sessé", in *Anales Jardín Botánico de Madrid,* X, vol. II, pp. 1–75, Madrid, 1952 in *Ciencia,* 12 (5–6), 169; Fernández del Castillo, F. "Historia de las Revistas médicas de México" in *Gaceta Médica de México,* LXXXIII (3), 229–244, México, 1953; "De lo que México debe al cirujano Francisco Javier Balmis", *Cirugía y Cirujanos,* (1953) XXI (9): 459–484; Alcántara Herrera, J. "Contribución a la Historia del Periodismo Médico en México", *Medicina,* XXXIV (691); 1–16 and (692): 39–46, México, 1954 in *Ciencia,* (1954) 13 (11–12), 300–301.

[44] Among his Publications related to Francisco Hernández that have appeared in Ciencia are the following: "Manuscrito firmado. Original del doctor Francisco Hernández aparecido en México", (1949), IX, (7–10), 209–210; "La partida de defunción del doctor Francisco Hernández", (1951), XI, (1–2), México, 50–52; "Nuevos manuscritos de Francisco Hernández aparecidos en Madrid", (1954), XIV, (4–6), 109–110.

[45] See News Section in *Ciencia* (1956), 15, (9–10), 210.

[46] On his scientific and educational work before and after his exile see Moreno Martínez (2021a, b).

[47] Bargalló (1955). It received an extensive review from José Joaquín Izquierdo in *Ciencia* (1956), 16 (1–3), 45–47.

he published important works on the history of mining in Mexico in the aforementioned publication.[48]

2.5 Conclusions

The dialogue between Spanish, Latin American and North American historians of science throughout the second half of the twentieth century reached its climax at the international congress "Science, Discovery and the Colonial World" held in Madrid between 25 and 28 June 1991. The interactions that took place at that conference between almost a hundred congress participants from various European and American countries was the result of the construction of a series of connections previously established within the Latin American Society for the History of Science and Technology, directed since its constitution in 1982 in Mexico by Juan José Saldaña.

When this society was founded, Eduardo Ortiz and Lewis Pyenson recalled that the first unanimous resolution adopted by its members was to send a telegram to José Babini, saluting "the great master of Argentina". In these pages I have been interested in highlighting that, alongside such relevant contributions as those of Babini to the historiographical field of the history of science and technology in Latin America, there are a number of Spaniards who carried out their historiographical work on the American continent in the middle decades of the twentieth century, either because they found academic accommodation there, as happened to José Babini's teacher Julio Rey Pastor, or because they were defeated in the Spanish Civil War, as happened to the Republicans Modesto Bargalló and Germán Somolinos. The latter, by the way, was one of the driving forces behind the creation in 1964 of the Mexican Society for the History of Science and Technology, which years later—after the death of Germán Somolinos, who died in 1973—would contribute to the founding of the Latin American Society for the History of Science and Technology.

In the prologue to the book Mundialización de la ciencia y cultura nacional José María López Piñero suggested that the historiographical impact of the congress "Ciencia, descubrimiento y mundo colonial" in June 1991 could be as important as the one he estimated that the IX Congress of the Académie Internationale d'Histoire des Sciences held in Barcelona and Madrid in September 1959 had on the Spanish historiographical sphere. Confirming or refuting this approach may be a question to be addressed in the future. On this occasion, it seemed appropriate to focus on highlighting the immediate and more distant antecedents of the 1991 Madrid congress, emphasizing that the dialogue between the historiographies of North American,

[48] These include the following: "Homenaje a Don Andrés Manuel del Río Fernández, en ocasión del primer centenario de su muerte (1849–1949)", *Ciencia*, X, (1950): 270–278; "Método de beneficio de los minerales de plata y oro por fundición, del mexicano D. Joseph Garcés y Eguía: cientocincuenta aniversario de su libro "Nueva téorica y práctica del beneficio de los metales de oro y plata", *Ciencia,* XII, (1952): 199–206.

European, and Latin American science was practiced in the pages of the journal *Ciencia*, the voice of the Spanish Republican scientific diaspora by prominent historians of science such as Germán Somolinos d'Ardois, Enrique Rioja and Modesto Bargalló.

References

Aguilera Ríos, S. (2000). *Quipu*: una revista latinoamericana de la historia de las ciencias y la tecnología. In *Biblio 3W. Revista Bibliográfica de Geografía y Ciencias Sociales* (p. 212). Universidad de Barcelona. On line. https://www.ub.edu/geocrit/b3w-212.htm

Argueta Prado, J. Q. (2010), *La revista Ciencia (1940–1975). Contribuciones a la ciencia mexicana del siglo XX.* Morelia, Universidad Michoacana de San Nicolás Hidalgo, Academia Mexicana de Ciencias.

Bargalló, M. (1955). *La Minería y la Metalurgia en la América Española durante la Época Colonial, con un Apéndice sobre la industria del hierro en México desde la iniciación de la Independencia hasta el presente.* D.F Fondo de Cultura Económica, edition sponsored by Compañía de Fierro y Acero de Monterrey.

Beltrán, E. (1984). La historia de la ciencia en América latina. Primera reunión latinoamericana de historiadores de la ciencia. Puebla, México, agosto 23-26, 1982. *Quipu, I*(1), 7–23.

Bigg, C., Reinisch, J., Somsen, G., & Widmalm, S. (2023). The art of gathering: Histories of international scientific conferences. *The British Journal for the History of Science, 56*, 423–433.

Canal, J. (2021). *25 de julio de 1992: la vuelta al mundo de España.* Taurus.

Capel, H., & Jori, G. (Ed.). (2010). Homenaje al profesor José María López Piñero (1933–2010). *Scripta Nova. Revista Electrónica de Geografía y Ciencias Sociales, 14* (343). [On line]. http://www.ub.es/geocrit/sn/ sn-343/sn-343.htm

Carapeto, C., Pulgarín, A., & Cobos, J. M. (2002). Ciencia. Revista hispano-americana de ciencias puras y aplicadas (1940-1975). *Llull, 25*(53), 329–368.

Cueto, M., & da Silva, M. A. D. (2020). Trayectorias y desafíos en la historiografía de la ciencia y de la medicina en América Latina. *Asclepio, 72*(2), 3–20. https://doi.org/10.3989/Asclepio.2020.21

Español, L. (2004). Mariano Hormigón Blánquez. In Memoriam. *Lllull, 27*, 502–505.

Geyer, M. H., & Paulmann, J. (Eds.). (2001). *The mechanichs of internationalism.* German Historical Institute.

Lafuente, A. (1986). La ciencia periférica y su especialidad historiográfica. In *El perfil de la ciencia en América,* Cuadernos de Quipu, n° 1. (Actas del simposio Historia y Filosofía de la Ciencia en América. Guadalajara, México 12 a 14 de noviembre de 1985). Sociedad Latinoamericana de Historia de las Ciencias y la Tecnología, pp. 31–40.

Lafuente, A., & Estrella, E. (1987). Scientific enterprise, academic adventure and drawing-room culture in the geodesic mission to Quito. In J. J. Saldaña (Ed.), *Cross Cultural Diffusion of Science: Latin America*, Cuadernos de Quipu, n° 2 (Symposium "Cross Cultural Transmission of Natural Knowledge and its Social Implications: Latin America". Vol V. Acts of the XVII International Congress of History of Science, Berkeley, Calif. 31 July-8 August 1985). Sociedad Latinoamericana de Historia de las Ciencias y la Tecnología, pp 13–31.

Lafuente, A., & Saldaña, J. J. (Eds.). (1987). *Historia de las ciencias.* Consejo Superior de Investigaciones Científicas.

Lafuente, A., & Saraiva, T. (2001). The savant's drama: Two cultures and two shores. In J. J. Saldaña (Ed.), *Science and cultural diversity. Filling a gap in the history of science* (pp. 97–121). Sociedad Latinoamericana de Historia de las Ciencias y la Tecnología.

Lafuente, A., Elena, A., & Ortega, M.L. (Eds.). (1993). *Mundialización de la ciencia y cultura nacional,* (Actas del Congreso Internacional "Ciencia, descubrimiento y mundo colonial"). Doce Calles.

Legg, S., Heffernan, M., Hodder, J., & Thorpe, B. (2022). *Placing internationalism: International conferences and the making of the modern world.* Bloomsbury.

López Piñero, J. M., & Pardo Tomás, J. (1994). *Nuevos materiales y noticias sobre la Historia de las plantas de Nueva España, de Francisco Hernández.* Universidad de Valencia-CSIC.

López Piñero, J. M., & Pardo Tomás, J. (1996). *La influencia de Francisco Hernández (1515–1587) en la constitución de la botánica y la materia médica modernas.* Universidad de Valencia-CSIC.

López-Ocón, L. (2014). Atlante en el exilio: actores y etapas de una editorial republicana hispano-americana. In A. Sánchez Cuervo & G. Zermeño Padilla (Eds.), *El exilio español del 39 en México. Mediaciones entre mundos, disciplinas y saberes* (pp. 63–100). El Colegio de México.

López-Ocón, L. (2015). Horacio Capel y el Instituto de Historia del Consejo Superior de Investigaciones Científicas: recuerdo de una larga relación. In L. Urteaga & V. Casals (Eds.), *Horacio Capel, geógrafo* (Barcelona ed., pp. 595–602). Publicacions i Edicions de la Universitat de Barcelona.

López-Ocón, L. (2023). *El cénit de la ciencia republicana. Los científicos en el espacio público (curso 1935–1936).* Silex.

Moreno Martínez, L. (2021a). Modesto Bargalló en España (1894-1939). Una biografía entre la historia de la educación y la historia de la ciencia. *Historia y Memoria de la Educación, 13,* 635–674.

Moreno Martínez, L. (2021b). Modesto Bargalló en México (1939-1981): exilio de un migrante entre la didáctica y la historia de las ciencias, *Saberes. Revista de historia de las ciencias y las humanidades, 4*(10), 144–163.

Ortiz, E. L., & Lewis, P. (1985). Eloge: José Babini, 11 May 1897-18 May 1984. *Isis, 76*(4), 567–569.

Peset, J. L. (Ed.). (1985). *La ciencia moderna y el Nuevo Mundo.* Consejo Superior de Investigaciones Científicas-Sociedad Latinoamericana de Historia de las Ciencias y de la Tecnología.

Peset, J. L. (Ed.). (1989). *Ciencia, vida y espacio en Iberoamérica.* CSIC, (colección "Estudios sobre la ciencia"). 3 vols.

Pimentel, J. (2020). *Fantasmas de la ciencia española.* Marcial Pons Historia.

Polanco, X. (Ed.). (1990). *Naissance et développement de la science monde. Production et reproduction des communautés scientifiques en Europe et en Amérique latine.* Éditions La Découverte.

Reinisch, J. (2016). Agents of internationalism. *Contemporary European History, 25*(2), 195–205.

Rey Pastor, J. (1942). *La ciencia y la técnica en el descubrimiento de América.* Espasa Calpe. (Fourth edition in 1970).

Rey Pastor, J., & Babini, J. (1951). *Historia de la matemática. De la antigüedad a la actualidad.* Espasa-Calpe Argentina.

Rey Pastor, J., & García Camarero, E. (1959). *La cartografía mallorquina.* CSIC (Departamento de Historia y Filosofía de la Ciencia "Instituto Luis Vives").

Sala Catalá, J. (1986a). La ciencia iberoamericana, entre su historia y su filosofía. In *El perfil de la ciencia en América,* Cuadernos de Quipu, n° 1 (pp. 19–29). Sociedad Latinoamericana de Historia de las Ciencias y la Tecnología.

Sala Catalá, J. (1986b). La localización de la capital de Nueva España, como problema científico y tecnológico. *Quipu, 3*(3), 279–297.

Sala Catalá, J. (1988). Ciencia y técnica en la metropolización de Lima. *Quipu, 5*(3), 389–412.

Sala Catalá, J. (1994). *Ciencia y Técnica en la Metropolización de América.* CSIC-Ediciones Doce Calles.

Sanz López, J. (2023). *1992. El año de España en el mundo.* Silex ediciones.

Somolinos d'Ardois, G. (1947). Estudio de la frecuencia del factor Rh entre los habitantes de México. *Revista Ciencia, 8(1), 13–19.*

Somolinos d'Ardois, G. (1960). *Vida y obra de Francisco Hernández* (Obras completas de Francisco Hernández) (Vol. I). UNAM.

Somolinos d'Ardois, G. (1962). Francisco Hernández, protomédico de las Indias. *Gaceta Médica de México, 92*(7), 609–616.

Somolinos d'Ardois, G. (1965–1966). Historia de la ciencia. *Historia Mexicana, 15*(2–3), 269–290.

Somolinos Palencia, J. (1985). Introducción general. In *Escritos varios. VI Obras Completas de Francisco Hernández*. UNAM.

Taton, R. (Ed.). (1963–1966). *A general history of sciences* (4 vols). Thames & Hudson.

Varey, S. (Ed.). (2000). *The Mexican treasury. The writings of Dr. Francisco Hernandez*. Stanford University Press.

Varey, S., Chabrán, R., & Weiner, D. B. (2000). *Searching for the secrets of nature. The life and works of Dr. Francisco Hernandez*. Stanford University Press.

Leoncio López-Ocón Cabrera Scientific researcher working in Madrid in the Department of History of Science at the Institute of History of the Higher Council for Scientific Research. He holds a PhD in Geography and History from the Complutense University of Madrid and a Master's degree in Social Sciences with a specialization in Andean History from the Quito branch of FLACSO. He is currently president of the Spanish Society for the History of Science and Technology (SEHCYT). His books include Breve historia de la ciencia española (2003) and El cénit de la ciencia republicana. Los científicos en el espacio público (curso 1935–1936) (2023).

Part II
The Latin American Society for the History of Science and Technology (LASHST): Foundation, Challenges and Perspectives

Chapter 3
Presence of Juan José Saldaña in the Latin American History of Science as Seen from Colombia (1981–2018)

Luis Carlos Arboleda

Abstract This paper discusses several episodes in the development of the Social History of Science and Technology in Latin America as seen from Colombia in which the participation of Juan José Saldaña was decisive as president of the Latin American Society for the History of Science, and later as director of the Internet Community for the History of Science in Latin America. Saldaña's presence is shown to have had an important impact on the institutionalization and professionalization of this field of studies in this country, specifically about problems of interest in the history of science in Latin America. The presentation of these episodes is testimonial and is based on experiences of a close and rich friendship and scientific collaboration between the author and Saldaña, which unfolded for more than 40 years in numerous scientific activities and projects in my country, in the region, and internationally.

Keywords History of Latin American history of science · Latin American society for the history of science and technology · Social history of science in Colombia · Construction of scientific communities in Latin America

3.1 Introduction

In the late 1970s, Juan José Saldaña and I advanced our doctoral training in Paris. Saldaña, a philosophy graduate from the Autonomous University of Nuevo Leon, Mexico, was working at the Institute of History and Philosophy of Science (IHPST for its French acronym) in rue du Four, directed by Suzanne Bachelard, daughter of the philosopher Gaston Bachelard, with a strong Husserlian orientation towards research in the philosophy of physics, particularly on the relations between reason and experience. Suzanne Bachelard had been preceded in the direction of the IHPST

L. C. Arboleda (✉)
Universidad del Valle, Cali, Colombia
e-mail: luis.carlos.arboleda@gmail.com

M. de la P. Ramos-Lara, L. C. Arboleda (eds.), *History of Science in Latin America: The Construction of an Intellectual Field (20th century)*,
https://doi.org/10.1007/978-3-031-75536-1_3

by Abel Rey, her father, and Georges Canguilhem. For my part, I had been trained as a graduate in mathematics and physics at the Universities of Valle and Santiago de Cali, Colombia, and I was preparing to obtain my doctorate at the Alexander Koyré Center on rue Colbert, founded in 1958 by Koyré as the Center for Research in the History of Science and Techniques. Upon Koyré's death in 1964, Taton took over the direction of the focal point together with his colleague and friend Father Pierre Costabel.

Taton was already well known for his work on various topics in the history of mathematics, including one that was very useful in my doctoral work, the genre of scientific correspondences as a resource for studying the life and work of scientists. His four-volume *Histoire Générale des Sciences* was by then an international reference work (Taton, 1957–1964). Taton also ensured, together with Suzanne Delorme, the direction of one of the most important periodicals in this field, the *Revue d'histoire des sciences*. He also stood out in France and abroad for his teaching activity in the discipline, in the guidance of theses and research projects at the rue Colbert seminar, and for his impulse to the institutional development of the international community of the History of Sciences. Saldaña's thesis, oriented by Bachelard at the University of Paris I (Panthéon-Sorbonne), focused on studying the notion of scientific revolution applied to the case of the introduction of wave theories in optics at the beginning of the nineteenth century. At the same time, I defended my thesis under Taton's tutelage at the *Ecole des Hautes Etudes en Sciences Sociales* (EHSS for its French acronym), on the introduction of general topology and functional analysis at the beginning of the twentieth century based on the study of the scientific correspondences of the French mathematician Maurice Fréchet, a line of research that I have maintained throughout my career.

On his return to Mexico, Saldaña endeavored to disseminate his research on the introduction of wave optics in Fresnel's work, in particular his work (Saldaña, 1983), which he gave in numerous lectures at different universities in Mexico, one of them at the National Congress of Physics in Jalapa in November 1982. This talk shows his concern to make known in his country his approach to the history of physics, which combined the analysis of epistemological and philosophical aspects with a history of the practice of knowledge production situated in specific social and political contexts. As stated in the abstract, he proposed to analyze "the political and scientific moment in which the wave explanation of the phenomenon of diffraction emerged and the opposition that the wave theory encountered among the supporters of the *théorie de l'emission*. An attempt is made to show the methodological relevance of the social history of the sciences in the historical explanation of these" (Saldaña, 1983).

The experience we had as students in France was a recurring theme in personal conversations and discussions with colleagues in the early 1980s. The opportunity to have been trained in two centers of excellence that were equally poles in the institutionalization of the field of history of science in France and internationally, and to have done so at a time of great effervescence of ideas and historical-epistemological approaches, allowed us to acquire a professional background as researchers and to familiarize ourselves early on with the criteria and procedures that distinguished the professional practice of the historian of science. These included the rigorous study of archives and documentary sources, seminar-based training, scientific exchanges with researchers from different backgrounds, participation in congresses, publication in specialized books and journals, and the creation of groups and societies. But this European academic tradition, so important in our professional activity as historians, would soon have a critical counterpart in the formulation, mainly by Saldaña, of an alternative program for the development of the history of science in Latin America.

As mentioned above, during the first years of his academic activity in Mexico and the region, Saldaña disseminated in conferences and publications the orientation of his thesis on history, epistemology as well as the philosophy of science and physics. At the same time, the inclination for new objects of study related to the history of Latin American science began to emerge.[1] He explains, as editor of the journal *Quipu*, how this change of orientation came about, based on the conviction that "Latin American science should be studied on its own, in its own right, and not only as an epiphenomenon of science carried out in other latitudes (mainly Europe)" (Saldaña, 2004a). He also recalls that: "This idea gradually made its way and programs such as the history of colonial science began to yield space to the history of Latin American and more generally regional science. And *Quipu* was, naturally, a pioneer in the expression of this change and one of its best-known voices" (Saldaña, 2004a).

Earlier I recalled some episodes from our academic environments as students in Paris. However, Saldaña and I did not meet personally at that time, but shortly thereafter, as participants in our first international congress as explained below. In the Paris years, we only had some slight news of each other through other Latin American colleagues who were free attendees of our respective seminars, among them the Argentinian physicist Carlos Galles and the Colombian physicist-mathematician Regino Martínez-Chavanz. This personal situation of disconnection between the young people who at the same time were being trained in our discipline in Europe and the United States also reflected the state of isolation in which the most significant undertakings in the history of science in the region found themselves and their disconnection about the theoretical and methodological developments of the discipline in the international academic focal centers (Fig. 3.1).

The situation began to change in the early 1980s, especially with the return to their countries of origin of the first foreign-trained PhDs in the History of Science, among them the two of us, thus foreshadowing a new stage in the professionalization of our field of studies in the region. This led to a close friendship and a rich

[1] A list of the lectures and courses he gave in Mexico upon his return from France can be found in his 2018 Curriculum vitae. It can be seen that little by little the epistemological, historical, and sociopolitical themes of physics and the sciences gave way to the social history of science in Mexico and Latin America. To get an idea of this transition, see a list of his lectures in 1981 and 1982 when he was a professor at the Universidad Autónoma Metropolitana-Iztapalapa, before joining the Universidad Autónoma de México (UNAM for its acronym in Spanish): https://www.uam.mx/sah/pre-pa/tema04/Rosado%2082/64filos.html

Fig. 3.1 Poster of the author's participation in the posthumous tribute to Saldaña in Mexico in 2022. (The photos belong to the archive of Juan José Saldaña and have been generously donated to the editors of the book by his family)

scientific collaboration that unfolded in numerous activities and projects in the area and internationally. This paper will only comment on the episodes of our cooperation that, in my opinion, were most significant for research in the Social History of Science and Technology in Colombia, its impact on problems of national interest such as Science and Cultural Diversity, Science and Nation, the contribution from the history of science to the improvement of science education and the use of the history of science in the design and implementation of public policies in Science, Education, and Development. First, it is necessary to place the programmatic context in which this collaboration of more than 40 years is established.

3.2 Our Encounter at the Bucharest Congress (1981) and the Emergence of an Innovative Academic Project, the SLHCT

In September 1981, seven historians of science from Latin America participated in the XVI International Congress on the History of Science held in Bucharest under the sponsorship of UNESCO and the National Academy of Sciences of the former Socialist Republic of Romania. In a first contact, Saldaña and I proposed to locate these participants and invite them to a meeting. That is how we met personally, Saldaña and I, with colleagues Elena Pennini de Vega (Argentina), Ático Vilas Boas de Mota (Brazil), José López Sánchez and Pedro Marino Pruna (Cuba), and Inés Harding (Chile).

By the end of the meeting, which lasted several hours, we agreed to sign the *Bucharest Declaration*[2] that set the course for the formation of an academic community, the Latin American Society for the History of Science and Technology (SLHCT for its Spanish acronym), which soon after would have the journal *Quipu* as its mobilization instrument. Before adding a brief commentary to those presented on the same document in several chapters of this book, it may be useful to provide some information on the profile of its signatories at that time.

Elena Pennini de Vega Elena (1908–1983) was a doctor in Biochemistry and Pharmacy with an outstanding career in teaching and research within the Chair of History of Medicine at the School of Medicine of Buenos Aires, directed by Argentino Jorge Landaburu, which also included Dr. Alfredo G. Kohn Loncarica who would maintain an active collaboration with the SLHCT. In 1974 Dr. Pennini chaired the Ateneo de la Historia de la Medicina and the VI Jornadas de Historia de la Medicina (1978). She was a member of numerous scientific associations. She founded the Argentine Association of Humanism of the Argentine Medical Association. He directed the magazine *Manual Farmaceutico of Buenos Aires*. He passed away in 1983. The following year his influential article *Darwin en Argentina* was published in the first issue of *Quipu* (Pennini, 1984).

Ático Vilas Boas da Mota (1928–2016) was, during the years of the Bucharest Congress, an outstanding Brazilian professor, poet, and linguist with a doctorate in Romance philology, and an expert in the study of Romance languages and literature. He was then recognized for his research on historical, linguistic, ethnographic, and folklore aspects characteristic of the culture of the state of Goiás, as well as for his studies on the relationship between territory and cultural diversity in popular medicine and the history of the gypsy roms community.

For his part, the medical epidemiologist and prominent political figure of the Cuban revolution, Professor José López Sánchez (1911–2004), was recognized at the congress as the dean of the historians of medicine in Cuba. Among his long list of publications on the history of science and medicine in the late 1970s, his 1967 book on *Tomás Romay and the origin of science in Cuba*, in which he gathered his long years of research on the life and work of the Havana health physician, precursor of the introduction of the vaccine and the scientific description of yellow fever

[2] Not to be confused with the declaration agreed upon at a symposium held simultaneously at the end of the same Bucharest Congress, but with a different character from the declaration we have been referring to (Saldaña, 1981). It is the appeal to world public opinion to stop the arms race, which men of science and personalities from 33 participating countries, including five Latin Americans, decided to launch at the symposium *Scientists and Peace*. Of those five Latin Americans, three had signed the *Bucharest Declaration* on the Latin American history of science: Juan José Saldaña, Ático Vilas Boas da Mota, Pedro Marino Pruna, and Luis Carlos Arboleda. As members of the International Initiative Committee, Saldaña and I took several actions in this direction upon our return to our countries. Saldaña published the translation of the text of the declaration in Spanish with an introduction calling on Mexican scientists to join this committee (Saldaña, 1982). For my part, I gave lectures on armament and the anti-war movement of the time based on the manuscript (Arboleda, 1982). For a review of this Bucharest symposium see: https://biblioteca-digitala.ro/reviste/dacia/26_dacia_revue-archeologie-historie-ancienne_SN_XXVI_1982_007.pdf

in the late eighteenth century, stood out. He also began his research on Cuban physician Carlos J. Finlay, famous for his work on the transmission of yellow fever.

The second of the Cuban signatories was the biologist Pedro Marino Pruna Goodgall (1943–2017), a researcher at the Institute of Zoology of Cuba who had also obtained his doctorate a year before the congress at the Institute of History of Natural Sciences and Technology of the Academy of Sciences of the USSR, about the formation of the hierarchical classification of plants and animals in the eighteenth and early nineteenth centuries. Pruna published an article in the first volume, number 3, of *Quipu* (1984) on the reception of Darwin's ideas in Cuba during the nineteenth century (Pruna, 1984).

Finally, another signatory was Inés Harding (deceased in 1996). Harding was a distinguished professor of mathematics and computer science at the Universidad Técnica del Estado (today Universidad de Santiago de Chile) who by the time of the congress had been removed from her position by the dictatorship. Before her dismissal, she had the project of writing the history of mathematics in Chile from the conquest until 1980. She was recognized by the Chilean mathematical community, in addition to her innovative performance in the reform of mathematics teaching, for being the first woman in the country to have completed doctoral studies at the Lomonosov University in Moscow, 10 years before the first doctorate in mathematics was formally awarded to a woman in Chile. At the Bucharest Congress, she participated in the symposium *Women in Science* in which the Commission on Women of the International Union in History and Philosophy of Science (Division of History of Science and Technology) was formed. Harding is listed in the records of the Commission as a founding member representing Chile and West Germany.

In the beginning, *the Bucharest Declaration* recognized a certain boom in the work of History of Science in Latin America but regretted its isolation in the region and to the international community. It also noted that this work was being carried out in different countries in precarious conditions with an incipient degree of institutionalization and professionalization. It concluded in the following terms (Saldaña et al., 1981):

> Faced with this state of our discipline, the undersigned participants in the XVI International Congress on the History of Science commit ourselves to undertake all our efforts to overcome the obstacles that hinder its development. The main agreement, to be advanced in the coming months, is the creation of an Informative Bulletin on the different activities being carried out in Latin America in the history of science, as well as the preparation of a Directory of Latin American Historians of Science (...) We have also welcomed as an objective of great interest, the proposal of the Mexican delegate to seek to hold next year in his country (or in another country that expresses its agreement and offers the necessary conditions), the First Meeting of Latin American Historians of Science (...) We have also accepted the proposal of the Mexican delegate to hold next year in his country (or in another country that expresses its agreement and offers the necessary conditions), the First Meeting of Latin American Historians of Science.[3]

[3]Frente a este estado de nuestra disciplina, los suscritos participantes en el XVI Congreso Internacional de Historia de las Ciencias nos comprometemos a emprender todos nuestros esfuerzos para superar los obstáculos que frenan su desarrollo. El acuerdo prioritario, a adelantar en los

3.3 The Program for the Construction of the SLHCT at the Puebla Meeting (1982)

This meeting was held on August 23–26, 1982 at the Autonomous University of Puebla, Mexico. It had the collaboration of historians from several countries committed to the *Bucharest Declaration*, and the generous support of this University. Saldaña made there an eloquent demonstration of his leadership and convening capacity to attract the efforts of different actors in the history of science and make them converge towers the construction of an academic community. Delegates from 16 countries of the Americas and Europe participated and presented nearly 100 papers in 15 scientific sections structured around the following six programmatic axes: (1) History of the scientific and cultural heritage of Latin American countries, (2) Latin American contributions to the History of Science, (3) Methodology of the History of Science, (4) Science and Society in Latin America, (5) Promotion of Science History Teaching in Latin America and (6) Sources and Historiography for the History of Science in Latin America.

Finally, the SLHCT was established at the Puebla Meeting under the presidency of Saldaña. Its statutory regime was governed by the following principles: (1) To contribute to the promotion of the History of Science and Technology and its recognition by the States and International Organizations, (2) To support scientific research, academic programs as well as university studies on the development of science and technology and their technological appropriations, and to contribute to the training of highly competent historians through the best resources available within the region, (3) To favor an exhaustive investigation of our cultural and scientific heritage, and of the evolution of our schools of knowledge, to contribute to the design of coherent scientific and cultural policies that are more in line with our historical and social reality, (4) To promote links with research centers and scientific societies, primarily in Latin America, so that our projects may benefit from the most

próximos meses, es la creación de un Boletín Informativo sobre las distintas actividades que se vienen adelantando en Latinoamérica en historia de las ciencias, así como la preparación de un Directorio de Historiadores Latinoamericanos de las Ciencias. (...) Igualmente hemos acogido como un objetivo del mayor interés, la proposición del delegado mexicano de buscar realizar el año próximo en su país (o en otro país que manifieste su acuerdo y ofrezca las condiciones necesarias), el Primer Encuentro de Historiadores Latinoamericanos de las Ciencias. (The translation and slight changes in quotations are mine, LCA). In view of this state of our discipline, we, the undersigned participants in the XVI International Congress on the History of Science, commit ourselves to undertake all our efforts to overcome the obstacles that hinder its development. The priority agreement, to be advanced in the coming months, is the creation of an Informative Bulletin on the different activities being carried out in Latin America in the history of science, as well as the preparation of a Directory of Latin American Historians of Science. (...) We have also accepted as an objective of great interest the proposal of the Mexican delegate to hold next year in his country (or in another country that expresses its agreement and offers the necessary conditions), the First Meeting of Latin American Historians of the Sciences. (The translation and slight changes in quotations are mine, LCA).

Fig. 3.2 Fragment of the *Bucharest Declaration* and scene of the Puebla meeting in 1982

advanced developments and experiences in other fields of knowledge, and reciprocally (Fig. 3.2).

With these principles and programmatic axes in mind, the SLHCT advanced in the following years an impressive agenda of novel initiatives that exceeded by far the expectations of the *Bucharest Declaration*, particularly in terms of the purposes of professionalization and regional institutionalization of the discipline. It became evident with this agenda that it was possible to break the isolation and private character that had prevailed until then in the activities and projects, through the publication of works and the association with a network of professionals in the history of science, as well as the amateurs who did it more occasionally and informally. Regarding the purpose of institutionalization, noticeable changes were beginning to take place in the introduction in several universities of chairs, seminars as well as teaching courses in the history of science or related subjects, and also in the realization of research projects and the publication of their results. In his opening lecture at a meeting of the SLHCT held in Cali, Colombia, in November 1984, Saldaña made the following comment on his achievements in the first 2 years of its creation, highlighting among them the publication of three issues of the first volume of the journal *Quipu* (Saldaña, 1986):

> In five countries of the region, the Society has promoted or catalyzed the establishment of national societies for the history of science and technology. In others, here in Colombia for example, this is expected to happen soon, and in some more, the revitalization of old associations is being stimulated. The different symposia and meetings promoted by the Society, its Latin American and international character, have allowed the constitution of a community of historians, the knowledge of the work being done in other countries, diverse forms of collaboration and, I think it is important to point out, friendships and reciprocal knowledge among those who otherwise, as was usual, had to go to Europe to "discover Latin America". The journal *Quipu* plays a significant role in this respect. No less significant is the role it is performing as a vehicle for making known outside Latin America the importance and variety of research we are doing, as well as the value and nature of Latin American

scientific practice and its tradition. This, indeed, is beginning to be recognized internationally.[4]

Later on, we will have the opportunity to refer to Saldaña's participation in the Cali meeting mentioned above. We must first highlight his presence, in November 1983, together with other members of the SLHCT from several countries, in an event that would prove to be strategic for the constitution of the field of history of science studies in Colombia.

3.4 The Bogotá Seminar on Methodology for the Social History of Sciences in Latin America (1983)

This seminar was decisive in the implementation of the Social History of Science Project in Colombia, which, in turn, would become the cornerstone for the establishment of this new field of study in the country. Within the framework of this project, which had begun in March 1983 with the support of Colciencias (today the Colombian Ministry of Science) and the Organization of American States-OAS, many professional, research, academic training, and dissemination activities were carried out over the next 10 years, all of which contributed to the demarcation of the discipline as an autonomous and legitimate field of study in the country.

The seminar papers first appeared in 1983 in the Colciencias journal *Ciencia, Tecnología y Desarrollo* (volume 7, no. 3, July to September quarter). A decade later they were compiled in the first volume (*Fundamentos Teórico-Metodológicos*) of the ten-volume collection *Historia Social de la Ciencia en Colombia* (Quevedo et al., 1993).[5] Since its first publication, they have been the subject of discussions in university groups interested in the history of science and among government officials involved in public policies on education, science, and society. The presence at

[4] En cinco países de la región la Sociedad ha promovido o ha servido de catalizador para la constitución de sociedades nacionales de historia de las ciencias y la tecnología. En otros, aquí mismo en Colombia por ejemplo, esto mismo se espera que acontezca próximamente, y en algunos más la revitalización de antiguas asociaciones está siendo estimulada. Los diferentes simposios y reuniones que ha promovido la Sociedad, su carácter latinoamericano e internacional, han permitido la constitución de una comunidad de historiadores, el conocimiento de los trabajos que se realizan en otros países, formas de colaboración diversas y, me parece importante señalarlo, amistades y conocimiento recíproco entre quienes de otra forma, como era habitual, debían salir a Europa para "descubrir a América Latina". La revista *Quipu* está desempeñando un papel significativo a este respecto. No menos significativo es el papel que está desempeñando como vehículo para dar a conocer fuera de Latinoamérica la importancia y variedad de investigaciones que estamos realizando, así como el valor y la naturaleza de la práctica científica latinoamericana y de su tradición. Lo cual, en efecto, empieza a ser reconocido internacionalmente.

[5] Both publications contain contributions to the Methodological Seminar by Eduardo Aldana, Carlos Vasco, Diana Obregón, Joseph Hodara, Juan José Saldaña, Ubiratan D'Ambrosio, Ruy Gama, Hebe Vessuri, Néstor Miranda, Emilio Quevedo, Antonio Lafuente and José L. Peset, Ernesto Yépes and Luis Carlos Arboleda.

this seminar of the international guests, in particular the SLHCT delegation headed by Saldaña, marked a decisive turn in the institutionalization and professionalization of the history of science in Colombia. As acknowledged in the *Presentation* of the aforementioned first volume (Quevedo et al., 1993), the contributions of the SLHCT guests had an important role in the debates of the research team, and their guidelines on the "new social history of science" were taken into account in many of the studies of the project (Fig. 3.3).

The theme of Saldaña's paper at the Bogotá seminar (*Towards a Historical, Theoretical and Methodological Critique of the Latin American Historiography of the Sciences*) (Saldaña, 1993a) was based on several considerations on the need to delimit and underpin the development of the field of studies in the early 1980s by the epistemological, social and cultural specificities of Latin America. The basic assumption of his critique was that science and technology in the peripheral countries had played a different role in their historical development compared to what they played in the central countries. Consequently (Saldaña, 1993a),

> the categories for *thinking historically* about the evolution of science and technology in Latin America cannot continue to be those resulting from the import—consciously or not—of "methodologies" coming from situations and scientific traditions different from those that have prevailed in our region. However, the absence to date of a theoretical approach to these issues makes it necessary to review and critique the available methodologies, even when they do not consider our specificity, to think through them in our past of science and technology.[6]

Fig. 3.3 Aspects of the Methodological Seminar on the Social History of Science in Bogota in 1983

[6] las categorías para *pensar históricamente* en la evolución de la ciencia y la tecnología en América Latina no pueden seguir siendo las que resultan de la importación -consciente o no- de "metodologías" provenientes de situaciones y tradiciones científicas distintas de las que en nuestra

In the final part, Saldaña formulates some of the characteristics of the new history of science in Latin America, which emerge from a historical, theoretical as well as methodological critique of the historiographical past. He does so by negatively arguing the role that the historical and epistemological categories and notions in vogue in international debates had in the reflection on the nature of our practices. Saldaña suggests that the new history in Latin America will not be internalist, continuist, or externalist. It will not be a history of contributions to the history of the "universal" sciences nor a positivist history of the sciences. It will not be a history detached from the epistemology and sociopolitics of the sciences. Nor will it be a recursive history.

To give content to the methodological characteristics of the new history, Saldaña includes as an appendix to his paper the guidelines of a research project in which this approach is applied to the case of the historiography of the sciences in Mexico (Saldaña, 1993a). These ideas were also presented in the lecture *Towards a Theory of the History of Science in Latin America* that Saldaña gave at the Faculty of Sciences of the Universidad del Valle during his visit in November 1984, at a time when a certain critical mass in teaching and research in the history of science and mathematics was being formed in this institution.

3.5 The Cali Seminar on Alternatives for the Teaching of History of Science and Technology (1984)

On February 21–25, 1983, the Advisory Board of the SLHCT, formed at the Puebla meeting the previous year with representation from Argentina, Brazil, Colombia, Cuba, Mexico, and Peru, met in Campinas. The organizational and administrative guidelines to ensure the take-off of the society included the establishment of commissions and sections specialized in strategic issues. Two academic sections were initially established. One section was headed by Ubiratan D'Ambrosio on the Methodology of the History of Science and the other by Luis Carlos Arboleda on the Promotion of the Teaching of the History of Science and Technology in the region.[7] In the following year, both defined their program agendas in meetings in Lima and Cali, with the participation of members of the society and international guests (Fig. 3.4).

región han prevalecido. Sin embargo, la ausencia hasta hoy de un planteamiento teórico sobre estas cuestiones hace necesaria una revisión y una crítica de las metodologías disponibles, aun cuando no consideren nuestra especificidad, para pensar a través de ellas en nuestro pasado de ciencia y tecnología.

[7] A few months later, the sections on the History of Institutionalization of Science in Latin America, the History of Ideas in Latin America, Iberoamerican Studies on the History of Science and Technology, and the History of Physics in Latin America were created. The Newsletter of the SLHCT published in several issues the working programs of these scientific sections.

Fig. 3.4 Aspects of the Latin American Seminar on Methodology of Science History Teaching in Cali in 1984

The seminar in Cali, held at the Universidad del Valle on November 4–10, 1984, was the first international meeting in Latin America devoted to the teaching of the history of science (Arboleda, 1986).[8] Having as a frame of reference the situation of this teaching in the international focal points represented at the meeting, the discussions of the symposia and workshops studied the experiences in this area at the regional level, and advanced the exchange of ideas on research programs and comparative studies in the following areas: (1) Teaching the history of science in science and engineering curricula, (2) Modalities of popularization of the history of science in the region, and (3) Training of Latin American researchers in the history of science. Three symposia and a series of conferences were also held to illustrate the variety of interests and new approaches in this field of study in the region.[9] Finally, advantage was taken of the presence of the leading Colombian historians at this meeting to create the Colombian Society for the History of Science and Technology.[10]

[8] Together with Saldaña, other international guests participated in their respective conferences: Education, Revolution and Cultural Evolution (F. Diaz de Avila-Pires, Brazil), The History of Science in Quebec (Jean-Claude Guédon, Quebec), On the State of the History of Science and Technology in the Federal Republic of Germany) and the History of Science and Physics at the University of Pavia (F. Bevilacqua, Italy). Celina Lértora Mendoza (Argentina), Ubiratan D'Ambrosio (Brazil), Baldomero Valiño (Cuba), Jorge Nuñez (Cuba), and Giulano Bellodi (Italy) also actively participated in the symposia and workshops. Additionally, 40 Colombian professors from different universities in the country.

[9] These symposia already foreshadowed some of the lines that will be fundamental in the orientation of research in the history of science in Colombia: Epistemology and the history of science, history of science and ethnoscience of the Prehispanic cultures, and science in Colombia at the time of the Botanical Expedition of José C. Mutis.

[10] Emilio Quevedo's chapter in this book studies this society, which unfortunately did not manage to consolidate itself.

In his inaugural lecture at the Cali Seminar, Saldaña made an overall assessment of the situation of the history of science in Latin America at that time (Saldaña, 1986). We have previously commented on some parts of this talk about the challenges[11] and achievements of the SLHCT since its creation and the role played by the journal *Quipu* in this respect. We should add that as a prelude to the November Seminar and to stimulate discussion among the participants, issue 2 of *Quipu* was published, devoted exclusively to the topic of the history of science and teaching.[12] In the following years, this issue and the memory of the seminar will become permanent reference documents in the design and execution of different academic initiatives to implement this strategy in the region, particularly in Colombia.

It is possible to recognize in Saldaña's talk some of the mobilizing ideas and suggestions for programmatic lines that have been guiding the initiatives of researchers and university groups in this strategy up to the present time. To Saldaña, the introduction of the history of science and technology in our educational systems was a task that the SLHCT had to face as something more transcendent than training scholars to simply satisfy an individual pleasure. It was a matter of given an identity formation by what we are and what we aspire to be. In other words, to teach history in such a way as to recover our past and achieve a more integrated vision of our culture. "It is not possible to continue training scientists who ignore the particular course taken by the science they study in their respective countries and Latin America. To inform the scientist and the technician, as well as the social scientist and the humanist, about the national and Latin American history of the sciences, is to teach them how we arrived at the present state, and to give them the means to contribute to its modification, that is, in the sense of reality and national projects".[13]

[11] In Cali, Saldaña refers to the challenges of the SLHCT based on the following anecdote: "Some years ago, in his apartment in rue Descartes in Paris, Professor René Taton told me that during the years in which he was preparing the Histoire Générale des Sciences (Taton, 1957–1964), he was unable to find someone to write the chapter on science in Latin America, in an overall view, since the Latin American historians and scientists whom he consulted were ignorant of the regional scientific process" (Saldaña, 1986). (Saldaña, 1986). ("Hace algunos años, en su departamento de la rue Descartes de París, el profesor René Taton me comentaba que en los años en que preparaba la *Histoire Générale des Sciencies*, no logró conseguir quien pudiera escribir el capítulo de la ciencia en Latinoamérica, en una visión de conjunto, pues los historiadores y científicos latinoamericanos a quienes consultó se mostraron ignorantes del proceso científico regional"). I also personally heard Taton refer to the same issue at the Seminar of the Koyré Center and on other occasions when the subject of Latin America was present.

[12] The following papers are included: Historia y enseñanza de las matemáticas (Luis Carlos Arboleda), La historia de las ciencias en la enseñanza (Tomás A. Brody), O ensino da história da técnica na facultade de arquitetura e urbanismo da Universidade de São Paulo (Ruy Gama), Papel de la historia de las ciencias en la integración curricular de la Escuela Colombiana de Medicina (Emilio Quevedo), Historia da ciência e o seu ensino na Universidade de São Paulo (Shozo Motoyama et al.).

[13] No es posible seguir formando científicos que ignoran el curso particular que tomó la ciencia que ellos estudian en su respectivo país y en Latinoamérica. Informar al científico y al técnico, como también al científico social y al humanista, de la historia nacional y latinoamericana de las ciencias, es enseñarle cómo llegamos al estado actual, y darle medios para que pueda contribuir a su modificación, es decir, en el sentido de realidad y proyectos nacionales.

Since this was an event on the pedagogical purposes of the history of science, one of the leitmotifs of the new history of science advocated by Saldaña and the leadership of the SLHCT could not be absent from the talk: history as a resource for forming citizenship for the common good. “It is through the history of science and technology that we will be able to understand the role of Latin American societies, their research institutions, their universities, their governments, their economies, etc., on science and technology, to build an endogenous, national and Latin American science and technology”.[14]

Saldaña and the direction of the SLHCT considered that the formation in the region of an endogenous capacity of knowledge in the history of science with these characteristics required concerted efforts by the countries of the region to advance the institutionalization of our field of studies to levels comparable with international centers. It was necessary to promote the creation, within a medium-term period, of a Latin American Research and Postgraduate Center. This center should have a specialized library and a bibliographic information system for the entire region, and would also be an excellent channel for maintaining and expanding cooperative relations with similar centers in other regions of the world. It could also serve as an advisor to governments in the design of scientific policies and the conservation of cultural and scientific heritage. These approaches, formulated in summary form at the 1984 Cali meeting, contain the major mobilizing ideas of Saldaña's program for the new Latin American history of science.

3.6 The IV Latin American Congress on the History of Science and Technology in Cali (1994)

Ten years after the 1984 Cali meeting, Saldaña returned to Colombia on the occasion of the IV Latin American Congress on the History of Science and Technology, which I was in charge of organizing at the Universidad del Valle in my role as president of the SLHCT. As described in the Introduction to the Memoirs of the Congress (Arboleda & Osorio, 1997) and in the report presented to Colciencias, one of the financing entities[15]:

[14] Es mediante la historia de las ciencias y de las técnicas como podremos comprender el papel de las sociedades latinoamericanas, de sus instituciones de investigación, de sus universidades, de sus gobiernos, de sus economías, etc., sobre las ciencias y las técnicas, a efecto de constituir una ciencia y una tecnología endógenas, nacionales y latinoamericanas.

[15] Seventy researchers from Argentina, Brazil, Chile, Costa Rica, Cuba, Spain, the United States, France, Mexico, Peru, Puerto Rico, Uruguay, and Venezuela, and more than 100 Colombians participated. The main topics dealt with the socio-historical analysis of disciplines within the framework of the constitution of national sciences. Also on the methodological approach to specific problems such as the comparative reception of Darwinism, the protagonists of national science, the development of public spaces for science, science projects in the 18th and 19th centuries, scientific societies, technological innovation in Ibero-American cities, change and continuity in the technological modernization of Latin America, education and science, history and science teaching. See:

> The main theme of the Congress, *Nationalism and Internationalism in the History of Science and Technology in Latin America*, provided a space for reflection and conceptualization on the role of science and technology in the dynamics of development in the region, their interaction with international trends and interests and with the processes of (historical) construction of nationalities in Latin America.[16]

During the years following the founding meeting in Puebla, this question, like no other, had attracted the interest of historians from the region and abroad, as is particularly evident in the proceedings of the previous congresses (Havana-1985, São Paulo-1988, Mexico-1992), in *Quipu, Cuadernos de Quipu* and other national and international publications of those years. Not only in the regional sphere but also at the international level, the SLHCT became an important laboratory of ideas and a pole of investigation of significant historical situations about that invisible interweaving of threads between nationalism and internationalism, from which our cultural formations were constituted. Among the situations of this type that preceded and influenced the organization of the Cali Congress on this subject, I remember the impact on us of two publications, both from 1992. The memoir of the congress *Science et Empires* (Petitjean et al., 1992),[17] and No. 4 of the *Cuadernos de Quipu* dedicated precisely to *Los orígenes de la ciencia nacional* (Saldaña, 1992a) (Fig. 3.5).

To contribute to the analysis of this complex historical interaction between national science and what he designated as its "external sector", Saldaña oriented his inaugural conference of the fourth congress to present the point of view that he had been maturing in his practice as a professional researcher and as a head of the Seminar of History of Science at UNAM. This was based on a rigorously documented study of the specific case of the institutionalization of ecology as a scientific discipline in Mexico. He showed that North American scientific philanthropy and Pan-Americanism or international cooperation among scientists were undoubtedly transcendent factors in the institutionalization of conservationist practices. But this international action only became viable and took place when sufficient and necessary conditions of the local context were present. These conditions were (Saldaña, 1997):

> The creation of the Mexican Institute of Renewable Natural Resources (1951), the existence of its main promoter and author of the project Enrique Beltrán; the existence of a Mexican biological community that was essentially constituted during the two previous decades (with foreign and local contributions), the participation of important Mexican social sectors that had also fought for its establishment. (...) The international component, although its existence is due to the very nature of science, seen historically, is an idealization. Therefore,

---

Consideraciones al IV Congreso Latinoamericano de Historia de las Ciencias y la Tecnología, at: http://repositorio.colciencias.gov.co/handle/11146/1988

[16] El tema central del Congreso, *Nacionalismo e Internacionalismo en la Historia de la Ciencia y la Tecnología en América Latina*, permitió un espacio de reflexión y conceptualización sobre el papel de las ciencias y la tecnología en las dinámicas de desarrollo de la región, su interacción con las tendencias e intereses internacionales y con los procesos de construcción (histórica) de nacionalidades en América Latina.

[17] See the study of this memory in Martha Cecilia Bustamante's chapter in this book.

Fig. 3.5 Aspects of the IV Latin American Congress on the History of Science in Cali in 1994

> and not for sentimental nationalistic reasons as has been supposed, it is affirmed in the historiography of Latin American science that the "injection" of science or "resuscitating action" cannot constitute the minimum historical explanation of science that takes place in the here and now.[18]

This same point of view had been presented by Saldaña in other publications of the time, based on general conceptual and methodological arguments, and documented in historical case studies of various kinds on Mexican science and that of other countries. Of these publications, I especially retain one that was very important in the preparatory activities for the Cali congress (Saldaña, 1992b). This fifty-page essay was a synthesis of the discussions held on this issue in the early 1990s in various areas of the SLHCT. Particularly in the exchanges with its students and the professors from abroad invited to its Seminar on the History of Science. These included Antonio Lafuente, Luis Carlos Arboleda, Xavier Polanco, Emilio Quevedo, José Sala Catalá, Milton Vargas, Joan Bromberg and Rachel Laudan.

[18] La creación del Instituto Mexicano de Recursos Naturales Renovables (1951), la existencia de su principal promotor y autor del proyecto Enrique Beltrán; pero también la existencia de una comunidad biológica mexicana que en lo fundamental fue constituida durante las dos décadas anteriores (con aportaciones foráneas y locales), la participación de importantes sectores sociales mexicanos que habían pugnado igualmente por su establecimiento. (…) El componente internacional, si bien obedece su existencia a la naturaleza misma de la ciencia, visto históricamente, es una idealización. Por ello, y no por sentimentales motivos nacionalistas como se ha supuesto, se afirma en la historiografía de la ciencia latinoamericana que la "inyección" de ciencia o "acción resucitadora", no puede constituir la mínima explicación histórica de la ciencia que tiene lugar en un aquí y en un ahora.

Based on an analysis of the state of the art of the question in regional and international literature, Saldaña establishes his points of view on different sensitive issues at that time such as the globalization of science, colonial science, national science, national science and colonialism, national science and national state. Then he applies his considerations to relevant historical situations of Mexican and Latin American science in the transition from the colony to the republic to give more foundation to his argumentation. As we mentioned before, in his inaugural lecture of the congress he exemplifies his reflection on national science with the systematic study of the institutionalization of Mexican conservationism.

During his stay at the Universidad del Valle on the occasion of the fourth congress of the SLHCT, Saldaña carried out several academic activities, among which I remember two very important ones. The first was his lecture *El interés de la historia social de las ciencias en América Latina*, organized by the History of Mathematics Group that had just been created in the Faculty of Sciences under my coordination. The talk of a pedagogical nature allowed Saldaña the opportunity to expand on some of the ideas previously presented in his article (Saldaña, 1993b). From the present perspective, this talk was very opportune at the time, since the professors who were then initiating the Seminar on the History of Science at the University—in connection with the Colciencias Social History of Science project—were able to have an overview of the recent evolution of our field of study in Latin America. The presence of Saldaña also contributed to legitimate the interest in addressing the socio-cultural aspects of the local context in the historical epistemological studies on the development of mathematical and scientific knowledge.

The second important academic activity of Saldaña in his 1995 visit to the Universidad del Valle was the completion and delivery of the consulting report contracted months earlier with the university administration, consisting of the proposal of an ambitious program of socio-historical studies of science and technology within the framework of the curricular reform that the university was then implementing, and the regional projection of several of these initiatives in two directions, the agenda of the Vice Rector's Office of Research under my charge and the emerging Program of Social Studies in Science and Technology. (Saldaña, 1996).

3.7 The International Seminar-Workshop on North-South and South-South Scientific Partnership in Cartagena de Indias (2001)

Six years after the Cali Congress, Saldaña returned to Colombia as a speaker at a workshop seminar on north-south and south-south scientific partnership for sustainable development (Saldaña, 2004b).

The meeting was held in Cartagena de Indias on November 28–30, 2001, and was attended by researchers from Latin America, Europe, Asia, and Africa. It was organized by the Cinara Institute of the Universidad del Valle—widely recognized for its research and development activities in environmental management with emphasis on water management—and the KFPE, Swiss Commission for Scientific Partnership with Developing Countries, among other sponsoring institutions. We sought to better understand the conditions under which the partnership was developing in the Latin American context, and how the participating institutions could orient their actions within appropriate partnership management schemes at the beginning of the new millennium. The concept of cultural diversity, present in all partnership initiatives, was among the characteristics of these new schemes for consideration by the participants. Particularly in the production of suitable scientific results and their appropriation in projects for sustainable development within the framework of the national science and technology systems that were being promoted in those years (Arboleda et al., 2004).

As an advisor to the Cinara Institute in the organization of the Cartagena Seminar-Workshop, the invitation to Saldaña was made in the framework of the recognition of the importance of the contributions of the community of historians of science in the reflection on science and cultural diversity, resulting from the initial dissemination of the International Congress on the History of Science held 4 months earlier in Mexico. I summarize the impact of Saldaña's lecture in the following terms (Arboleda, 2004):

> I believe that the participants agreed in appreciating the initiative of the organizers to include at the beginning of the Workshop a lecture on Science and Cultural Diversity from the perspective of the history of science in Latin America. Some of the central issues of this intervention served as inputs to feed the discussions at the thematic tables: The deculturation implied in the notion of diffusion and progress. The experiences of resistance of Creole scholars and professional scientists to this trend of deculturation. Criticism of the vision of peripheral science. The valuation of native scientific productions. The emergence and international legitimization of new scientific fields such as ethnomedicine, ethnomathematics, ethnoastronomy, etc. The need to abandon any point of view of claim or blame in Latin America's scientific relations with the North, and to replace these attitudes with the elaboration of professional histories of the processes of cultural symbiosis. The suggestion is that we assume ourselves within a world of cultural interdependence, leaving aside provincial visions and the local particularisms that unfortunately still predominate in our countries. The request is that we not be afraid to expose ourselves to the experience of frequenting and taking advantage of diverse cultures. That we should be vigilant in recognizing and criticizing the residues of imperial approaches to Science and Technology that feed certain diffusionist, developmental, and progress ideologies... Finally, the call made to us researchers in this communication, to contribute to the development of studies on the issue of cultural diversity in our countries, and to disseminate and take advantage of this socio-cultural information in our partnership projects, all to make research objectives more viable and scientific relations with our peers abroad more equitable.[19]

[19] En mi opinión, los participantes estuvieron de acuerdo en apreciar la iniciativa de los organizadores de incluir en el inicio del Taller una conferencia sobre Ciencia y Diversidad Cultural en la perspectiva de la historia de las ciencias en Latinoamérica. Algunas cuestiones centrales de esta intervención sirvieron de insumos para alimentar las discusiones en las mesas temáticas: La desculturación implicada en la noción de difusión y progreso. Las experiencias de resistencia de eruditos criollos y científicos profesionales a esta tendencia de desculturación. La crítica a la visión de ciencia periférica. La valoración de las producciones científicas autóctonas. La emergencia y

3.8 CIHCYTAL's Presence in Colombia Between 2013 and 2017

In the depletion phase of the SLHCT, Saldaña's efforts were oriented towards building an Internet Community for the History of Science and Technology in Latin America (CIHCYTAL for its acronym in Spanish). In this period, around 300 products in the history of science and technology were produced, represented in a Blog, interviews, seminars (webinars), online courses, e-books, and even a news agency. All of this was aimed at realizing Saldaña's dream of using information and communication technology resources and social networks to give new impetus to the institutionalization of our field of study. Let us now look at some relevant aspects of what Saldaña called "our digital adventure".

I particularly remember our interaction in organizing five of the 25 webinars held during this period in which we participated as panelists, alone or in the company of other colleagues from the region. The first of the series was on *Recursos de Internet para la investigación e Historia de las Ciencias* and was held on January 31, 2014. A second webinar was held on April 9 of the same year on *Las revistas de historia de las ciencias en América Latina en el siglo XXI: vigencia y perspectivas*. One more took place on April 8, 2015, on *Ver nacer y crecer a la Historia de las Ciencias en América Latina*. On June 9 and September 24, 2015, two seminars were devoted to presenting to the CIHCYTAL community our work and that of other experts on the life and work of Francisco José de Caldas. He was one of the enlightened scholars of colonial and republican science whom Saldaña and his students at the Mexico Seminary studied recurrently over the years. The two seminars were broadcast from the Bogota headquarters of the Academy of Exact, Physical, and Natural Sciences (ACCEFYN for its acronym in Spanish) on the occasion of the national commemoration of the bicentennial of the death of Caldas (1768–1816) (Fig. 3.6).[20]

In addition, videos and audios of 14 interviews were produced, which were available for years both on the CIHCYTAL website mentioned above and on several

legitimación internacional de nuevos campos científicos como la etnomedicina, la etnomatemática, la etnoastronomía, etc. La necesidad de abandonar todo punto de vista de reclamación o de culpa en las relaciones científicas de América Latina con el Norte, y reemplazar estas actitudes por la elaboración de historias profesionales de los procesos de simbiosis cultural. La sugerencia de que nos asumamos dentro de un mundo de interdependencia cultural, dejando de lado visiones provinciales y los particularismos locales que desafortunadamente aún predominan en nuestros países. La solicitud de que no temamos exponernos a la experiencia de frecuentar y aprovechar diversas culturas. Que. estemos vigilantes a reconocer y criticar los residuos de enfoques imperiales de Ciencia y Tecnología que alimentan ciertas ideologías difusionistas, desarrollistas y del progreso... En fin, la convocatoria que se nos hizo a los investigadores en esta comunicación, para que contribuyamos al desarrollo de los estudios sobre la cuestión de la diversidad cultural en nuestros países, y difundamos y aprovechemos en nuestros proyectos de partenariado esta información socio-cultural, todo ello con el propósito de hacer más viables los objetivos de las investigaciones y más equitativas las relaciones científicas con nuestros pares del extranjero.

[20] The CIHCYTAL and ACCEFYN webinars kicked off a series of commemorative activities that concluded with the publication of the work (Guerrero, 2020).

La **Comunidad en Internet de Historia de la Ciencia y la Tecnología en América Latina**

y la **Academia Colombiana de Ciencias Exactas, Físicas y Naturales,**

trabajando en conjunto en la difusión de la historia de la ciencia y la tecnología,

le **INVITAN** a la

SEGUNDA JORNADA DE CONMEMORACIÓN DE LA VIDA Y OBRA DE DON FRANCISCO JOSÉ DE CALDAS (1768-1816) notable científico y héroe de la independencia colombiana

a través de un Webinario abierto a todo público que llevaremos a cabo el Jueves 24 de Septiembre a las 11:00 hrs.
(Hora de las Ciudades de México y de Bogotá, -6 GMT)

CIHCyTAL

La **CIHCyTAL** le **INVITA**

a su próximo Webinario que se llevará a cabo el Miércoles 8 de Abril a las 11:00 hrs.
(Hora de la Ciudad de México, -6 GMT)

dando inicio a la serie: **CONVERSACIONES**, con la participación de los doctores

Luis Carlos Arboleda y Juan José Saldaña
(Universidad del Valle, Cali, Colombia /Universidad Nacional Autónoma de México), con el tema:

"Ver nacer y crecer a la Historia de las Ciencias en América Latina"

Fig. 3.6 Announcements of two CIHCYTAL webinars in Colombia in 2014–2015

community channels open on other platforms such as YouTube. The first of them was made by Saldaña to the Chilean historian Zenobio Saldivia and Arboleda on the occasion of our participation in the symposium *La ciencia y la tecnología en la historia de la América Latina* within the framework of the 39th Congress of Anthropology and History, held in late February 2014 at the University of Sonora in Hermosillo, Mexico. One month later, the interview that Saldaña and his student Libertad Díaz did with me on *La historia de las matemáticas en América Latina* was published in the Blog. Finally, I took part in two virtual courses on the history of mathematics that began with Euclidean arithmetic and analytic geometry, as a starting point of a plan that could have been developed in courses on other historical moments if the CIHCYTAL project had been consolidated.

On June 2015, Saldaña presented to the direction of the UNAM's Faculty of Philosophy and Letters a complete proposal for teaching the history of science using digital resources based on conversations that we had previously held through the web, he and some of his collaborators (Saldaña, 2015a). Together with the other virtual devices of the CIHCYTAL, these courses were intended to extend and unite the community, giving greater reach and social projection to the history of science in the region through teaching. At CIHCYTAL, said the proposal, "We currently have the appropriate means to make an educational offer such as the one we propose, such as the support of a work team that includes specialist professors in Mexico and Latin America and technical personnel with capacity and experience in the handling of diverse information and communication technologies; an important didactic collection of specialized and digitized reading materials; a significant group of followers and contacts in various countries with people interested in our subject matter; and, experience in the combined handling of face-to-face and online methods in conferences, courses, and seminars" (Saldaña, 2015a).

The operational plan of the proposal to the Faculty of Philosophy and Letters was accompanied by a methodological design of interactive teaching supported by personal readings, guidelines for continuous evaluation, and recommendations so that the professor's presentations took into account the particularities of face-to-face and distance teaching. Three decades after the 1984 meeting in Cali, this methodology for teaching the discipline, using robust and effective resources whose potential and scope we could not have imagined at the time, could not find the institutional support to continue its implementation. The courses and other initiatives of CIHCYTAL were carried out intermittently until Saldaña's resources and the time available to several colleagues allowed it. On April 11, 2017, I received his e-mail with the income and expenditure statement with an accumulated deficit of $31,000 between 2013 and 2017. Referring to the factors that led to the closure of the academic enterprise, Saldaña wrote the following:

> As you well know, this project was a good and pioneering initiative, but like others we have carried out in the past, it was a failure from a financial point of view. Our community, I conclude once again, does not have an appropriate culture to take on projects of this nature, even if it uses them with manifest benefit. And we still haven't been able or haven't been able to remedy that. In short, the attempt was made with absolute dedication and the desire to leave a path marked for the future when others with better means or conditions can continue it. I cannot end this note without thanking you, dear Luis Carlos, for all your support in making possible this enterprise and your company of so many years to produce and publicize the history of science in Latin America.[21]

3.9 Conclusion

In another e-mail dated April 2019, 2 years after the one mentioned above, Saldaña clarifies that despite having closed CIHCYTAL, he has kept the *Quipu* consultation open to the public free of charge. Saldaña writes at the end of the message: "I am about to give some life to this project again and I will keep you informed as I expect your collaboration". Never again did we discuss the matter in our communications,

[21] Como bien sabes este proyecto fue una buena y pionera iniciativa, pero como otras que hemos llevado a cabo en el pasado, desde el punto de vista financiero, un fracaso. Nuestra comunidad, concluyo una vez más, no tiene una cultura apropiada para para hacerse cargo por sí misma de proyectos de esta naturaleza aunque se sirva de ellos con manifiesto provecho. Y eso aún no hemos podido o no hemos sabido remediarlo. En fin, el intento se hizo con absoluta entrega y el deseo de dejar marcado un camino para el futuro cuando otros con mejores medios o condiciones lo puedan continuar. No puedo terminar esta nota sin agradecerte, querido Luis Carlos, todo tu apoyo para hacer posible esta empresa y tu compañía de tantos años en el propósito de producir y dar a conocer la historia de la ciencia de América Latina.

which became more and more sporadic,[22] so I assumed that the negotiations in this sense had failed.

His steadfast attitude to persevere in the virtualization of the HCT field in the region—in a refractory institutional and professional context and even at the cost of the health problems he was already beginning to manifest—was in no way due to personal interest. Because of his career as a university professor at UNAM and because he was an insightful observer of international dynamics like no other, Saldaña had a comprehensive vision of the impact of new technologies on the practices of scientific communities and their forms of organization and international communication. For example, as Secretary of the DHS/IUHPS Executive Committee, he proposed and implemented several initiatives in this area. He was persuaded that the CIHCYTAL community could contribute to integrating and enhancing the professional work that in the early 2010s was being carried out in several lines of the history of science by numerous groups and networks in our countries, relying on the appropriation and use of new technologies.

This chapter cannot be concluded without referring to two undertakings that resulted from our collaboration with Saldaña, which in the last decade have had a positive impact on the development of our activity in Colombia in teaching, research, and researcher training. The first was the publication in the journal *Quipu* (second stage)[23] of several papers by research groups linked to the Inter-institutional Doctorate in Education in Colombia.[24] Some of these on the history of mathematics education in Colombia appeared in volume 14 of 2012 of *Quipu*, under my editorial coordination at the request of the journal's director. One other on the history of physics education in Colombia was published in volume 17 of 2016. Even though these volumes of *Quipu* are no longer available on the web and are of restricted distribution, this body of work has been an important input in the academic and research activities of history and mathematics education, one of the most dynamic communities of history of science in Colombia and the Ibero-American region (Fig. 3.7).

[22] The last visit of Saldaña to Colombia was to participate in the Third Global Forum of the Institute of Mexicans Abroad (IME for its acronym in Spanish) in Bogota from October 18 to 20, 2019. In her letter of invitation the Ambassador of Mexico in Colombia writes: "Being you a high specialist in science and technology in Mexico and Latin America, I would like to extend you a kind invitation to be part of our Forum, in a Master Conference whose theme is *Los retos de la ciencia en América Latina*." (Galeana, 2018). About the Forum, see: https://www.gob.mx/ime/prensa/tercer-foro-global-del-ime-en-bogota-colombia?idiom=es

[23] Regarding the publication of works by Colombian historians of science in Quipu, see Quevedo's chapter in this book.

[24] The institutionalization and professionalization of the History of Sciences in Colombia has much to do with its insertion in the field of Education and Pedagogy. For example, the first doctoral research in the History of Mathematics, the History of Education, and the History of Mathematics and Mathematics Education in Colombia was carried out at the beginning of the new millennium within the framework of the Interinstitutional Doctorate in Education formed by the Universidad del Valle, the Universidad Distrital de Bogotá, and the Universidad Pedagógica Nacional. In general, this situation remains the same today.

Fig. 3.7 Saldaña's 2017 lecture on Science and Nation in Mexico (1917–1929)

The other accomplishment of our most recent cooperation with Saldaña was his essay *La ciencia en la conformación de la nación*, published in a collective work on the projects of nation in Colombia and Latin America in the nineteenth century (Saldaña, 2015b), one of the lines of research and doctoral training of the interdisciplinary studies group *Nación, Cultura y Memoria* of the Universidad del Valle. Some of the members of the group had become interested in Saldaña's ideas in the book *Las revoluciones políticas y la ciencia en México* (Saldaña, 2010), whose two volumes he had generously sent us to Cali. We consequently invited him to write an initial chapter that would offer, from the Mexican perspective, a conceptual framework to support the historical studies of the other chapters on state-nation relations, particularly regarding the training of a scientific elite capable of managing such relationships. My essay ended up focusing equally on examining the relationship of scientific elites with the establishment and organization of the republican state through mathematical knowledge (Arboleda, 2015).

From the standpoint presented in this article, a review of Saldaña's presence in Colombia reveals what is already recognized in other latitudes: his tireless effort to contribute to the institutionalization and the professionalization of the field of Latin American History of Science and Technology Studies, to its visibility at the regional and international level and to the formation of endogenous research capacities. Throughout the years, Saldaña supported important initiatives in the country to organize knowledge networks, groups, and scientific communities. Either with his presence in congresses, seminars, and other events or by making available to these groups and networks suitable instruments of communication and dissemination of their intellectual production. Colombian historians of science have benefited and we hope that they will continue to benefit from their original contributions to the theoretical construction of the field in several areas of our interest: Social History of Science and Technology, Science and Cultural Diversity, Science and Nation, as well as in the application of the history of science to the design and enforcement of public policies in Science, Education and Development.

References

Arboleda, L. C. (1982). El armamentismo y el movimiento anti-guerra en la situación actual. Manuscrito, 20 p. Archivo personal de Arboleda.

Arboleda, L. C. (Ed.). (1986). *Seminario Latinoamericano sobre Alternativas para la Enseñanza de la Historia de las Ciencias y la Tecnología*. Universidad del Valle, Cali, 4–10 de noviembre de 1984. Instituto Colombiano para el Fomento de la Educación Superior.

Arboleda, L. C. (2004). Conferencia de síntesis: El partenariado científico Norte-Sur y Sur-Sur. Balance y Perspectivas. EN: (Arboleda et al. 2004).

Arboleda, L. C. (2015). Élites, medidas y Estado en Colombia en la primera mitad del siglo XIX. Orden republicano y sistema métrico decimal. EN: (Quiceno, 2015), 177-–30.

Arboleda, L. C., & Osorio, C. (Eds.). (1997). *Nacionalismo e Internacionalismo en la Historia de las Ciencias y la Tecnología en América Latina.* Memorias del IV Congreso Latinoamericano de Historia de las Ciencias y la Tecnología (Cali, del 24 al 27 de enero de 1995). Facultad de Humanidades-Universidad del Valle.

Arboleda, L. C., García, M., & Pérez, H. (Eds.). (2004). *El partenariado científico para el desarrollo sostenible: aproximaciones y experiencias en América Latina.* Memorias del Seminario-Taller. Cartagena, Colombia, 28–30 de noviembre de 2001. CD-ROM.

Galeana, P. (2018). Carta de la Embajadora de México en Colombia con la invitación a Juan José Saldaña como conferencista del *Foro Global punto de encuentro para la comunidad mexicana.* Bogotá, 24 de septiembre del 2018. Manuscrito. Archivo personal de Saldaña.

Guerrero Pino, G. (Ed.). (2020). *Estudios Caldasianos, Ciencia y Nación. A 250 años del natalicio de Francisco José de Caldas*. Programa Editorial Universidad del Valle.

Pennini de Vega, E. (1984). Darwin en Argentina. *Quipu, Revista Latinoamericana de Historia de la Ciencia y la Tecnología, 1*(1), 119–132.

Petitjean, P., Jami, C., & Moulin, A. M. (1992). *Science and empires: Historical studies about scientific development and European expansion*. Springer.

Pruna, P. M. (1984). La recepción de las ideas de Darwin en Cuba, durante el siglo XIX. *Quipu, Revista Latinoamericana de Historia de las Ciencias y la Tecnología, 1*(3), 369–389.

Quevedo, E., et al. (1993). *Fundamentos Teórico-Metodológicos. Historia Social de la Ciencia en Colombia, tomo I.* Instituto Colombiano para el Desarrollo de la Ciencia y la Tecnología Francisco José de Caldas.

Quiceno Castrillón, H. (Ed.). (2015). *La nación imaginada: ensayos sobre los proyectos de nación en Colombia y América Latina en el siglo XIX*. Programa Editorial Universidad del Valle.

Saldaña, J. J. (trad.). (1982). *Declaración de Bucarest.* Simposio Internacional "Los científicos y la paz". Bucarest, septiembre 4 y 5 de 1981. Archivo personal de Saldaña.

Saldaña, J. J. (1983). La introducción de la óptica ondulatoria: la explicación fresneliana de la difracción. *Revista Mexicana de Física, 29*(3), 419–447.

Saldaña, J. J. (1986). Balance general de la Historia de las Ciencias en América Latina. Conferencia Inaugural. EN: (Arboleda, 1986), 3–11.

Saldaña, J. J. (Ed.). 1992a. Los orígenes de la ciencia nacional. *Cuadernos de Quipu,* n° 4. SLHCT-UNAM.

Saldaña, J. J. (1992b). *Acerca de la historia de la ciencia nacional.* EN: (Saldaña, 1992), 9–54.

Saldaña, J. J. (1993a). *Hacia una crítica histórica, teórica y metodológica de la historiografía latinoamericana de las ciencias*. En: (Quevedo, 1993), 163–178.

Saldaña, J. J. (1993b). Nuevas tendencias en la historia latinoamericana de las ciencias. *Cuadernos Americanos*, 2 (1993), n° 38, 68–91.

Saldaña, J. J. (1996). Los Estudios Socio-Históricos de la Ciencia y la Tecnología y la Reforma Curricular de la Universidad del Valle. *Política y Gestión Universitaria, 25*, 7–53.

Saldaña, J. J. (1997). El sector externo y la ciencia nacional. El conservacionismo en México (1934–1952). En: (Arboleda, 1997), 13–33.

Saldaña, J. J. (2004a). Mi experiencia como editor de *Quipu. Boletín Mexicano de Historia y Filosofía de la Medicina, 7*(2), 60–65.

Saldaña, J. J. (2004b). Conferencia plenaria de apertura. EN: (Arboleda et al, 2004).
Saldaña, J. J. (2010). *Las revoluciones políticas y la ciencia en México*. Tomo I: Ciencia y política en México en la época de la Independencia, Tomo II: Ciencia y política en México de la Reforma a la Revolución Mexicana. CONACYT.
Saldaña, J. J. (2015a). Propuesta para el empleo de tecnologías digitales en la enseñanza de la historia de la ciencia y la tecnología y de otras disciplinas en la Facultad de Filosofía y Letras de la UNAM. Manuscrito. Comunicación personal en email del 15 de junio de 2015.
Saldaña, J. J. (2015b). La ciencia en la conformación de la nación: el caso de México. EN: (Quiceno, 2015), 57–81.
Saldaña, J. J., et al. (1981). *Declaración de Bucarest*. Publicada inicialmente en el sitio web actualmente inexistente de la CIHCYTAL. Reproducido en. https://issuu.com/cihcytal/docs/declaracion_de_bucarestb
Taton, R. (Ed.). 1957–1964. *Histoire générale des sciences*. En tres tomos y cuatro volúmenes. Presses Universitaires de France. Reeditado en 1966–1983.

Luis Carlos Arboleda is Emeritus Professor at the Universidad del Valle, Cali, Colombia. For a little more than half a century he has taught the history of mathematics and science at this university, where he was also Vice Rector of Research. From 1991 to 1994 he was subdirector of Colciencias (today Ministry of Science), an entity that designated him as Emeritus Researcher in 2018. Since 2017 he has been a full member of the Colombian Academy of Exact, Physical and Natural Sciences. He was awarded the *Sábato Medal* of the Inter-American Commission on Mathematical Education (CIAEM for its acronym in Spanish) in 2023 in recognition of his services in history and mathematics education.

Chapter 4
Establishment of the First Latin American Community on History of Science

María de la Paz Ramos-Lara and Luis Carlos Arboleda

Abstract In 1981, during the XVI International Congress of History of Science (ICHS), a group of Latin American historians of science came together to establish the Latin American Society for the History of Science and Technology (LASHST) (1982) and the journal *Quipu* (1984). Juan José Saldaña (1944–2022) was the first president of the former and director of the latter. The formation of an academic network and the activities carried out contributed to the professionalization and institutionalization of this discipline in Latin America. The emerging Latin American community succeeded in incorporating new local topics into the mainstream of the history of science worldwide. LASHST worked on the search for methodologies specific to the region, organized a large number of Latin American and international events to give visibility and recognition to its field, and established specialized study programs for the training of new generations. Due to its outstanding performance Mexico hosted the XXI ICHS in 2001, for the first time in Latin America. In 2017, Brazil hosted the same event. Saldaña played a crucial role in these activities and in establishing close collaboration with international organizations. Consequently, he became the first Latin American to hold the position of General Secretary at the IUHPST/DHST, and the founder of IASCUD, Ceromayalist and CIHCyTAL.

Keywords Latin American Society for the History of Science and Technology (LASHST) · *Quipu* · International Congress of History of Science (ICHS) · XXI ICHS Mexico City 2001 · CIHCyTAL · Ceromayalist

M. de la P. Ramos-Lara (✉)
Centro de Investigaciones Interdisciplinarias en Ciencias y Humanidades, Universidad Nacional Autónoma de México, Mexico City, Mexico
e-mail: ramoslm@unam.mx

L. C. Arboleda
Universidad del Valle, Cali, Colombia

M. de la P. Ramos-Lara, L. C. Arboleda (eds.), *History of Science in Latin America: The Construction of an Intellectual Field (20th century)*,
https://doi.org/10.1007/978-3-031-75536-1_4

4.1 From the Eurocentric Model to the Search for Latin America's Models

In the first half of the twentieth century, the field of History of Science was professionalized and institutionalized in Europe and the United States. It was developed by establishing educational and research institutions, specialized societies and journals, international congresses, and publishing houses that facilitated the publication of books and journals.[1] In methodological terms, several currents in the history of science—the internalist (associated with positive science) and externalist currents—have influenced the study of the history of science globally, including in Latin America.

Latin American historians, both amateurs and professionals, tended to follow Eurocentric models in their research, which affected the direction and conclusions of their studies. It's understandable that an approach focused on Europe and the Europeans led them to closely scrutinize local scientific materials in search of evidence of participation in "positive science" by local historical figures. As a result, they ended up historicizing European science in Latin America rather than focusing on Latin American scientific practice (Saldaña, 1989, 342, 345). The lack of local contributions to "positive sciences" led to disillusionment, resulting in academic and governmental institutions losing interest in financing projects in this field (Aguilera, 2000). It was clear that a methodology for social history of science and technology in Latin America was needed (Saldaña, 1989, 362).

In terms of education, the French schools considerably influenced the first Latin American students to earn a doctorate in the History of Science. The Eurocentric model was so deeply ingrained in the curricular plans that they completed their studies solely within that framework. In other words, they selected research topics related to the history of European science, as very little was known about local contributions in other regions. Saldaña shared that René Tatón (1915–2004), a renowned French historian of science, expressed his interest in including chapters on scientific developments from regions outside of Europe in his famous work called *Histoire Générale des Sciences*. However, he faced difficulty finding someone to provide him with the necessary information. As a result, he only managed to include some scattered information he had available (Saldaña, 2004, 3).

It became evident that the mechanical transmission model of European science to Latin America did not consider the historical development of scientific practices and cultures in the region. LASHST members explore other alternatives in their

[1] The German Society for the History of Medicine, Natural Sciences, and Technology, also known as Deutsche Gesellschaft für Geschichte der Medizin, Naturwissenschaft und Technik, was established in 1901. This society is considered the first to focus on the history of science. In 1924, the History of Science Society (HSS) was founded, and now it is the leading professional society for the academic study of the history of science. On August 17, 1928, the Académie Internationale d'Histoire des Sciences/International Academy for the History of Science was established. In 1926, the International Committee of Historical Sciences/Comité International des Sciences Historiques (ICHS/CISH) was created in Geneva.

search to break this methodological and epistemological opposition in the objects and lines of research. We will see below that, during the 1980s, *Quipu* published articles on non-positivist approaches to understanding how Latin American countries produced local knowledge and received and adopted European science. The goal was to show how scientific practices were formed in different socio-cultural contexts.[2]

4.2 The Bucharest Declaration (1981) and the Establishment of an Association of Latin American Historians

In 1981, the XVI International Congress of History of Science (ICHS)[3] held in Bucharest (Romania) played a significant role in the professionalization of the History of Science in Latin America. A group of Latin American researchers, some of whom had obtained doctorates in History of Science from European universities and centers (mainly in France), registered to present their research findings at this Congress.[4] Despite their predominantly Eurocentric presentations, it was surprising to see little or no participation in Latin American history of science topics. During informal conversations at the Congress, they expressed the need to approach this field professionally in Latin America. They decided to hold a formal meeting to create a standardized professionalization process for countries that joined the project. The meeting's agreements were condensed into the "Declaration of Bucharest." The document highlighted the current state of science in Latin America and emphasized the need to institutionalize this field in countries with favorable cultural, political, and economic conditions (Saldaña, 1985, 3–4). In this context, the Congress served as a cornerstone for the growth and development of the field in Latin America.

The Latin American group developed a program to professionalize the field in Latin America, recognizing the need for doctoral training in the history of science and realizing that there were very few PhDs in this field in their respective countries (Ramos-Lara & Arboleda, 2024a). They followed the European model in terms of academic training and creating a society that would bring together individuals

[2] In the same decade, the conditions were being prepared for creating science and technology systems and building endogenous capacities in science and technology. In this context, some governments in the region have implemented various initiatives to support studies on the development of local science: seminars, publications, trips, regional exchanges, congresses, etc. It was done considering that the history of science could be a resource for the rationalization of science, technology, and innovation policy and management.

[3] Currently International Congress of History of Science and Technology (ICHST).

[4] Elena Pennini de Vega, History of Medicine, Universidad de Buenos Aires, Argentina. Atico Vilas-Boas da Mota, Cultural Historian, Brazil. Juan José Saldaña, History and Philosophy of Physics, Universidad Nacional Autónoma de México, Mexico. Luis Carlos Arboleda, History of Mathematics and its Teaching, Universidad del Valle, Colombia. José López Sánchez, History of Medicine, Academia de Ciencias de Cuba. Pedro Marino Pruna, History of Biology, Academia de Ciencias de Cuba.

and groups within the regional community. They looked to the International Union of History and Philosophy of Science and Technology (IUPHST) as an example, as it is responsible for organizing international congresses, such as the one in Bucharest, which they attended.[5] While discussing, they came up with the idea of creating a specialized journal with regional and international circulation. This journal would publish research results from the region and serve as a resource to promote forming an academic community. Similar to the journal *Isis,* which has been the most important worldwide in the History of Science since its foundation in 1912.[6]

The group designed a comprehensive program that includes formal postgraduate instruction, organizing Latin American congresses and symposia on specific topics, and establishing research centers and institutes. The initiative consisted of three main guidelines:

1. *Providing professional training*, mainly through postgraduate seminars at universities.
2. *Integrating an academic community and communicating results* through events, publications, and journals.
3. *Conducting research with specific projects* that spanned across the participating countries.

One of the most successful projects was focused on colonial studies, which was developed in the same regional context and had standard links to Iberian countries. This project strengthened the bond with Spanish historians of science to the point where they participated in the same events and researched related topics.[7]

This chapter provides an overview of the creation of the Latin American Society for the History of Science and Technology (LASHST), its activities in the first years of its establishment, and the launch and eventual discontinuation of its publication, *Quipu Latin American Journal of History of Science and Technology*. These efforts were crucial in reducing the use of a Eurocentric approach and promoting a shift from amateurism to professionalism in the field of History of Science in the region. They contributed to giving international visibility to the historical-scientific

[5] The creation of the IUHS took place in 1947 due to a UNESCO program to establish a body or "Institute of History of Science" attached to the International Council of Scientific Unions (ICSU). Its precursor, the International Academy of the History of Science—created in 1927 by an elite group of scholars focused on history—organized the first International Congresses of the History of Science (ICHS). The first was held in Paris (France) in 1929 and the second in London (England) in 1931. The latter was the most famous due to the presence of a Soviet delegation that presented a social approach to the history of science that had a considerable influence not only on many scientists but also on the world (including Mexico). This project led to the beginning of the journal Archives Internationales d'Histoire des Sciences. Years later, a P was incorporated, and although the name of the Union changed to the International Union of History and Philosophy of Science (IUPHS), its objectives were maintained (Petitjean, 2006b, 81–82). The Technology term was recently added, leaving the acronym IUPHST as it is today.

[6] Its founder and editor, George Sarton (1884–1956)—considered the father of the History of Science—devoted himself to it for 40 years (Garfield, 1985).

[7] In the previous section, Leoncio López-Ocón discussed this topic.

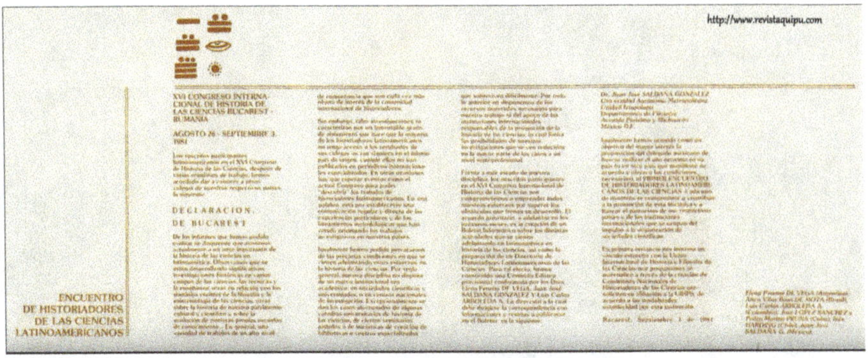

Fig. 4.1 The Bucharest Declaration, 1982. (Photographic archive Juan José Saldaña González)

material produced in the region and proposed a local methodology that takes into account both endogenous and exogenous factors (Saldaña, 2006, 10–11). The topics of the history of science and technology in Latin America became so crucial in the international community that it was agreed to hold one of its international congresses for the first time in a Latin American country, Mexico, in 2001, under the organization of Saldaña himself. Sixteen years later, another international congress XXV ICHS was held in Rio de Janeiro, Brazil (Fig. 4.1).

4.3 Creation of the LASHST in the City of Puebla (Mexico, 1982)

The LASHST was established in Puebla, Mexico, in 1982, at the First Latin American Meeting of Historians of Science organized by Juan José Saldaña (Saldaña, 1985, 4). One of its primary objectives was “to recover the historical memory of the Latin American peoples in a new context, which recognized scientific heritage as an integral part of the culture, economy, politics, and society of the Latin American people” (Aguilera, 2000). This success marked the beginning of institutionalizing this discipline in the region.[8] Another goal was to disseminate the research results of Latin American historians of science worldwide, transcending local efforts. The society created the journal *Quipu* to achieve it, as explained in the following section (Fig. 4.2).

In 1982, Saldaña became a part-time professor at the Universidad Autónoma Metropolitana Iztapalapa, a year after graduating in Paris. Despite not having institutional support, he managed to create a society with the help of his connections with Mexican and Latin American academics, some of whom held stable positions

[8] Another critical step was the creation of a Research Seminar in the Graduate School of the Faculty of Philosophy and Letters of the UNAM in 1985 to train professionals in this field. María de la Paz Ramos-Lara discusses this subject at greater length in the last section of this book.

Fig. 4.2 Installation of the Puebla Meeting, 1982. In the center, Saldaña and Alfonso López Pliego, rector of the University of Puebla. (Photographic archive Juan José Saldaña González)

in their universities. The Puebla meeting was made possible by the support he received from Carlos Contreras Cruz and other historians from the University of Puebla, who arranged for the event to be held at the university facilities in the capital of Puebla (Ramos-Lara, 2024).

During that meeting, 200 academics from 16 countries presented 120 papers. The Latin American Council and the Executive Committee members were elected to establish the Society to achieve the following objectives (Saldaña, 1985, 4):

- Bring together those engaged in research and teaching of the History of Science and Technology and other related disciplines in Latin America.
- Promote research and teaching of the History of Science and Technology.
- Encourage comprehensive research on Latin America's scientific, technical, and cultural heritage.
- Stimulate the preservation and rescue of Latin American scientific and technical documentation.
- Establish links and collaboration with similar and related associations in other countries, especially with the IUPHST/DHST.[9]
- Foster the institutionalization and professionalization of the discipline.

Juan José Saldaña was appointed president of the LASHST and also of the Latin American Council. The council was made up of four regional secretaries, namely Brazilian Ubiratan D'Ambrosio for the Southern Cone, Colombian Luis Carlos Arboleda for the Andean Zone, Cuban Pedro M. Pruna for Central America, and Spanish Antonio Lafuente for the Iberian Zone (Saldaña, 1985, 7).

The LASHST constituted an academic organization defined by thematic sections. Each one was headed by a president in charge of promoting them. All of them were specialists with doctorates in their fields, as shown in Table 4.1 below.

[9] Division of History of Science and Technology (DHST).

Table 4.1 List of chairperson of the 14 thematic divisions of the LASHST in 1985 (Saldaña, 1985, 22–24)

Chairperson	Country	Thematic division
Eduardo L. Ortíz	England/Argentina	History of Mathematics
Shozo Motoyama	Brazil	History of Physics
Pedro M. Pruna	Cuba	History of Biology
Rubén Ardila	Colombia	History of Psychology
Ernesto Yépes	Peru	History of Social Sciences
Simão Mathias	Brazil	History of Chemistry
Emilio Quevedo	Colombia	History of Medicine
Juan José Saldaña	Mexico	Documentation
Luis Carlos Arboleda	Colombia	History of Science Teaching
Ubiratan D'Ambrosio	Brazil	Methodology
Hebbe Vessuri	Venezuela	History of the Institutionalization of Science
Horacio Cerutti	Mexico	History of Ideas
Ruy Gama	Brazil	History of Technology
José Luis Peset	Spain	Iberoamerican Studies of the History of Science

The LASHST quickly formed a regional network of affiliated national societies from Argentina, Brazil, Colombia, Costa Rica, Cuba, Chile, Ecuador, Perú and Mexico. By 1984, three of these societies had become affiliated: the Sociedade Brasileira de História da Ciência,[10] the Grupo Argentino de Historia de la Ciencia,[11] and the Sociedad Ecuatoriana de Historia de las Ciencias y la Tecnología (*Quipu* vol. 1, no. 3).[12] A year later also joined, the Sociedad Colombiana de Historia de las Ciencias y de las Técnicas,[13] the Asociación Costarricense de Historia y Filosofía de la Ciencia, the Sociedad Cubana de Historia de la Ciencia y la Técnica,[14] the Grupo

[10] The Brazilian society was founded on December 16, 1983, in São Paulo, 1 year after the LASHST. https://www.sbhc.org.br/site/capa (accessed 28 February 2024).

[11] This Argentine group was created in 1933 by the Spanish mathematician Julio Rey Pastor (1888–1962) and the Italian industrial chemist Umberto Giulio Paoli (1876–1953). They intended to install the Italian historian of science Aldo Mieli (1879–1950), founder of the Académie Internationale d'Histoire des Sciences (Paris) in Argentina. Once settled in that country, in 1939, Mieli was appointed director of the Institute of History and Philosophy of Science, created a year earlier at the National University of Litoral (Babini y De Asúa, 2003).

[12] The bylaws of the Ecuadorian company were established in 1985.

[13] In 1984, the Colombian Society for the History of Science and Technology was established as part of the Latin American Seminar on Alternatives for Teaching the History of Science and Technology, held at the Universidad del Valle in Cali, Colombia. For more information, please refer to Emilio Quevedo's chapter in the fourth section of this book.

[14] The Cuban Society was founded in 1985 within the framework of the activities celebrating the First Congress of the LASHST, held in Havana. See the work of Alamino and Aguilar in the fourth section.

Chileno de Historia de la Ciencia, and the Sociedad Mexicana de Historia de la Ciencia y la Tecnología (SMHCT) (Saldaña, 1985, 24).[15] In 1986, the Sociedad Peruana de Historia de la Ciencia y la Tecnología and the Grupo Venezolano de Historia y Sociología de la Ciencia also became affiliated. Within 4 years, LASHST had affiliated with all the regional societies that would collaborate with it on its projects.

The establishment of the LASHST was highly appreciated at both regional and international levels. One of its notable outcomes was the formation of the "International Commission for the Conservation of the Scientific and Cultural Heritage of the Ibero-American Countries" (CICPCCPI).[16] By 1984, the commission was already functional with its members comprising of Arturo Alcalde Mongrut (Peru), José Babini (Argentina), Enrique Beltrán (Mexico), Rodrigo Fierro Benítez (Ecuador), Fernando Ernesto García Camarero (Spain), Pedro Laín Entralgo (Spain), José López Sánchez (Cuba), Simão Mathias (Brazil), Eduardo L. Ortíz (Argentina), Lewis Pyenson (Canada), Nathan Reingold (USA), Marcel Roche (Venezuela), Juan José Saldaña (Mexico), Francisco de Solano (Spain), Dirk J. Struik (USA), René Taton (France), Juan Vernet Ginés (Spain) (*Quipu* vol. 1, no. 1, 157–158).

Since its inception, the LASHST has encouraged collaboration among historians of science across Latin America, as well as with historians from other nations. This vibrant and ongoing exchange has greatly enriched the field of science history in the region, resulting in a range of positive outcomes such as:

- Enthusiasm was generated to organize events in other cities and disseminate them among academic communities, which forced the financing of a field that was beginning to have a presence in academic and political circles.
- This environment motivated individuals to form local organizations and reactive ones.
- Various undergraduate, master's, and doctoral programs began teaching the history of science chairs at the curricular level.
- Over time, colleagues from various countries developed collaborative relationships with each other.
- Interest in financing collective projects with local and international support emerged.
- Some of them have assumed positions at the highest international levels.

[15] The Mexican Society was founded in 1964 by the Mexican biologist and historian of science Enrique Beltrán (1903–1994) but declined very soon. See Ramos-Lara work in the forth section.

[16] Saldaña was the Executive Secretary of the commission created as part of the LASHST. The commission was active from 1983 to 1988.

4.4 LASHST Achievement from 1982 to 2004

The LASHST was very active in its first years, organizing 11 events in various countries of Latin America, Spain, and the United States of America. Table 4.2 provides a summary of these events (Fig. 4.3).

We refer to a meeting organized in Bogota where different perspectives were exchanged regarding the methodology of the Social History of Science in Latin America. This meeting aimed to discuss new theoretical and methodological alternatives to enrich the Latin American history of science. In this forum, Saldaña suggested the importance of examining Latin America's local traditions to develop a unique theory and methodology of historiographic practice. According to Saldaña, Latin American history of science should not be seen as internalist, continuist, or externalist, nor as a history of contributions or universals, nor as positivist or recursive. Instead, it should be seen as a history that considers the epistemology and sociopolitics of science. Other speakers were Ubiratan D'Ambrosio (Brazil), Ruy Gama (Brazil), Hebe Vessuri (Venezuela), Antonio Lafuente (Spain), Ernesto Yépes

Table 4.2 List of events organized by the LASHST from 1983 to 1985. (Saldaña, 1985, 5–6)

Año	City/Country	Meeting themes
1983	Caracas, Venezuela	Institutionalization of the sciences
1983	Bogota, Colombia	Methodology of the social history of science
1984	Mexico	History of science and scientific and technological foresight
1984	Cali, Colombia	Teaching the history of science
1984	Lima, Peru	Problems of the history of science
1984	Chicago, USA	New directions in the history of science
1985	Mexico	Documentation for the history of science
1985	Puebla, Mexico	History of social sciences
1985	Havana, Cuba	First Latin American congress on history of sciences and technology
1985	Berkeley, USA	The intercultural diffusion of sciences in Latin America

Fig. 4.3 View of participants in the Seminar on Documentation for the History of Science and Technology, 1985. (Photographic archive Juan José Saldaña González).

(Peru), and Colombians Eduardo Aldana, Carlos Eduardo Vasco, Nestor Miranda, Emilio Quevedo, Diana Obregón and Luis Carlos Arboleda (De Ríos, 1984).[17]

After these meetings, the LASHST organized others, of which the Latin American congresses stood out as the most relevant and exciting for specialists in the region. The First Latin American Congress on the History of Science and Technology (CLHCT) was held on July 21–25, 1985, in Havana (Cuba) and was sponsored by the Cuban Academy of Sciences. This event constituted a considerable advance in the purposes of this society in terms of communication, teaching, and research, as well as consolidating the processes of institutionalization and professionalization of the History of Science and Technology in Latin America (Saldaña, 1985, 9).[18]

As previously mentioned, the LASHST played a significant role in forming new national associations or societies of the History of Science in various South American countries, such as Brazil,[19] Colombia,[20] Costa Rica,[21] Cuba, Chile, Ecuador, Peru, and Venezuela. It also helped revitalize existing ones, such as those in Argentina and Mexico. The LASHST committed to arranging a Latin American congress every 3 years. Consequently, following the Puebla congress, the I CLHCT took place in Havana in 1985, the II in São Paulo in 1988, the III in Mexico City in 1992, the IV in Cali in 1995, the V in Rio de Janeiro in 1998, and the VI in Buenos Aires in 2004. Also, it organized (on its initiative or jointly with other associations) meetings, colloquia, symposiums, and congresses on specific topics in Mexico, Latin America, and other cities such as Paris, Madrid, and Chicago (Saldaña, 2004, 6).[22]

---

[17] This event was sponsored by Colciencias, now the Ministry of Science of Colombia, and the OEA. It was part of the Program for the elaboration of a collective work on the Social History of Science in Colombia (De Ríos, 1984). It made an essential contribution to the theoretical and methodological foundation of the interdisciplinary project on the Social History of Science in Colombia, which, after 10 years of work would result in the publication of a great piece entitled Social History of Science in Colombia, consisting of 10 volumes. It was financed by Colciencias and edited by Emilio Quevedo (Quevedo, 1993). Quevedo and Arboleda expand on this subject's considerations in their respective chapters.

[18] As will be discussed in the fourth section.

[19] According to Simão Mathias, a Brazilian chemistry historian, the Sociedade Brasileira de História da Ciência was established on December 16, 1983, and subsequently became affiliated with the LASHST. In his work, Mathias expresses gratitude to the LASHST and Saldaña for their dedication in promoting the philosophy and history of science not only in Brazil but also in other countries (Mathias, 1986).

[20] For details on the influence of the SLHCT in the introduction and development of this field in Colombia, see the chapters by Arboleda and Quevedo in this book.

[21] According to Ruíz Gaytán (2003), the Asociación Costarricense de Historia y Filosofía de la Ciencia (ACOHIFICI) was established in 1983 and was connected to the LASHST from the very beginning. He notes that this association emerged from another one that was formed in the late 1970s but did not function effectively. In contrast, ACOHIFICI, linked to the LASHST, started organizing a series of congresses in which many members of the LASHST, including Saldaña himself, took part.

[22] It is described in more detail in Sect. 4.4.

Several events were organized in Europe to promote the internationalization of the history of science and technology in Latin America, particularly in Spain and France, as we saw in the previous section. Additionally, the Latin American delegation attended the XVII International Congress at the University of California, Berkeley (USA) in 1985. For the first time in that Congress, a symposium specializing in the history of science and technology in Latin America was held. The papers discussed the issue of intercultural diffusion of science in the region. Saldaña published it as an integral part of the congress proceedings in *Cuadernos de Quipu* no. 2 (Saldaña, 1987).[23] The topic was significant in the Berkeley congress scenario, and the *Cuadernos de Quipu* no. 2 received immediate reception within and outside the region. It paved the way for discussions on the same topic from different theoretical perspectives, which became more dynamic in the following decade. It explains the strong participation of the academic community in the region in the Mexico Congress on Science and Cultural Diversity (refer to Fig. 4.4).

During his tenure as President of the SMHCT (1988–2008), Saldaña held several international positions and took on additional responsibilities as a lecturer and organizer. He actively participated in academic and administrative tasks for specific commissions in the history of science and related fields. For instance, he was elected Vice President of the International Scientific Council for Science and Technology Policy Development (UNESCO) in 1991. From 1989 to 1997, he also served as an advisor to the Executive Committee of the IUHPST/DHST (ICSU-UNESCO). From 1986 to

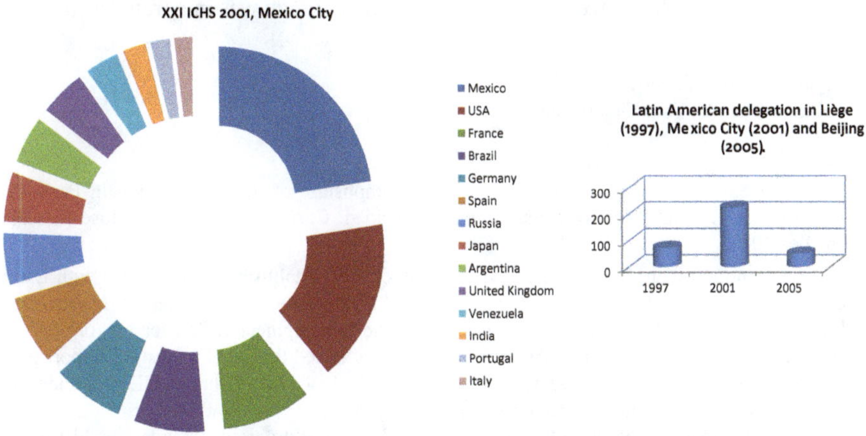

Fig. 4.4 Largest delegations at the XXI IXHST 2001, Mexico City (left). Right is the total number of Latin American participants in the 1997 (Liège) and 2005 (Beijing) ICHSs. (Own elaboration)

[23] The *Cuadernos de Quipu* collection was another publication of LASHST under Saldaña's direction. The first issue of this series is *El perfil de la ciencia en América* (1986); the second, *Cross-Cultural Diffusion of Science: Latin America* (1987); the third, *Descubrimientos de Carbón Mineral y Petróleo en México* (unpublished text by J. A. Alzate) (1988); the fourth, *Los orígenes de la ciencia nacional* (1992); and the fifth, *Science and Cultural Diversity. Filling a gap in the History of Science* (2001).

1994, he was a member of the Natural Sciences Committee of the National Commission of the United Mexican States for UNESCO. Between 1992 and 1997, he was a member of the Advisory Council of the Society for the History of Technology (SHOT) in the United States. In 1999, he was the first Executive Secretary of the International Association for the Study of Science and Cultural Diversity (IUHPST/DHST) in Bologna.[24] Additionally, he served as a Member of the Board of Directors of the International Committee for the History of Technology (ICOHTEC) from 2001 to 2005, among other positions (Ramos-Lara & Arboleda, 2024b).[25]

The LASHST's increasing establishment in Latin America and Saldaña's active participation in international organizations brought well-deserved recognition to the organization. Saldaña saw an opportunity to take the ICHS to Latin America for the first time, an idea he had considered for some time. In 1989, he unsuccessfully proposed Mexico City as ICHS site. Finally, in 1997, during the XX ICHS held in Liège (Belgium), Saldaña collaborated with some Mexican colleagues (as delegates)[26] to make Mexico the first Latin American venue for the most important congress of the International Community of Historians of Science and Technology. On November 28–29, 2000, Saldaña organized an international symposium on Science and Cultural Diversity, preparatory to the 2001 Congress. Within the framework of this symposium held in Mexico (Pachuca, Hidalgo) was created the International Association for Science and Cultural Diversity (IASCUD) on November 30, 2000.[27]

The IUPHS/DHS General Assembly approved the integration of IASCUD as a new section. It also approved its Executive Council, which was integrated as follows in 2000:

Paulus Gerdes, President
Karine Chemla, First Vice President

---

[24] Juan José Saldaña González organized the first symposium of this commission in Pachuca, Mexico in 2000. Academics from ten countries attended. Curriculum vitae Juan José Saldaña González 2018.

[25] It is worth noting that UNESCO had expressed interest in establishing connections with Latin America since 1947 when a meeting was held in Mexico City. A year later, in Montevideo (Uruguay), the Panel of Experts in Latin America on the Development of Science was organized. About 15 scientists from ten Latin American countries attended the panel to design support programs for the region's scientific research development. It included establishing UNESCO scientific cooperation offices in Latin America (Petitjean, 2006a, 71–72).

It was not surprising that decades later, UNESCO's support programs included the history of science through the IUHPST/DHST. Several authors in this work refer to essential initiatives supported by UNESCO and organized by the LASHST or in which it played an important role, for example, the 1990 Paris Congress on "Sciences et Empires" or the XXI ICHS in Mexico 2001. Saldaña's Curriculum Vitae is a revealing record of the importance of UNESCO in the development of her academic career and, through it, in the institutionalization and international recognition of the Society.

[26] Among them, María de la Paz Ramos-Lara, co-author of this paper.

[27] http://iuhpst.org/pages/inter-division-commissions/iascud.php. Accessed in May 2024. Olival Freire, Michel Paty and Ana María Carrillo refer to IASCUD's beginnings and role in the next section.

Irfan Habib, Second Vice President
Luis Carlos Arboleda, General Secretary
Martine Duquesne, Assistant Secretary General
Wesley Stevens, Treasurer

With approximately 30 members from five continents, IASCUD officially began its journey in the History of Science 24 years ago and continues to work with the following aims "coordinate and promote research on the history and philosophy of science, contributing to the unity of science in general and establishing links between different branches of knowledge".[28]

The XXI ICHS was a significant event for Latin America, held in Mexico City in 2001, with almost a thousand papers presented. Brazil hosted the congress again in 2017, after Mexico. Saldaña's efforts in promoting the internationalization of Latin American history in science and technology opened doors for other historians from the region to take relevant positions in international organizations in our scientific field. Marcos Cueto, who is now the President of the IUHPST, is one such historian with a brilliant academic background.

During the XXI ICHS, the Mexican delegation was the largest among all the participating countries. Similarly, the Latin American delegation was also the biggest compared to the previous and subsequent congresses, i.e., XX ICHS conducted in Liège (Belgium) in 1997 and XXII ICHS conducted in Beijing (China) in 2005, as indicated in Fig. 4.4.

It is indisputable that the Mexican and Latin American communities were the primary beneficiaries of this conference. The international community took part in various symposia related to scientific practices in this region (Fig. 4.5). To provide

Fig. 4.5 During the XXI ICHS Mexico City, a group of congress participants were photographed in front of the Palace of Fine Arts downtown. (Photographic archive Juan José Saldaña González)

[28] https://uia.org/s/or/en/1100004812. Accessed in May 2024.

the reader with an idea of the Latin American themes, we present the following list as they appeared on the program[29]:

Ciencias agrícolas y cultura en América Latina en el siglo XX
Ciencia y tecnología en el México Antiguo
La materia médica novohispana del siglo XVI
Evolucionismo darwinismo y las ciencias biológicas y sociales en América Latina
Social history of medicine in developing countries
La historia de las matemáticas en países latinoamericanos
Archive of science in Latin America
Ciencias, salud y poder en América Latina y el Caribe
Comparative studies in the rise of research in modern physics in Latin America and the Pacific Rim Countries
Los precursores de la enseñanza y la investigación en la historia de la ciencia en Latinoamérica

In addition to the specific symposia, Latin American historians of science participated in various thematic panels, including those with Western or other cultural content. It is worth noting that the Spanish language, as the congress's official language, also increased the number of speakers from the region. The Proceedings published many lectures in Spanish, too (Fig. 4.6).

Fig. 4.6 Program and proceedings cover of the XXI ICHS 2001 Mexico City

[29] A specific table on methodology was presented "Methodology and sources for the history of science and technology".

4.5 *Quipu, Revista Latinoamericana de Historia de la Ciencia y la Tecnología*[30]

The LASHST, like all scientific societies, started publishing a specialized quarterly journal called *Quipu, Revista Latinoamericana de Historia de la Ciencia y la Tecnología* in 1984. The journal's founders chose this symbolic name, which refers to the information processing, communication, calculation, and computing systems used by the Inca people of the Andean Cordillera, to highlight their commitment to preserving, studying, and appreciating the scientific and technical history of Latin America's people. *Quipu*'s main goal was to support and spread research on Latin American science and technology history. Its role was crucial in the surge of interest in the field during the 1980s and 1990s. *Quipu* challenged the prevailing cultural and professional norms by demonstrating that Latin American science deserved to be studied on its terms rather than viewed exclusively through a Eurocentric lens. The journal presented the scientific advancements made in the region, including contributions from notable individuals, the establishment and functioning of institutions and laboratories, the study of textbooks and teaching methods, technological applications, archives, and specialized bibliography. The journal's unique and significant subject matter drew the attention of librarians from various parts of the world, including Europe, the United States, Japan, and Australia, in addition to numerous others (Saldaña, 2004, 5–9).

The first issue of *Quipu* (Vol. 1, No. 1, pp. 5–6) was launched with these words:

> Quipu constituye el primer esfuerzo editorial para consolidar los significativos avances que se están operando en América Latina en la comunicación entre historiadores de las ciencias y de la tecnología, y en la profesionalización de su actividad. La historia de las ciencias y de la tecnología en América Latina no había constituido hasta años recientes, una disciplina regularmente cultivada en los medios académicos, universitarios o gubernamentales dedicados a la promoción de la actividad científica y tecnológica en los países de la región. Desde luego, esta disciplina tuvo en el pasado y en diferentes países latinoamericanos, cultores eminentes que contribuyeron con sus investigaciones a descubrir el importante y rico pasado científico y técnico de sus respectivos países. El carácter privado y el aislamiento de esas investigaciones, fueron sus rasgos distintivos. Sería hasta una época reciente, cuando en diversos países se iniciara una investigación histórica regular [...].[31]

The journal publishes articles in Spanish, Portuguese, English, and French. Initially, the Editorial Board consisted of the director and founder, Juan José Saldaña, eight

[30] *Quipu, Latin American Journal of the History of Science and Technology.*

[31] The Quipu initiative is the first editorial effort to consolidate the significant advances taking place in Latin America regarding the communication between historians of science and technology, as well as the professionalization of their activity. In recent years, the history of science and technology in Latin America has become a discipline regularly cultivated in academic, university, or governmental circles dedicated to promoting scientific and technological activity in the region's countries. Although eminent researchers in different Latin American countries have contributed to discovering the essential and rich scientific and technical past of their respective countries, the private character and the isolation of these investigations were their distinctive features. However, in recent times, regular historical research has begun in several countries (authors' translation).

EDITOR IN CHIEF
Juan José Saldaña González

EDITORIAL BOARD

José Babini (Argentina)	José Luis Peset (Spain)
Simão Mathias (Brazil)	Enrique Beltrán (Mexico)
Luis Carlos Arboleda (Colombia)	Arturo Alcalde Mongrut (Peru)
José López Sánchez (Cuba)	Marcel Roche (Venezuela)
Rodrigo Fierro (Ecuador)	Dirk J. Struik (USA).

ASSOCIATE EDITORS

Venezuela	**Brazil**	**Argentina**	**Spain**
Hebe Vessuri	Shozo Motoyama	Ana Celina Lértora M.	Antonio Lafuente
Peru	**Colombia**	**Ecuador**	**Costa Rica**
Ernesto Yépez	Emilio Quevedo	Eduardo Estrella	Angel Ruíz

Fig. 4.7 Editorial Board of Quipu Magazine in 1984 (vol. 1, no. 1)

other Latin American personalities, and two prominent international figures in our discipline from Spain and the United States of America (Fig. 4.7).

In 1994, the journal's Editorial Board was reformed with historians from both inside and outside the region, reaffirming its research vocation, international presence, and professional development achievements: a I. B. Cohen (USA), Marcos Cueto (Perú), Roy MacLeod (Australia), Michel Paty (Francia), Lewis Pyenson (Canadá), Thomas F. Glick (USA), Silvia F. Figueirôa (Brasil), Carlos Galles (Argentina), Teresa Rojas Rabiela (México), además de Luis Carlos Arboleda (Colombia), Antonio Lafuente (España), Pedro M. Pruna (Cuba), Hebe M. C. Vessuri (Venezuela) and Juan José Saldaña (México).[32] All of them, including the regional editors and directors of scientific sections, authored a significant number of articles published in various volumes of Quipu.

The magazine was circulated in print form during its early years, which was the norm then. It had both advantages and disadvantages. The advantage was that individuals and institutions interested in the magazine could access the physical collection or be informed about the publication of an article of their interest through international databases. They could then consult the article or request it be sent to specific libraries. However, the major disadvantage was the distribution process from the place of production (UNAM, Faculty of Philosophy and Letters, Mexico City) to inside and outside Latin America. This process involved costs such as bank transfer and shipping of copies. Therefore, the editor and his team had to put great

[32] *Quipu* 1994 (vol 11, num 1).

Fig. 4.8 Saldaña, founder and director of Quipu magazine. (Photographic archive Juan José Saldaña González)

effort into taking copies to all the countries the editor traveled to congresses, meetings, and international events (Fig. 4.8).

The journal survived for many years thanks to a group of subscribers and libraries. However, more than subscriptions alone were needed to cover the editorial production costs, which required institutional support. In order to obtain this support, special issues were produced with the help of affiliated national societies and dedicated to the history of science in countries such as Brazil, Cuba, Mexico, Colombia, and Venezuela. The journal published articles until 2000 but experienced financial difficulties in its final years. While the Consejo Nacional de Ciencia y Tecnología de México (CONACYT) provided funding for some time,[33] the journal's budget had to be supplemented with contributions from other institutions and universities.[34] Unfortunately, financial problems ultimately led to the closure of the journal. During its initial stage (1984–2000), the editorial process was almost entirely done by hand due to a limited budget that did not allow for the payment of professionals for editing, publishing, and sales (Saldaña, 2004, 11–12).

The journal's closure was not related to the quality of its articles, which were of a very high standard. Both the authors and the referees, responsible for strict and objective reviews, were part of the community of professional historians of science with a high level of work. The main problems were the economic crisis in Mexico in 1994 and the temporary and insufficient suspension of CONACYT support in 1993, which caused irregular publication of the journal. Two years later, CONACYT

[33] National Council of Science and Technology of Mexico. Recently renamed Consejo Nacional de Humanidades, Ciencias y Tecnologías (CONAHCYT), Mexico.

[34] For example, Colciencias-Minciencias of Colombia, the Academy of Sciences of Cuba, CNPq of Brazil, CONICIT of Venezuela, the Instituto Politécnico Nacional (Mexico) (Aguilera 2000, 7), the Universidad del Valle (Colombia) and in Mexico the FFyL-UNAM and the Universidad de Puebla.

conditioned the budget to the journal being registered in international indexes, but the distribution problems seemed insurmountable (Aguilera, 2000, 8).

The arrival of the internet brought about the transition to a digital magazine format, which added to the already difficult circumstances that led to the magazine's closure. However, later, it was seen as an opportunity to reopen the magazine under the name Quipu-Segunda Época, with the original production categorized under Quipu-Primera Época. The second phase of the magazine began in 2012 as a fully digital publication, utterly separate from the LASHST. Its administration, financing, production, and copyright belonged to Saldaña, the director, until 2014.[35] In this second phase, authors paid a moderate fee to publish their work, and readers did the same.

Starting in 2014, the journal, along with other virtual resources such as webinars, interviews, courses, the Ceromayalist, and access to the database of volumes since its inception, was included in a new project conceived by Saldaña. The goal of the project was to contribute to and support the development of our field of study within the new institutional and professional realities of the global scientific world: la *Comunidad en Internet de Historia de la Ciencia y la Tecnología de América Latina* (CIHCyTAL). The project gained attention from the Latin American community, but, unfortunately, the journal could not continue publication due to a lack of resources and a suitable academic and administrative management team. Eventually, Saldaña finished the CIHCyTAL project and had close access to its resources and services. We hope the platform will be reactivated soon and made available to the academic world by some entity.

It is worth noting that the Latin American community played an active role in the *Quipu* journal, even during difficult financial times. They contributed by submitting articles for publication, serving as referees, writing reviews, and collaborating on bibliographic essays. Those who knew Juan José Saldaña knew that the journal was the most cherished possession of the LASHST, and he tried to keep it going even in the worst circumstances. However, not publishing issues for a few years would mean the end of the journal because authors need their publications to validate their academic work and cannot wait for a journal to recover from a relapse. In the case of *Quipu*, it was evident that the supportive community was crucial for the journal's continuity as it was their only collective publishing project. The chart below shows the countries of origin of authors who published the most articles in *Quipu* from 1984 to 2000 (Fig. 4.9).

[35] The *Quipu* Journal is available at: https://search.app/ZbYmvvAWekuypJuU6 https://web.archive.org/web/20220804093919 https://www.revistaquipu.com/

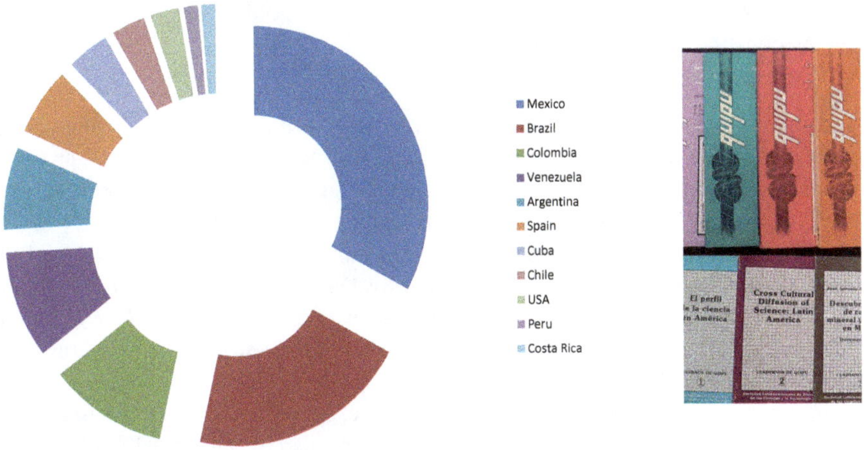

Fig. 4.9 Participation of authors in *Quipu* categorized by their country of origin. (Own elaboration)

4.6 Final Remarks

We regret the slow extinction of the LASHST, its journal *Quipu*, and the CIHCyTAL Community project despite their significance for the professionalization of the history of science in several Latin American countries. These initiatives played a vital role in establishing new local societies and revitalizing old ones. Moreover, they contributed to the identification of crucial themes for the development of the epistemological, cultural, and territorial identity of the history of Latin American science, which have now become lines of research and training in many countries. IASCUD, for example, is a still active institution that has made significant contributions to the fields of history of science and cultural history.

The Bucharest Declaration's requirements have been met mainly across various countries, although some differences remain. The community of historians of science in Latin America is now rightfully integrated into the global community. In the region, there are scientific journals, national societies, archives, databases, specialized libraries, institutes, centers, and research groups that offer training programs for historians along strategic lines. Although there are still limitations to overcome, progress has been made, for example, in the study and valuation of our scientific heritage and in how we use and appropriate the history of science in science education. Communication between the countries in the region and with countries in both hemispheres is fluid. Scientific mobility resources include networks, collaborative projects, and participation in events and publications. Academic efforts that were once heroic are now becoming commonplace as the following:

- The involvement of leading historians of science in Latin America in formulating and executing science policies.
- Their involvement in orienting research programs in universities and national science and technology systems.

- Demonstration of commitment and leadership by holding senior management positions in our field or related areas.
- Proposals to organize international congresses in Latin America.

The authors know that Saldaña approached several institutions and publishing companies with proposals to recommence *Quipu* as an academic business project. Before his death, he was considering at least one proposal from abroad. Unfortunately, none of these proposals were successful in reaching an agreement. Currently, *Quipu*, the *Cuadernos de Quipu*, and other valuable documents of the LASHST are part of the legacy that Saldaña left behind for his family. During the *I Jornada Internacional de Historia de la Ciencia y la Tecnología en América Latina. Juan José Saldaña González In Memoriam* held on March 23 and 24, 2023, at UNAM,[36] several academics expressed their support for reviving *Quipu* in different ways. However, this matter is still pending. The authors of this chapter hope that their contribution will help create better conditions for its realization. They also trust that the links between the Latin American community and the history of science will continue to be strengthened.

Acknowledgments We would like to extend our heartfelt thanks to Dra. Clara Isabel Carpy Navarro, the Saldaña's widow, for generously providing us with photographs from the Photographic Archive Juan José Saldaña González. We would also like to express our gratitude to Luz Marina Duque for sharing the document related to the establishment of the LASHST with us.

References

Aguilera Ríos, S. (2000). Quipu: una Revista Latinoamericana de la Historia de las Ciencias y la Tecnología. *Biblio 3W. Revista Bibliográfica de Geografía y Ciencias Sociales* 212. https://biblat.unam.mx/es/revista/quipu/

Alzate y Ramírez, J. A. (1988). Descubrimientos del Carbón Mineral y Petróleo en México (Documento inédito de J. A. Alzate). *Cuadernos de Quipu 3*. LASHST.

Babini, N., & De Asúa, M. (2003). La Historia de la Ciencia en Argentina en el último cuarto de siglo. *Llull: Revista de la Sociedad Española de Historia de las Ciencias y de las Técnicas, 26*(56), 731–738.

De Ríos, D. (1984). Memorias del Seminario sobre Metodología de la Historia Social de las Ciencias en América Latina. *Quipu, Revista Latinoamericana de Historia de las Ciencias y la Tecnología, 1*(3), 469–472.

Garfield, E. (1985). The life and career of George Sarton: The father of the history of science. *Journal of the History of Behavioral Sciences, 21*, 107–117.

Mathias, S. (1986). Reminiscências. *Quipu, Revista Latinoamericana de Historia de las Ciencias y la Tecnología, 3*(2), 245–250.

Petitjean, P. (2006a). The 'periphery principle' crossing borders. Contributing to the development of science in Latin America. In *Sixty years of science at UNESCO 1945–2005* (pp. 71–72). UNESCO. https://unesdoc.unesco.org/ark:/48223/pf0000149137

[36] I International Conference on the History of Science and Technology in Latin America.

Petitjean, P. (2006b). Social aspects of science. Meetings of minds. UNESCO and the creation of the International Union of History of Science. In *Sixty years of science at UNESCO 1945–2005* (pp. 81–82). UNESCO. https://unesdoc.unesco.org/ark:/48223/pf0000149137

Quevedo, E. (Ed.). (1993). *Historia Social de la Ciencia en Colombia*. Colciencias.

Quipu, Revista Latinoamericana de Historia de las Ciencias y la Tecnología. 1984. vol. 1, no. 1 and 3.

Ramos-Lara, M. P. (2024). Juan José Saldaña en la Historia de la Ciencia y la Tecnología en México. *Saberes. Revista de Historia de las Ciencias y las Humanidades, 7*(15), 87–104. https://www.saberesrevista.org/ojs/index.php/saberes/article/view/301

Ramos-Lara, M. P., & Arboleda, L. C. (Eds.). (2024a). *La cátedra que profesionalizó la Historia de la Ciencia en México. Contribución de la FFyL-UNAM a América Latina*. México, Universidad Nacional Autónoma de México (UNAM).

Ramos-Lara, M. P. & Arboleda, L. C. (2024b). Eloge: Juan José Saldaña González (1944–2022). *ISIS Journal, 115(2)*. https://www.journals.uchicago.edu/ https://doi.org/10.1086/730306

Ruíz Zúñiga, Á. (2003). La profesión de la historia de las ciencias y las matemáticas en Costa Rica en los últimos veinticinco años. *Llull: Revista de la Sociedad Española de Historia de las Ciencias y de las Técnicas, 26*, 701–730.

Saldaña, J. J. (1985). *¿Qué es la Sociedad Latinoamericana de Historia de las Ciencias y la Tecnología? ¿What is the Latin American Society for the history of science and technology?* Grupo Edición.

Saldaña, J. J. (1989). Marcos conceptuales de la historia de las ciencias en Latinoamérica. Positivismo y economicismo. In *Introducción a la Teoría de la Historia de las Ciencias* (pp. 337–363). Universidad Nacional Autónoma de México. (second edition, first edition 1985).

Saldaña, J. J. (2004). Mi experiencia como editor de la Revista Quipu. *Boletín Mexicano de Historia y Filosofía de la Medicina, 7*(2), 60–65.

Saldaña, J. J. (2006). Introduction. The Latin American scientific theater. In *Science in Latin América* (pp. 1–27). University of Texas Press.

Saldaña, J. J. (Ed.). (1986). El perfil de la ciencia en América (XI Congreso Interamericano de Filosofía. Actas del Simposio Historia y Filosofía de la Ciencia en América, Guadalajara, Jalisco, Mexico, November 12–14, 1985). *Cuadernos de Quipu 1*. LASHST.

Saldaña, J. J. (Ed.). (1987). Cross cultural diffusion of science: Latin America (Symposium "Cross Cultural Transmission of Natural Knowledge and its Social Implications: Latin America", Acts of the XVII International Congress of History of Science, Berkeley, California, 31 July to 8 August 1985). *Cuadernos de Quipu 2*. LASHST.

Saldaña, J. J. (Ed.). (1992). Los orígenes de la ciencia nacional (monografías del Seminario de Investigación y Tesis sobre Historia de la Ciencia y la Tecnología de la División de Estudios de Posgrado, Departamento de Historia de México, Facultad de Filosofía y Letras, UNAM). *Cuadernos de Quipu 4*. LASHST and Facultad de Filosofía y Letras (UNAM).

Saldaña, J. J. (Ed.). (2001). *Science and cultural diversity. Filling a gap in the history of science*. LASHST.

María de la Paz Ramos-Lara is a researcher at CEIICH-UNAM, where she founded the History of Science Department and two book collections: *Bibliotheca Mexicana Historiae Scientiarum* and *Science and Technology in the History of Mexico*. Head of the Laboratory for Interdisciplinary Research and Complexity Sciences, and the Science and Technology Department. She is also a Philosophy of Science Postgraduate Program professor at UNAM and a member of the National System of Researchers (SNI-Conahcyt, Mexico). She has authored three books, published 16 more, and written a hundred chapters and articles on the History of Science in Mexico. In recent years, she has adopted the theoretical and methodological framework of complex systems to study the history of science.

Luis Carlos Arboleda is Emeritus Professor at the Universidad del Valle, Cali, Colombia. For a little more than half a century he has taught the history of mathematics and science at this university, where he was also Vice Rector of Research. From 1991 to 1994 he was subdirector of Colciencias (today Ministry of Science), an entity that designated him as Emeritus Researcher in 2018. Since 2017 he has been a full member of the Colombian Academy of Exact, Physical and Natural Sciences. He was awarded the *Sábato Medal* of the Inter-American Commission on Mathematical Education (CIAEM for its acronym in Spanish) in 2023 in recognition of his services in history and mathematics education.

Chapter 5
Notes on the Contributions of the LASHST to the Development of the Historiography of Science in Latin America

Maria Amélia Mascarenhas Dantes

Abstract The text follows the performance of the Latin American Society for the History of Science and Technology (LASHST) in the 1980s and analyzes its contributions to the integration and professionalization of the community of Latin American science historians. It also deals with the theoretical and methodological debates that it implemented, and that contributed to the consolidation of the institutionalization of the History of scientific and technical activities in Latin American countries and the implementation of contextualized approaches.

Keywords Science · Latin America · Historiography

5.1 Introductory Text

As an introduction to this text, first I want to record my memories of the early 1980s, when I had contact with a group of Latin American historians, led by the Mexican Juan José Saldaña, on a visit to the University of São Paulo. The group represented the newly created Latin America Society for the History of Science and Technology (LASHST). During these years, me and several Brazilian colleagues began to dedicate ourselves to studies on the History of Brazilian sciences and, from this first contact, I had the opportunity to join the community of historians that was forming in the region. I was also able to take part in debates that were being implemented by the Society and which focused on a central theme for my research: the lack of recognition of scientific and technical past of peripheral regions by world historiography.

In this text, in addition to following the creation and performance of the LASHST in its first decade, my interest is especially focused on the debates on the

M. A. M. Dantes (✉)
Dep.de História – FFLCH-USP, Universidade de São Paulo, São Paulo, Brazil
e-mail: mamdantes@usp.br

M. de la P. Ramos-Lara, L. C. Arboleda (eds.), *History of Science in Latin America: The Construction of an Intellectual Field (20th century)*,
https://doi.org/10.1007/978-3-031-75536-1_5

methodology of the History of Sciences in Latin America that were taking place in these years. And which, in my opinion, were fundamental for the institutionalization and recognition acquired by the studies on the History of science and technology in the countries of the region.

5.2 The Creation of LASHST and the Context of the History of Science in the World and Latin America

The creation of LASHST, in 1982, took place in a context of great expansion in the field of History of Science worldwide, with the formation of communities of professionals that spread across several continents.

Since the post-war period, the area has undergone transformations. From a methodological point of view, from the 1960s and 1970s onwards, social studies of science expanded. Also, as the Australian historian Roy MacLeod recalls, with the reorganization of international relations and the end of colonial systems, these years were marked by the debates about the role of science and technology in the development of former colonies and peripheric regions, and by the implementation of studies about the presence of science in European expansions (MacLeod, 2001). The article by the North American historian George Basalla was in this research line and analyzed the process of worldwide diffusion of modern Western science. Also encouraged the studies on the development of sciences outside Europe (Basalla, 1967).

In these years, debates about the relationship between science, technology and development were also widespread in Latin American countries. Among the many books written by renowned scientists, we can mention "*Ciência e libertação*", by the Brazilian José Leite Lopes (Lopes, 1969) and "*Ciencia y política en América Latina*" by the Argentine Amilcar Herrera (Herrera, 1971). They were texts that sought to value the role that scientific and technological research could have for the economic development of our countries. In this context, studies on the scientific and technological past in Latin America gained prominence for the contribution they could bring to scientific and technological policies.

Thus, from the 1970s onwards, with the first generation of professional Latin American science historians—trained in their countries or abroad—the national sciences gained prominence as a research topic. It was in this context that the LASHST was created in 1982.

5.3 The Creation of LASHST

According to the testimonies of Luis Carlos Arboleda and Juan José Saldaña, the project to create a Latin American association was prepared in 1981, by some young researchers trained in Europe—like Arboleda and Saldaña—and some mentors from the History of Science in Latin America. This group drafted a platform of commitments that aimed to change the conditions of the discipline, bring together the community of scholars—professionals and amateurs—then active, and contribute to the institutionalization and professionalization of History of Science in Latin America (Arboleda, 1984; Saldaña, 1983, 1984).

This group was responsible for the first meeting on the History of Sciences held in Latin America, which took place in the city of Puebla, Mexico, in 1982. At this meeting, which saw the significant participation of representatives from 16 countries in America and Europe, the LASHST was created, being chosen as first president Juan José Saldaña and as members of the Latin American Council: José López Sánchez, for Cuba; Luis Carlos Arboleda, for Colombia; Ruy Gama, for Brazil; Elena Pennini de Vega, for Argentina and Arturo Alcalde Mongrut, for Peru.

The society's objectives were:

1. Support for the rescue and preservation of scientific and technological heritage of Latin America (by holding debates and actions aimed at the rescue and preservation of documents).
2. Support and promotion of historical research (by creating favorable institutional conditions for activities in the area: chairs, courses, seminars, financing).
3. Organization and dissemination of the work of Latin American historians (through publications and meetings).
4. Promotion of teaching History of Science and Technology (by meetings and disseminating existing experiences).

But, as we will see, the LASHST, in addition to contributing to the professionalization of the History of Science, also played a fundamental role in the institutionalization of a new area of research, that of historical studies on the sciences and technologies developed in the countries of the region.

5.4 LASHST in Action. Realization of Events and the Integration of Latin American Researchers

Seeking to fulfill its objectives, the LASHST developed intense activities in its first years. In the 1980s, it held two Latin American Congresses: the first, in 1985, in Havana, Cuba; and the second in São Paulo, Brazil. In the following years, Congresses were held every 3 years.

Also, in the early years, it organized several events about different questions: methodology of History of Science, historical sources, and teaching of History of Science. In addition to these Latin American congresses, the Society took part in the organization of two international congresses in 1985. It was responsible for a symposium on Latin America at the XVII International Congress on the History of Science, held in Berkeley, United States (Saldaña, 1987). And it also took part in the organization of a congress that took place in Madrid, with Latin American,

Portuguese, and Spanish researchers, about the theme "*Science in the New World*" (Peset, 1985).

I consider it important to note that the Latin American Congresses in these early years had many participants and were spaces for meetings and debates for the Latin American researchers. They also generated texts that circulated across the continent and other countries, disseminating research that was being carried out and debates on theory and methodology of the History of Science. I had the opportunity to take part in several of these events—in Mexico, Colombia, and Brazil—where I also met, in addition to Mexican and Colombian colleagues, historians from Argentina, Chile, Venezuela, Peru, Ecuador, Cuba, Costa Rica, but also from Spain and Portugal and the United States and Canada.

Thus, they met one of the concerns of the society's founders: the lack of integration among researchers in the region.

As the Mexican Enrique Beltran emphasized in an article published in the first issue of *Quipu* journal, the contacts among Latin American intellectuals were deficient, and they were little informed about what was happening in the sister republics, despite their common origin, language, and customs (Beltran, 1984).

With these events, this situation began to change.

5.5 LASHST in Action. The Publications

5.5.1 Quipu

Quipu journal deserves special mention. A quarterly publication, it began to be published in 1984 and had Juan José Saldaña as its editor in the 1980s and 1990s, with regular publication until 1992 and some issues published from 1996 to 2000. The journal contributed greatly to the dissemination of scientific production which, in these years, was being carried out by Latin American science historians. The name of the journal—*Quipu*, an accounting system used by the Incas—also met one of the objectives of the LASHST, the recognition of the scientific and technological past of the people of Latin America.

The analysis of published texts shows us that there were already a significant number of historians of science active in the various countries and that the major theme of the moment were studies on aspects of the implementation of scientific and technical activities in Latin America, for the colonial period and for independent nations. Among the topics covered were the diffusion, reception, adaptation, and institutionalization of scientific theories; scientists and institutions—schools, academies, commissions, expeditions, research institutes, publications; etc. Theoretical and methodological issues and other topics related to the discipline were also published—such as the special volume on the teaching of History of Science in Latin America.

Several numbers of the journal were dedicated to the historiography produced in some countries. Thus, there are volumes on the History of Science and Technology in Mexico, Colombia, Argentina, Venezuela, Brazil, and Cuba.

In a 1993 evaluation, Luis Carlos Arboleda, then president of LASHST, in addition to characterizing *Quipu* as a laboratory of trends and investigative strategies, emphasized the journal's contribution to the dissemination of more up-to-date references of scientific production between our historians of science and for the international recognition of their work. And he concludes by stating that "En buena medida *Quipu* es el resultado de la producción científica de una comunidad científica que dificilmente se habría construído sin el instrumento de la revista".[1]

5.5.2 The "Cuadernos de Quipu" (Quipu's Notebooks)

It is also important to record the two volumes published as "*Cuadernos de Quipu*" and which were fundamental for the dissemination of the debates that were implemented by the LASHST about the historiography of sciences of peripheral regions and specially of Latin America.

Our next topic is precisely about these debates.

5.6 LASHST and the Debates on the Historiography of Latin American Sciences

As we saw, when analyzing the volumes published by the *Quipu* journal, in the 1980s, there was already a significant number of researchers who were dedicated to studying the scientific and technological practices developed in the various Latin American countries. However, these themes were still hardly recognized by the world historiography of science.

In this context, the debates implemented by the Society on methodological questions in the History of Science in Latin America were fundamental.

In November 1985, the Society held in Guadalajara, Mexico, the symposium "*Historia y filosofia de las ciencias en America*", which resulted in a collection of texts published as the First *Cuadernos de Quipu* with the title "*El perfil de la ciência en America*". On the back cover of the book, the editor explains that the publication sought to contribute to the significant efforts that were being made to disseminate studies on the sciences and techniques of Latin America. And poses as central questions of the book: What were the American contributions for the development of science? and what are its historiographical and philosophical approaches? (Saldaña, 1986a).

[1] Arboleda, 1984, p. 7.

Articles from researchers who were very active in the Society are presented in the book. Latin Americans Juan José Saldaña, Hebe Vessuri, Luis Carlos Arboleda, Xavier Polanco, Ubiratan d'Ambrosio and Celina Lértora; and the Spaniards Antonio Lafuente and José Sala Catalá. The book also had a text by the American historian Sally Kohstedt.

In this book we see that the term "peripheral science" was used differently by the authors. For some, what mattered was the question of the scientific periphery, because in the global system of knowledge production, there are centers—with a decisive role in the paths of this production—and peripheries, with a secondary position (Polanco, 1986). Other authors sought to integrate the scientific question with the economic one. Like Hebe Vessuri who, in her text, deals with the history of sciences in underdeveloped or developing countries (Vessuri, 1986).

The Second *Cuadernos de Quipu* also deserves our attention. It was published in 1987 with the title "*Cross Cultural Diffusion of Science: Latin America*" and presented the proceedings of the symposium organized by LASHST at the XVII International Congress of the History of Science, held in Berkeley, United States. In addition to Juan José Saldaña, Hebe Vessuri, Ubiratan d'Ambrosio and Antonio Lafuente, the researchers Regis Cabral, Rafael Chabran, Eduardo Estrella, Desiderio Papp and Rosaura Ruiz were present in this book. It is a volume also dedicated to the History of Science of Latin America, more specifically to the topic of the diffusion of sciences from the centers to the so-called scientific peripheries, a topic that was then widely debated in the line of studies on "Sciences and Empires". According to the book's editor, the various texts proposed a new conceptualization of the transmission process which, in his view, should be understood "not only as a mechanism for the diffusion from cultural centers, but as a complex process of epistemological, social and cultural nature" (Saldaña, 1987).

From the analysis of these two publications, we can say that the interest of Latin American historians was that, until then, the History of Science had privileged the past of the countries considered leaders in the international scientific production. And that, in this picture, Latin American countries were absent. They sought to understand this historiographical characteristic—which they considered to have a theoretical and methodological background—since, in their view, sciences had always been present in the history of Latin American countries. They also questioned the models for diffusion of modern science—with emphasis on the model proposed by George Basalla—which, in their view, did not consider the complexity of the process of implementing science in peripheral regions.

We can distinguish two movements in these debates: the criticism of the traditional historiography of science; and proposals for a new methodology for the History of Science in Latin America.

5.6.1 Historiographical Criticism and New Proposals

The book "*El perfil de la ciencia en America*" brings a variety of approaches for historiographical criticism.

The text written by Saldaña, "Marcos conceptuales de la historia de las Ciências en Latinoamérica: Positivismo y economismo" (Saldaña, 1986b) is greatly elucidative, presenting a historiographical analysis. The author analyzes how the history written by Latin Americans about the science of their countries has changed since the end of the nineteenth century, when a positivist historiography was disseminated, until the 1950s, when externalist histories were written in various countries. It is an analysis that is situated in the internalism–externalism debate, then very present in the historiography of science.

Initially, he analyzes works by Mexican and Argentine authors from the end of the nineteenth century, concluding that they developed a positivist, internalist and Eurocentric history, that identified European science as a paradigm of universal science. In other words, as they didn't incorporated the action of social factors in shaping scientific activity, these authors just considered Latin American contributions to universal science. For them, Latin American science was a simple copy of European science and had contributed little to global scientific development.

Saldaña also analyzes Mexican and Cuban authors, who followed the externalist approach, which had as a reference the book *Science in History* by John Desmond Bernal, published in 1954. He concludes that, despite considering the action of social variables, they fell into a reductionist economism that linked very directly scientific development to the demands of the productive sector and didn't consider that, in fact, science is a social practice in the broadest sense, that is, a cultural, political, and ideological practice. In conclusion, he considers that the positivist and economistic approaches were not suitable for the History of Science in Latin America: "Ambas concepciones no permiten aprehender la especificidade de la práctica científica latinoamericana y conducen, en cambio, a um mimetismo metodológico. El mimetismo es siempre una alineación. Imitar en historia es, por tanto, una pérdida de la propia identidad".[2]

The authors of the other articles in the book also present a criticism that is close to Saldaña: for them, traditional historiographical production was not suitable for the study of peripheral regions. Hebe Vessuri, based on the concept that science is always integrated with culture, argued that historiographical production should focus on studying the cultural roles of science in different social contexts (Vessuri, 1986). Antonio Lafuente proposes the replacement of the word "science" by "scientific activity" because science might be seen as a social, concrete activity, performed by people and with determined institutional spaces. He emphasizes that this conceptual change would contribute to making historiographical production more connected with the social reality in focus (Lafuente, 1986.)

[2] Saldaña, 1986b, p. 79.

Thus, for the theoretical-methodological proposals, in summary, we can say that the different authors defended the use of a History of Science that treated science as a social activity whose development would be influenced by several factors, economic and political, but also cultural and ideological; and that considered the values and meanings attributed to scientific practices in different social contexts. The analysis of LASHST publications in these years—*Quipu* journal and *Cuadernos de Quipu*—shows that these new references were beginning to be used by the Latin America researchers.

The texts published in *Cuadernos de Quipu* n. 2, "*Cross Cultural Diffusion of Science: Latin America*" (Saldaña, 1987), for example, presented new readings on the theme of the diffusion of scientific theories and practices from central to peripheral regions, until then seen as simple receivers of external actions. The new studies, following the precepts of sociological studies of science, began to analyze in a more differentiated way the action of local variables in the process of domestication of theories.

The *Quipu* journal was also a fundamental vehicle for disseminating the research that was being carried out and which sought to contribute to the recognition of the Latin American scientific and technological past, with its specificities.

Then, some of these texts had great repercussion, due to the innovative way in which they worked and became references in debates on methodological questions of the History of Science in our countries. They also presented extensive research that, in addition to being integrated with the most current references in the Social History of Science, were based on an extensive survey of primary sources.

Initially, I want to remember two articles published in 1987 that deal with events that occurred at the end of the colonial period in different regions of Spanish America.

Luis Carlos Arboleda's article, "Acerca del problema de la difusión científica en la periferia. El caso de la física newtoniana en la Nueva Granada", published in volume 4(1) of the journal *Quipu* (Arboleda, 1987), as the title indicates, focuses on the dissemination of Newton's work in a peripheral region, far from the main European centers. However, the text shows the existence of an active enlightened community in New Granada, which closely followed the debates taking place in scientific centers and which identified Newtonianism with experimental physics. The text also questions the duality center—periphery because New Granada was not in disadvantage in relation to the metropolis and the first Castilian edition of Newton's Principia was published in the colony, through the action of José Celestino Mutis, a Spaniard living there.

And the article by Mexican historian Patricia Aceves Pastrana "La diffusion de la ciencia en la Nueva España en el siglo XVIII: la polémica en torno a la nomenclatura de Linneo y Lavoisier", published in volume 4 (3) of *Quipu* (Pastrana, 1987). This text also shows the existence of an active illustrated community in New Spain at the end of the eighteenth century. The emphasis is the performance of the criollo priest José Antonio Alzate who, together with a group of enlightened criollos from the university and protomedicate, opposed the Spanish Crown's project to introduce the theories of Linnaeus and Lavoisier in the colony. The author shows how various factors were present in the actions of these intellectuals. On one hand, its alignment with the independence movements. But also, the way they integrated pre-Columbian knowledge with European scientific traditions spread during the colonial period.

Also, the article by Peruvian historian Marcos Cueto "Nacionalismo y ciencias Médicas: los inicios de la investigación biomédica en el Perú: 1900-1950", published in volume 4(3) of the journal (Cueto, 1987) was much widespread in those years. There, Cueto analyzes the specific forms of the process of implementation of bacteriology and physiology in Peru, from 1900 to 1950. Regarding bacteriology, he shows how the spread of positivism and nationalism led researchers to choose the study of a national disease, the Peruvian Wart, and how this choice gave international prestige to the Peruvian Alberto Baron. About physiology, he also analyses how indigenous and nationalist ideologies marked the actions of scientists Carlos Medrano Monge and Alberto Hurtado in the study of effects of altitude on the human body and were present in the controversy they maintained with foreign scientists. The author also questions the duality center-periphery, when he emphasizes how these episodes shows the concrete possibility of the implementation of areas of scientific excellence in an underdeveloped country like Peru.

Finally, I consider it important to emphasize that these texts show that already in the 1980s a new historiography of science in Latin America began to be constructed.

The work of the LASHST was essential for this. The debates it implemented were fundamental for Latin American science historians to stay in tune with the most up-to-date trends in the Social History of Science which, in these years, began to define science as a contextualized practice, both for the central regions and for the peripherical ones (Pestre, 1995).

We can say that a turning point in this trend was the publication, in 1992, of the book "*The Scientific Revolution in National Context*" which, instead of focusing on one Scientific Revolution, presented studies about the multiple Scientific Revolutions that had taken place in the various countries on the European continent (Porter & Teich, 1992).

5.7 Developments of LASHST'S Performance: Latin American Dialogues and International Projection

The multiple activities implemented by LASHST led to the consolidation of academic relationships between Latin American researchers. An example of the construction of more solid and fruitful exchanges was the implementation of the network "Historia de la Salud en América Latina y Caribe—HISPALC" in 2000.

We can also mention the proliferation of books and articles that brought together researchers from different countries and that, for example, contributed to carrying out a comparative analysis of the implementation of scientific theories in the different contexts. Like the books published about the dissemination of Darwinist theories: "*A recepção do Darwinismo no Brasil*" (Domingues et al., 2003) and "*El darwinismo en España e Iberoamérica*" (Glick et al., 1999).

We can also remember, as a later development, the book "*Medicine and Public Health in Latin America. A History*" recently published (Cueto & Palmer, 2015).

The international projection soon achieved by the LASHST also contributed to the integration of Latin American historians into the international community. Thus, in the late 1980s and early 1990s, we were actively taking part in international meetings on the relations between Sciences and Empires and on the globalization of sciences, which were quite prestigious at the time. Examples of these events were: the congress held in Paris in 1990, the proceedings of which were published in the book *Science and Empires. Historical studies about scientific development and European expansion* (Petitjean et al., 1992); and the event held in Spain in 1991 to commemorate the 400th anniversary of Columbus's voyage, with proceedings published in the book *Mundialización de la Ciência y cultura nacional* (Lafuente et al., 1992).

We also cannot forget the commitment of LASHST's main leaders to ensure that the XXI International Congress on the History of Science was held in Mexico City, in 2001. This event was the entity's first congress in a developing country and clearly demonstrates the international recognition already achieved by the production of History of Science in Latin America.

5.8 Conclusion

Finally, I consider that it is important to keep in mind how far we are from what happened in the 1980s. Today, as we know, the Historiography of Sciences of our countries is institutionalized. Regarding Brazil, which I know best, now we have several generations of historians who are dedicated to the study of scientific practices in various historical periods and different regions of the country. At an international level, we can say that the scientific past of our countries today is perfectly integrated into the Historiography of Science.

It is also interesting to observe how the themes of ICSTH congresses changed and adapted to new trends. Thus, if the XXI International Congress on the History of Science, held in Mexico City, in 2001, had as its central theme "Science and Cultural Diversity", the XXV Congress held in Rio de Janeiro, in 2017, had the theme "Science, Technology and Medicine—between Global and Local".

Acknowledgments To Malu Dantes for her continued support.
To Marcos de Oliveira for reviewing the English text.

References

Arboleda, L. C. (1984). De como construir uma comunidad cientifica en la periferia. El caso de la Sociedad Latinoamericana de Historia da las Ciencias y la Tecnologia (SLHCT). In S. Figueiroa & M. Lopes (Eds.), *Geological sciences in Latin America. Scientific relations and exchanges (Anais do XVIII INHIGEO congress)* (pp. 3–7). UNICAMP.

Arboleda, L. C. (1987). Acerca del problema de la difusión cientifica en la periferia. El caso de la física newtoniana en la Nueva Granada. *Quipu, 4*(1), 7–30.

Basalla, G. (1967). The spread of Western science. *Science, 156*, 611–622.

Beltrán, E. (1984). La Historia de la Ciencia en America Latina. *Quipu, 1*(1), 7–23.

Cueto, M. (1987). Nacionalismo y ciencias médicas: los inicios de la investigación biomédica en el Perú: 1900-1950. *Quipu, 4*(3), 327–355.

Cueto, M., & Palmer, S. (2015). *Medicine and public health in Latin America: A history*. Cambridge University Press.

Domingues, H. M. B., de Sá, M. R., & Glick, T. (Eds.). (2003). *A recepção do Darwinismo no Brasil*. FIOCRUZ.

Glick, T. F., Ruiz, R., & Puig-Samper, Á. (Eds.). (1999). *El darwinismo en España e Iberoamérica*. UNAM/CSIC/Doce Calles.

Herrera, A. O. (1971). *Ciencia y política en America Latina*. Siglo XXI Editores.

Lafuente, A. (1986). La ciencia como ficción. Historia y contexto. In J. J. Saldaña (Ed.), *El perfil de la ciencia en América* (pp. 31–40). SLHCT.

Lafuente, A., Elena, A., & Luiza Ortega, M. (Eds.). (1992). *Mundialización de la ciencia y cultura nacional*. Doce Calles.

Lopes, J. L. (1969). *Ciência e libertação*. Editora Paz e Terra.

MacLeod, R. (2001). Introduction. *Osiris*, vol. 15, *Nature and empire. Science and the colonial enterprise*, pp. 1–13.

Pastrana, P. E. A. (1987). La difusión de la ciencia en la Nueva España en el siglo XVIII: la polémica en torno a la nomenclatura de Linneo y Lavoisier. *Quipu, 4*(3), 357–385.

Peset, J. L. (1985). *La ciencia moderna y el Nuevo Mundo*. CSIC.

Pestre, D. (1995). Pour une histoire sociale et culturelle des sciences: Nouvelles définitions, nouveaux objects, nouvelles pratiques. *Annales Histoire, Sciences Sociales, 50*(3), 487–522.

Petitjean, P., Jami, C., & Moulin, A.-M. (Eds.). (1992). *Science and empires. Historical studies about scientific development and European expansion*. Kluwer Academic.

Polanco, X. (1986). La ciencia como ficción. Historia y contexto. In J. J. Saldaña (Ed.), *El perfil de la ciencia en América* (pp. 41–56). SLHCT.

Porter, R., & Teich, M. (Eds.). (1992). *The scientific revolution in national context*. Cambridge University Press.

Saldaña, J. J. (1983). La primera reunion latino-americana de Historia de las Ciencias. In *Boletín de Historia de las Ciencias* (numero 1, pp. 3–4). Ed. Universidad del Valle.

Saldaña, J. J. (1984). Balance general de la Historia de las Ciencias en America Latina. In *Anais do Seminario Latinoamericano sobre Alternativas para la enseñanza de la Historia de las Ciencias y la Tecnologia* (pp. 3–11). Universidad del Valle.

Saldaña, J. J. (Ed.). (1986a). *El perfil de la ciencia en América*. SLHCT.

Saldaña, J. J. (1986b). Marcos conceptuales de la historia de las ciencias en Latinoamérica. Positivismo y economicismo. In J. J. Saldaña (Ed.), *El perfil de la ciencia en América* (pp. 57–80). SLHCT.

Saldaña, J. J. (Ed.). (1987). *Cross cultural diffusion of science: Latin America*. SLHCT.

Vessuri, H. M. C. (1986). Los papeles culturales de la ciencia en los países subdesarrollados. In J. J. Saldaña (Ed.), *El perfil de la ciencia en América* (pp. 7–17). SLHCT.

Maria Amélia Mascarenhas Dantes, Retired Professor—Dep. of History-FFLCH-USP. Area of Activity—History of Science, History of Sciences in Brazil. Organizer, among others, of the book *Espaços da ciência no Brasil*, published in 2001 by Editora FIOCRUZ, Rio de Janeiro. She also published dozens of articles and book chapters and supervised around 50 master's and doctorate degrees on topics in the History of Science in Brazil.

Chapter 6
The VI Congress of the LASHST (Buenos Aires)

Carlos D. Galles and Norma Isabel Sánchez

Abstract In the Mexican city of Puebla, in August 1982, the First Latin American Meeting of Historians of Science took place, attended by 200 researchers who presented 120 communications for analysis, representing 16 countries. In that enthusiastic atmosphere, the constitution of the Latin American Society for the History of Science and Technology (LASHST) was agreed upon by the attendees. At the Constituent Assembly, a Latin American Council and an Executive Committee were elected to implement the bold and imaginative project that the LASHST represents and a figure of consensus to lead it emerged: Juan José Saldaña. Over the years, five Latin American congresses organized by the LASHST were held in the following venues: Havana (1985), Mexico City (1988), São Paulo (1991), Santiago de Cali (1994), and Rio de Janeiro (1998). All these congresses were characterized by camaraderie, friendship, and the sensible work of the committees examining articles. This allowed attendees to become acquainted with universities, cultural centers, libraries, magnificent buildings, as well as landscapes, culinary tastes, and customs. It is worth mentioning that these meetings had been held north of the Tropic of Capricorn. The next congress was destined for the Argentine Republic, considerably south of the American continent. One of the signatories of this article, Carlos Galles (C.G.) was elected at the Rio de Janeiro meeting as president of the LASHST. This president had been a study partner of Juan José Saldaña in Paris.

Keywords VI Congreso · LASHST · Buenos Aires

C. D. Galles (✉)
Independent Researcher, Rosario, Argentina
e-mail: carlos.galles@gmail.com

N. I. Sánchez
Institute of History of Science, Sociedad Científica Argentina,
Buenos Aires, Argentina

M. de la P. Ramos-Lara, L. C. Arboleda (eds.), *History of Science in Latin America: The Construction of an Intellectual Field (20th century)*,
https://doi.org/10.1007/978-3-031-75536-1_6

6.1 On the Preparation of the Congress

The beginning of the twenty-first century was a time of political and economic turmoil for Argentines. The government fell, and there were changes in the National Executive Power. For the then head of the Latin American Society for the History of Science and Technology (LASHST), Carlos Galles (C.G.), without political affiliation, it wasn't easy to obtain support from the institution where he taught: the National University of Rosario. He encountered various difficulties and obstacles. We detail a few: Kohn Loncarica, our friend, expressed concern more than once about the likelihood of its realization, not due to lack of dedication, but due to the particularities the country was going through. One of these was the almost impossible task of finding economic resources; he knew the topic well because he held two positions at the time: he was the general secretary of the Faculty of Medicine and president of Sociedad Científica Argentina (SCA). He sought help from the university; he found little response; not because of undervaluation of the meeting, but for other reasons. A similar outcome was the result of his efforts before the Rectorate of the University of Buenos Aires.

The possibility remained of obtaining collaboration from the SCA, which held two fundamental characteristics: the institution committed to providing the use of its halls and offered the cooperation of part of the administrative staff. Those interested in holding the congress did not give up and added Celina Lértora Mendoza, who could take charge of dissemination, as she was an active member of the LASHST society from the early days, with solid experience in organizing congresses and founder of Foundation for the Study of Argentine and Ibero-American Thought (FEPAI), with rapid communication with interested parties from all over the country, Latin America and Europe. Additionally, Pablo Lorenzano, secretary of the LASHST, doctor of philosophy from the Free University of Berlin, professor at the National University of Quilmes, and director of the magazine *Redes*, accepted to participate. Thus, a trio was formed, which collaborated with Carlos Galles and finally, the celebration was a success. It was wonderful to see the delegates fraternizing, making friends, chatting about their academic ventures, and remembering those who had passed away. The central theme addressed was as follows: 20 years of Historiography of Science and Technology in Latin America.

Due to the number of attendees, more than one meeting place had to be enabled. The Sociedad Científica Argentina was the main venue for the event. It has a majestic building and an excellent location; easy for travelers: from Ezeiza airport to lodging near Santa Fe Avenue 1145, in the heart of the city of Buenos Aires. The current building of the institution was inaugurated in 1934. On the ground floor, it has halls (where the different committees met), named: Carlos Darwin, Luis F. Leloir, Francisco P. Moreno and Estanislao S. Zeballos. On the first floor is located the spacious Florentino Ameghino Auditorium, where the General Assembly was held. On the third floor, is the magnificent and valuable Domingo F. Sarmiento library (which recalls the educator and president of the Republic between 1868 and 1874). It was founded in 1872, by a group of young students from the Department of Exact Sciences of the University of Buenos Aires, determined to promote mathematical, physical, and natural studies, and their applications. A great promoter of this idea was Estanislao S. Zeballos (born in Rosario, lawyer and later a prominent jurist and promoter of science in a broad sense). It was accompanied by Luis A. Huergo, as the first president, who had graduated as an engineer from the University of Buenos Aires just 2 years earlier (the first with such a title in the country).

From the outset, it was considered convenient to promote its own publication, which up until today is maintained under the name of *Anales de la Sociedad Científica Argentina.* This journal had its heyday in the last decades of the nineteenth century and the first of the next, a peculiarity that expresses the epochal taste for geology and geography, astronomy, the recognition of national territory, paleontology, etc. There were also articles on physics, chemistry, and mathematics. At the same time, it turned to the inquiry about the Patagonian soils, the southern seas, and even Antarctica, and studies of law, linguistics, history, bioethics, and many more. This task was completed by sponsoring exploration and reconnaissance expeditions, with an emphasis on the Andes mountain range and thoughtful studies to advance in setting boundaries with the neighboring republic of Chile. It has sponsored several congresses. Two of them, held in Argentina, are fondly remembered: the one in 1898, known as the First Scientific Congress, which had representatives from 14 nations and the presentation of 121 communications; and the one in 1910, the year commemorating the Centenary of the May Revolution (a milestone that marks the beginning of the Rio de la Plata's project of independent life, with legal-political separation from Spain).

In this opportunity, attendees exceeded a thousand with several hundred papers, featuring international and national researchers. Since then, many renowned figures have given lectures in its halls: the German physicist Albert Einstein; the Swedish Otto Nordenskjöld (in the time of his Antarctic expeditions); the Italian philosopher Rodolfo Mondolfo, and the three Argentine Nobel laureates in science: Bernardo A. Houssay, Luis Federico Leloir and César Milstein. It is located in the historical, cultural, and commercial center of Buenos Aires. Its management is open to cultural, educational, and academic activities, and it is part of national and international organizations dedicated to promoting science, such as the Argentine Association for the Advancement of Science, whose origins lie with the members of the Society. In sum, the Sociedad Científica Argentina turned out to be the appropriate place for the celebration of the VI Congress of the LASHST. The success of the congress, which had as its motto: "Twenty Years of Historiography of Science and Technology in Latin America", is due to all the collaborators, mainly its directors and the academic staff of its commissions, and especially to one of its active members, Mariano Asla, responsible for the excellent services provided, from gastronomy to accommodation, and transportation. We are grateful and have no other way to acknowledge it but by saying: "Thank you".[1]

[1] https://cientifica.org.ar/historia-de-la-sociedad-cientifica-argentina/

6.2 The Personality of Alfredo Kohn Loncarica

Alfredo G. Kohn Loncarica was the president of the Sociedad Científica Argentina when the VI Congress took place. Who was this professional who supported the organization of the congress with such dedication? To begin with, he was a doctor who liked history, so he cultivated it in two ways: international connections and the past/present of science with a determined orientation towards the history of medicine. He was the director of the Library of the National Congress, a post he competed for. In the Faculty of Medicine, University of Buenos Aires (UBA), he went through all the steps, from "vaccine assistant" to full professor of History of Medicine. He directed the Department of Medical Humanities (with its institutes, library, archive, Athenaeum, museums, publications, chairs, etc.). He received study and improvement scholarships; some of them were fulfilled at the Consejo Superior de Investigaciones Científicas (CSIC, History area) of Spain, where he intensified ties with those who worked there. He shared lengtly conversations with Pedro Laín Entralgo and Francisco Guerra. His last position was that of general secretary of the Faculty of Medicine (UBA).

In all these places, he cultivated very good friends, and when he proposed the convenience of adding the subject of Bioethics to undergraduate courses, he was supported. In the SCA, he also had a long career as member of the Board of Directors, secretary, responsible for *Annals* and President (2001–2005). Under his leadership, outstanding honorary members were admitted, and "in folder" others were treasured. Unfortunately, he did not have enough time left in his life. When he assessed the VI Congress, he was very satisfied. It was his farewell to friends from the continent and Spain. A few months later, he passed away.[2] He felt great respect for Juan José Saldaña and always commented on the good he had achieved by promoting from Mexico, so distant and yet so beloved, the studies that the LASHST was carrying out, and by realizing the degree of isolation of most Latin American researchers (to the point that some ignored each other's productions). He partially solved this problem by implementing a simple bulletin, *Quipu,* the subsequent *Quipu Notebooks* (*Cuadernos de Quipu*, original Spanish name) and by promoting face-to-face meetings.

6.3 The Other Meeting Point

Additionally, commissions met at the Postgraduate Unit of the National University of Tres de Febrero and at the Borges Cultural Center, located at the Galerías Pacífico, in the downtown area of the city. The two venues are separated by only a few blocks.

[2] "Vamos al rescate de la historia de la medicina argentina (bibliografía y encuentros)", por Norma Isabel Sánchez; en: (publicación virtual) *Revista de Historia de la Medicina y Epistemología Médica*. Facultad de Medicina (UBA); vol XI, n° 1; 1° semestre de 2019 (ISSN 1852–6152).

This classic building (by Argentine standards) houses magnificent murals in the 450 square meters of its central dome, painted by Argentine artists: Antonio Berni, Juan Carlos Castagnino, Lino E. Spilimbergo, Demetrio Irruchúa, and the Spaniard Manuel Colmeiro.

6.4 Main Activities During the 4 Days

More than two hundred papers were presented; their authors mainly represented Argentina, Brazil, Mexico, Colombia, and Venezuela while there were also contributions from Uruguay, Peru, Cuba, Chile, Costa Rica, El Salvador, USA, France, and Great Britain.

Five plenary sessions were held, along with ten symposiums, two conferences, a round table, and the presentation of a project.[3]

The following details these activities:

(a) **Plenary Sessions**

- The first, led by Juan José Saldaña, titled: "From the Private to the Public in Science: The First Institutionalization of Science in Mexico".
- The second, led by Luis Carlos Arboleda, "Topics in History and Mathematical Education".
- The third, by Hans-Jörg Rheinberger, "History of Science and the Practices of Experiment".
- The fourth, by Ubiratam D'Ambrosio, "The Sciences in the Cultural Dynamics of Encounters: Towards a History of Ethnoscience".
- The fifth, at the closing, with the lecture by Eduardo L. Ortiz, "Science, Nationalism, and Territory in Argentina in the 1930s: The Commission of the Meridian Arc".

(b) **Symposiums**

- History and Teaching of Sciences, coordinated by Mercedes Ayala.
- History of Public Health and Socio-Sanitary Medicine in Latin America, coordinated by Norma Isabel Sánchez and Alfredo G. Kohn Loncarica.
- Evolution of Ideas in Physics in Latin America through Institutions and Their Protagonists, coordinated by Cecilia von Reichenbach, Guillermo Bibiloni, Leandro Andrini.
- The Articulation of Science-Technology-Industry in Mexico in the nineteenth Century and up to 1940, coordinated by Juan José Saldaña.
- International Relations of Science in Latin America, coordinated by Eduardo L. Ortiz and Diego H. de Mendoza.

[3] https://fundacionazara.org.ar/portfolio_page/vi-congreso-latinoamericano-de-historia-de-las-ciencias-y-la-tecnologia/

- Builders of Latin American Science: In Honor of Marcel Roche, coordinated by Yajaira Freites, María Rachel Fróes da Fonseca, and Natalia Priego.
- History of Health, coordinated by Álvaro L. Casas Orrego.
- History of Basic Biomedical Scientific Research in Latin America, Alfredo G. Kohn Loncarica.
- Theories of Inheritance, eighteenth to twentieth centuries, coordinated by Pablo Lorenzano.
- Technological Devices of Psychology in Argentina, coordinated by Ana María Talak.

(c) **Conferences**

- Physics and Society in Brazil in the Immediate Post-World War II Period, by Antonio Passos Videira.
- Physics, Astrophysics, and the Rockefeller Foundation in Argentina (1925–1945), by Eduardo Ortiz.

(d) **Round Table** On the topic "Linkage between Physics and Technology in Latin America", coordinated by Gabriel Bilmes, with invited panelists: C. Martínez Vidal, J. Albano, A. A. Passos Videira, María de la Paz Ramos Lara, Yajaira Freites.

(e) **Presentation of the project** for the worldwide scientific bibliography online (ICSU-International Council of Scientific Unions), by its general coordinator, Juan José Saldaña.

6.5 The Symposium: The Articulation of Science-Technology-Industry in Mexico in the Nineteenth Century and up to 1940

The coordination was led by Juan José Saldaña.

Ten papers were presented, nine of which were from professionals at UNAM and one from the University of Puebla. It is presumed that the majority of the presenters are directly or indirectly disciples of Saldaña. The master's guiding-hand is evident in the careful writing, which is not uncommon in texts from Mexico and also in the meticulous bibliographic citations. In general terms, these writings deal with the incorporation of European science and technology into a country with its traditions in goods production. Core topics include water, nutrition, the development of studies in civil engineering, and the beginnings of the domestic pharmaceutical industry. For example, the importance of the National School of Agriculture in introducing new ways of farming, such as the incorporation of fertilizers into ancestral tasks, is highlighted. Gradually, some sciences like Physics, especially through engineering graduates, are becoming part of general knowledge. The significant development of mining during the period under consideration, which has a previous basis in earlier centuries, is also discussed.

Meanwhile, Saldaña's paper, "From the Private to the Public in Science: The First Institutionalization of Science in Mexico," is a mature proposal, written by someone who knows the history of his country very well and therefore does not need to provide details. Hence, his ability to insert the new into a historical past that he dominates: the break from Spain, the work of the "insurgents," the new State, and certain politicians who "leave their mark"; and, most importantly, how, once legal-political independence was achieved, cultural independence was also necessary.

6.6 Our Tribute to Juan José Saldaña

- I (Carlos Galles) met him in Paris, at the elegant and thought provoking seminar of Madame Tonnelat, 4 Rue du Four. We continued to see each other, and when he launched his idea of a society that would bring us together, I supported him. I recall a small anecdote: on one occasion (specifically, at the XX International Congress of History of Science -Liege/Belgium- 1997), the idea of having the next meeting in New York arose, and noticing that Juan José was leaning towards Mexico, I supported him (along with others), and it was successful. As the temporary president of the society, I always had his support, and he never disappointed me as a friend. His immense capacity for work was remarkable.
- I (Norma Isabel Sánchez) met him at congresses; I was deeply impressed by how much he knew about Argentine history and how he valued our Nobel laureates in science, making frequent comparative comments with Spain. On one occasion, he invited me to write a brief note about our mutual friend, Kohn Loncarica, and I did; he published it. I have valued his effort to maintain a publication that brings us together.

Appendix

The following are the minutes of the LASHST members' meeting.

Minutes

On March 18, 2004, at 9:30 pm, in the Florentino Ameghino Auditorium, of the *Sociedad Científica Argentina*, on the occasion of the VI Latin American Congress of the History of Sciences and Technology (VI LACHST),[4] the members present of the Latin American Society of the History of Sciences and Technology gathered for

[4]VI Congreso Latinoamericano de Historia de la Ciencia y la Tecnología (VI CLHCT) in Spanish.

an assembly to discuss the future of the organization. The session was chaired by the outgoing President, Carlos D. Galles.

Firstly, Juan José Saldaña took the floor and presented a plan for the renewal of the LASHST. His comprehensive proposal, endorsed by former presidents of the LASHST, was disseminated through the CEROMAYA list and essentially comprises the following points:

The new LASHST must be a modern organization to fulfill its ongoing tasks and the new ones that are also emerging. We will need to develop electronic means for, at least, the following tasks:

1. Instant and open communication with colleagues interested in Latin American science and technology, regardless of their location on the planet: Listserv with all the possibilities it offers (information, coordinated debates, CVs, etc.).
2. Internet portal with information and documents (as well as interesting links, etc.) that give international visibility and full transparency to the activities carried out by the LASHST.
3. Contribute to the elaboration of "History of Science in Latin America on the Internet: Bibliography and Documentary Sources," as part of the global project in the field carried out by IUHPS/DHS, and drawing on national works on the subject.
4. Publish an electronic journal.

(The above tasks are mentioned indicatively, and others may be conceived and carried out.)

Next, Alfredo Kohn Loncarica expressed his support for the proposal and emphasized the need for the existing National Groups to take an active part in the Society and the imperative to reach imaginative solutions. He also proposed the creation of a Council composed of the five former presidents, with the power to be expanded.

Subsequently, Roberto de Andrade Martins, in addressing the issue of the next LASHST congress, pointed out the need to consider the problem of its realization. Carlos Galles, referring to this issue, commented that the application to host the congress is open to all, without any pronouncements being made in this regard.

Ariel Barrios Medina expressed his support for the motion presented by Juan José Saldaña.

Yajaira Freites then stated that the LASHST must be inclusive, inviting all Latin American researchers to participate, especially the members of its former Latin American Council as a governing body.

Luis Carlos Arboleda supported the motion presented by Juan José Saldaña, emphasizing the usefulness of virtual networks.

Ubiratan D'Ambrosio highlighted the benefit of promoting initiatives from different groups, fostering the Society's encouragement towards research, with a federalist vocation. He indicated that the Society is alive, and its actions are transparent. He also noted that since the former presidents are the only ones present, and they are part of the Council as provided in the Statutes, it is up to them to temporarily assume the responsibility of leading the transitional stage of the LASHST.

Porfirio García de León expressed his agreement with what D'Ambrosio said, indicating that the prompt designation of the venue for the next congress is of crucial importance.

Yolanda Texera stated that in the past, she was Venezuela's representative to the Council, but that was many years ago and her mandate was never renewed. Therefore, she supports the motion for the former presidents to lead the society until its life is normalized, as it corresponds to them legitimately.

Pablo Lorenzano expressed his support for the re-establishment of the LASHST, indicating the need to formulate a precise plan of action, involving as many researchers as possible. He also noted the usefulness and benefit for every one of holding congresses, as their convening capacity has been demonstrated.

Sara Aguilera Ríos pointed out the need to incorporate young researchers into the LASHST and to give them a real say in the decisions of the institution.

After this exchange of opinions, the motion presented by Saldaña and expanded by Kohn Loncarica, to re-establish the LASHST using the possibilities offered by modern communication technologies, and temporarily administered by a Council composed of former presidents and those whom they consider appropriate to invite, with the mission of finding an appropriate venue for the next congress, was submitted to the vote of the Assembly and was approved by consensus. As a result, the LASHST will be administered by a provisional Board of Directors consisting of the five former presidents of the Society, who will be coordinated for 6-month periods in a rotating manner by each of them: Juan José Saldaña, Ubiratan D'Ambrosio, Luis Carlos Arboleda, Silvia F. Figueirôa, and Carlos D. Galles.

The former presidents present accepted to be part of this Provisional Board of Directors, and it was decided to invite Silvia Figuerôa, who was absent, to confirm her decision to join this Board.

The assembly was adjourned at 10:50 pm.

Certified by:

C. D. Galles

Carlos D. Galles. Graduated in physics from Universidad Nacional de Rosario. Magister in History and Philosophy of Science for the Université de Paris I (Panthéon-Sorbonne). Ex president of the LASHST. Assisted at the seminars of professors Marie-Antoinette Tonnelat and René Taton in Paris.

Norma Isabel Sánchez. Master in History. Consulting Professor at the UBA. Professor Emerita at the UCES. Member of the board; director of the Library and the Institute of History of Science (SCA).

Chapter 7
LASHST Female President and the V Congress in Rio de Janeiro. Silvia Figueirôa Interview

María de la Paz Ramos-Lara

Abstract The Latin American Society for the History of Science and Technology (LASHST) was founded in 1982. For over two decades, LASHST provided a fundamental platform for historians of science and technology to come together, share their knowledge, and work towards a common goal. The society was led by distinguished individuals such as Juan José Saldaña from Mexico, Ubiratan D'Ambrosio from Brazil, Luis Carlos Arboleda from Colombia, Silvia Fernanda de Mendonça Figueirôa from Brazil, and Carlos D. Galles from Argentina. This chapter features an interview with Silvia Figueirôa, the only female president of LASHST who served from 1995 to 1998. She organized the V Latin American Congress on the History of Science and Technology (LACHST) in 1988. She is currently leading the field of history of geosciences in Brazil and Latin America.

Keywords LASHST · Silvia Figueirôa · V CLAHCT (Río de Janeiro) · Female president · Interview

7.1 Silvia Figueirôa's Curriculum Summary

Geologist from the University of São Paulo (1981), master's degree (1987), and doctorate (1992) in Social History from the University of São Paulo, both in the specialty of History of Sciences. She obtained her professorship in 2001 at the State University of Campinas and became a full professor in 2006. Her postdoc was at the Centre Alexandre Koyré d'Histoire des Sciences et des Techniques (France, 2002). From 1987 to 2013, she was a professor at the Institute of Geosciences at UNICAMP, where she held the position of Director (2009–2013). In 2014, she joined the Faculty

M. de la P. Ramos-Lara (✉)
Centro de Investigaciones Interdisciplinarias en Ciencias y Humanidades, Universidad Nacional Autónoma de México, Mexico City, Mexico
e-mail: ramoslm@unam.mx

M. de la P. Ramos-Lara, L. C. Arboleda (eds.), *History of Science in Latin America: The Construction of an Intellectual Field (20th century)*,
https://doi.org/10.1007/978-3-031-75536-1_7

of Education at UNICAMP. She has experience in the area of History, with emphasis on the History of Sciences, as well as (Geo)Science Teaching and the subject of scientific archives, working mainly on the following topics: History of Sciences and Geosciences, with emphasis on Brazil; relations between History of science and teaching; scientific/technological documentation. She works at the undergraduate and graduate levels, guiding scientific initiation, masters and doctoral research, and supervising postdoctoral studies. He participates in the following scientific associations: Member, since 1987, of the International Commission for the History of Geological Sciences (INHIGEO IUPHS/IUGS), serving as President from 2008 to 2012; President of the Latin American Society for the History of Science and Technology (SLHCT) (1995–1998); Councilor (1997–1999 and 2015–2017), President-elect (1999–2001), and President (2001–2002) of the History of Earth Sciences Society (HESS); President of the International Commission "Science & Empires" (IUPHS-DHS) (2001 to 2005). Corresponding Member of the International Academy of the History of Sciences (IAHS) (from 2019) (Fig. 7.1).[1]

7.2 Interview Held in April 2024

Tell us about your interest in the History of Science studies. When and how did you become fascinated in this particular field?

My interest emerged near the time of my graduation from college, with a major in Geology, when I attended a course on Philosophy of Science by Professor Pablo Mariconda. Nothing was said about the geosciences, only about Physics and Mathematics—and I even translated some original texts by Pierre Duhem (1861–1916) –but this lack of discussion on geological themes prompted me to search for the literature available on this topic. It was much less than today since this happened in 1981. From then onwards, my interest and fascination grew steadily, *pari passu* I made acquaintance with paramount writings and learned more about the subject.

Some Latin American countries do not offer graduate programs in History of Science. Can you provide information on graduate programs in this field in Brazil?

The field of History of Science and Technology is now consolidated in Brazil. The Brazilian Society for the History of Sciences (SBHC) completed 40 years in 2023. The Society was born around a year after the foundation of the larger Latin American Society on the History of Science and Technology (LASHST), in a meeting attended by about 50 persons and benefiting from a local and international propitious moment. From that point on, the Brazilian Society grew steadily: It currently

[1] https://bv.fapesp.br/pt/pesquisador/88725/silvia-fernanda-de-mendonca-figueiroa/ (accessed on April 15, 2024).

Fig. 7.1 Cover of the program for the V LACHST event held in Rio de Janeiro in 1998. Silvia Figueirôa's personal archive

has 397 members from almost all Brazilian states—a national representation that has also been reflected in the composition of the Boards. Until the beginning of the 1980s, there were practically only two postgraduate programs: at the University of São Paulo (USP), where Shozo Motoyama, Maria Amélia Mascarenhas Dantes, and Simão Mathias were located, and at the University of Campinas (UNICAMP), more in tune with the Logic and Philosophy of science, with History present but with less emphasis. Since then, the number of postgraduate programs has grown a lot, and it is common for History of Science and Technology (HST) to be present in programs linked to the area of Science Teaching or Scientific and Technological Policy. At the IG/UNICAMP, for example, this is the case: linked to both themes in two different programs.

In certain Latin American countries, scientists trained in the History of Science are not employed by scientific institutes because this field falls under the humanities category. I am curious about your personal experience with the Geosciences Institute. Considering other scientific fields (e.g., physico-mathematical), do you believe Brazil provides more opportunities for historians of science to work within scientific institutions?

My entry into the Institute of Geosciences was due to the peculiarity of its institutional design. Since its foundation in 1979, the idea of its creator, Professor Amílcar Herrera, was not to compete with existing institutes and to create something innovative. Thus, the four initial nuclei involved Mineral Politics and Economics, Scientific and Technological Policy, Metallogenesis and Geochemistry (a cutting-edge approach at the time), and Education in Geosciences, where I connected. HST was also present in the Scientific and Technological Policy group, and Professor Hebe Vessuri was hired to implement it. Other places also valued and still value HST, but today, this presence is more substantial in groups linked to Science and Mathematics Teaching. However, I think Brazil has more space than other Latin American countries.

Could you provide insights into your initial interaction with Juan José Saldaña and LASHST?

As soon as LASHST was created, we soon found out in the group at USP, where I was doing my postgraduate studies, as professors Shozo, Maria Amélia, and Simão Mathias, in addition to Ubiratan d'Ambrósio who also traveled there, were in complete contact with to LASHST. Some even participated in its founding. Soon, the first copies of the journal *Quipu* and the thematic series *Cuadernos de Quipu* arrived. From then on, I discovered Saldaña as a relevant bibliography. 1986, I met him personally when I took a course he offered at USP. From then on, we remained in constant contact at events, including when he was Secretary of the History of Science Division of the International Union of Philosophy and History of Science (IUHPST/DHST).

Could you share with us how you became the president of LASHST?

We were at IV LACHST in Cali, Colombia, and there were some possibilities for candidatures, but some countries were facing difficulties. Brazil was a large country with a significant community and an association in the area that existed for more than a decade. These were relevant criteria for choosing someone from Brazil.

What were the main opportunities and difficulties you faced as LASHST president?

The most significant opportunities were the growing use of digital technologies for communication. This fact greatly facilitated the management of LASHST, which had members throughout Latin America, the USA, Canada, and Europe. We created an electronic discussion list called the "Finlay List", which was set up and managed by our colleague Emilio Quevedo from Bogotá. It represented, in addition to agility, also savings on postal costs. The most considerable difficulties lay in financing. There were many difficulties in receiving annual fees and restrictions in Brazil at the time for the administration of financial resources from abroad following the scandals and impeachment of President Fernando Collor de Mello. Another difficulty concerned the secretariat infrastructure, as IG/UNICAMP did not have many employees with whom I could ask for support. My solution was to pay an extra amount to one of the IG secretaries, using my resources from my salary.

Tell us about the organization of the V LACHST, held in Rio de Janeiro in 1998. Please include the advantages and disadvantages of the congress and any problems you have noticed.

The organization of V LACHST was laborious in all events of this type and involved much dedication from the Organizing Committee. Nevertheless, on balance, it was a success: we received a very significant number of colleagues from Latin America and other countries, in addition to many Brazilians. The most significant positive points were 2, in my opinion: (1) great integration of participants, especially Brazilians, in research and discussion networks within Latin America; this was favored by the arrangement of the scientific sessions, which purposely mixed participants from different countries on the same themes; (2) the chosen theme, "Gender and Science in Latin America" was still emerging, and allowed stimulating these studies and bringing in researchers who worked internationally in the area, such as Londa Schiebinger (USA) and Elizabeth Bortolaia Silva (UK). Getting appropriate financial support was, of course, the main issue.

What was your main contribution as president of LASHST, and what did that experience mean to you?

I think that my most outstanding contribution as president of LACHST was to keep it alive, very active, and increasingly inserted at the international level. These actions were the continuity of the excellent work done previously by the former presidents. For instance, the XXth International Congress of History of Science and Technology (Liège, 20–26 July 1997) had a record participation of Latin Americans, and we managed to bring, through supporting the colleagues from Mexico and Saldaña himself, this congress for the first time to Latin America, which took place in Mexico City in July 2001.

Would you like to add anything else as your final comment?

I have nothing else to add. Just congratulate the editors of this book for the relevant, significant initiative. Many thanks for including me in the project!

María de la Paz Ramos-Lara is a researcher at CEIICH-UNAM, where she founded the History of Science Department and two book collections: *Bibliotheca Mexicana Historiae Scientiarum* and *Science and Technology in the History of Mexico*. Head of the Laboratory for Interdisciplinary Research and Complexity Sciences, and the Science and Technology Department. She is also a Philosophy of Science Postgraduate Program professor at UNAM and a member of the National System of Researchers (SNI-Conahcyt, Mexico). She has authored three books, published 16 more, and written a hundred chapters and articles on the History of Science in Mexico. She was awarded the Dr. Enrique Beltrán Prize, the Sor Juana Inés de la Cruz, and José Joaquín Izquierdo Medals.

Part III
Latin American Science Policy and Participation in the International Community: Incorporating Latin American History of Science into a World Wide Agenda

Chapter 8
Geometry, Counter-Disciplinarity, and the Path to Cubism and Relativity in Buenos Aires, 1895–1918

Lewis Pyenson

Abstract The Argentine embrace of modern art and modern physics was prefigured in indifferently-received lessons by two university professors active in the Rio de La Plata region during first decades of the twentieth century, the geometer-architect Claro Cornelio Dassen (1873–1941) in Buenos Aires and the artist-geometer Emilio (Emile Bonnet) Coutaret (1863–1949) in La Plata. The demands of Technology suppressed interest in their Non-Euclidean geometry, even with the recognition that Technology made use of geometry. Successful engineers and designers in Argentina resisted purchasing avant-garde painting, and until popularization of the eclipse results of 1919 the bourgeoisie there scarcely acknowledged a universe of curved space. During the early decades of the twentieth century, artists and mathematicians in Argentina engaged counter-disciplinary enterprise at the expense of cultivating their field of expertise, unlike the specialist dedication that guided Picasso's easel-painting and Einstein's theoretical physics. My aim is to clarify the counter-disciplinary currents that were inhospitable to Cubism and Relativity.

Keywords Cubism · Relativity · Geometry · Drawing · Architecture · Argentina · Counter-Disciplinarity

8.1 Introduction

Recent opinion maintains that the study of mathematics in Argentina remained in an anemic state until the arrival from Spain of Julio Rey Pastor in 1917, and that specialist research about the theories of relativity began only in the mid-1920s (Schubring, 2002; Fernández, 2002, 2011; Ortiz & Rubinstein, 2009; Ortiz, 2011;

L. Pyenson (✉)
Department of History, Western Michigan University, Kalamazoo, MI, USA
e-mail: lewis.pyenson@wmich.edu

M. de la P. Ramos-Lara, L. C. Arboleda (eds.), *History of Science in Latin America: The Construction of an Intellectual Field (20th century)*,
https://doi.org/10.1007/978-3-031-75536-1_8

Gangui & Ortiz, 2015). Furthermore, art historians find little attention to the European non-figurative avant-garde in Argentina until the 1930s (Baldasarre, 2020; Cortés, 2016; Herrera, 2014). Here I examine this learned consensus by focusing on two university professors active in the Rio de La Plata region during the early decades of the twentieth century, the geometer-architect Claro C. Dassen in Buenos Aires and the artist-geometer Emilio Coutaret in La Plata, whose indifferently-received lessons anticipated later enthusiasm for modern art and modern physics.

Under the Neo-Idealist project of Modernity, which they touted, Argentine intellectuals should have welcomed both Picasso and Einstein. A social current, which I call Counter-Disciplinarity, prevented this enthusiastic embrace. Counter-Disciplinarity is the simultaneous engagement of a number of intellectual specialties or fields, which were known in the twentieth century as disciplines, without an attempt to define a new area of learned inquiry. Far from being hostile to specialization, Counter-Disciplinarity as a way of assembling knowledge depends on it. Counter-Disciplinarity differs from the interdisciplinary (or even as Paul Forman argues, the antidisciplinary) climate of Postmodernity (Forman, 2012). It bears some resemblance to the Counter-Modernism identified by Herbert Mehrtens in the early twentieth-century intuitionist mathematics of L. E. J. Brouwer, which Mehrtens contrasted to the modern mathematics of David Hilbert, although from my point of view, both Brouwer and Hilbert were Neo-Idealist moderns. Counter-Disciplinarity is more closely related to Transdisciplinarity, a task-force model of scientific research advocated in Switzerland at the beginning of the twenty-first century.[1] Unlike Transdisciplinarity, however, which encourages temporary and transactional groupings to achieve a particular goal, Counter-Disciplinarity serves a larger social vision. In Argentina, the larger vision gave a prominent place to ideas and material structures from late-nineteenth-century Western Europe.

Diana B. Wechsler has emphasized the limited impact of Cubism in Argentina through the middle 1920s, with particular reference to Emilio Pettoruti and his advocate Xul Solar. Wechsler sees Cubism in Latin America, when it appeared, as a special intellectual alloy on the western side of the Atlantic (Wechsler, 2011). If that is so, Cubism would have been received differently from Relativity. An argument could be that, while Argentina does not appear within Picasso's horizon of experience, Einstein, before his final formulation of general relativity, was in contact with two Argentine academics, physicist Jacobo Laub (1882–1962) and astronomer Charles Dillon Perrine (1867–1951). Picasso never saw Argentina, but in 1925 Einstein lectured there. I argue that these personal circumstances are best understood in larger, social terms, which help account for the reception of both Cubism and Relativity.

Jacobo Laub, born and raised in Polish Rzeszów (at the time, part of Austria-Hungary), studied physics at Vienna and Kraków before traveling to Göttingen and attending a mathematical seminar led by David Hilbert on the electron theory, the

[1] (Pyenson, 2021), 357 for Mehrtens, Brouwer, and Hilbert; 89 for Transdisciplinarity. (Mehrtens, 2004, Pohl & Hirsch, 2010).

theory in the context of which Einstein formulated special relativity. Laub took a doctorate with Wilhelm Wien and a postdoctoral assistantship with Philipp Lenard—two early Nobel laureates. Between 1907 and 1911, Laub worked closely with Einstein on special relativity, coauthoring several papers. In 1911, he accepted a professorship of geophysics at the University of La Plata. Shortly afterwards he moved to a teachers college in Buenos Aires, where he announced and likely taught a course on Relativity and also conducted experiments with X rays.[2] He took Argentine citizenship, became friendly with the government of Hipólito de Irigoyen, and spent more than 20 years in the Argentine diplomatic corps in Europe. He died in Switzerland (Barrachina, 2003; Pyenson, 1988). Laub certainly applied theoretical physics to diverse problems, as Einstein did, but unlike Einstein, Laub was counter-disciplinary in his career choices.

Charles Dillon Perrine, self-educated with an honorary doctorate in 1905 from Santa Clara College in California, rose to the position of Astronomer at Lick Observatory in California, whence he photographed solar eclipses around the world before in 1909 becoming director of the Córdoba Observatory, succeeding two previous directors from the US. Encouraged by colleagues at Lick and at the Berlin Observatory to look for the "Einstein Effect" of light bending around the Sun, Perrine sent expeditions to photograph solar eclipses in 1912 (Brazil), 1914 (Ukraine), and 1916 (Venezuela). No useful images resulted. Nor were photographs suitable that Perrine had taken at a 1909 eclipse to detect a hypothesized planet close to the Sun for explaining the disparities between the observed orbit of Mercury and its orbit predicted according to Newton's inverse-square formulation of gravitational force. Perrine could not obtain funding to photograph the crucial eclipse of 1919 in Brazil. He accumulated cultural capital from corresponding with astronomers in California and Germany, and he spent more than 40 years in Argentina. I have not found wider Argentine interest to continue his early attempts at measuring light-bending during eclipses.[3] Perrine turned his telescopes to diverse celestial objects, but he was a disciplinary paragon of observational astronomy.

Laub and Perrine each cuts a distinct figure in the world of science. Laub was a difficult person, in both professional and personal life, who, after being close to Einstein, drifted from one scientific domain to another before beginning a

[2] The year before Laub arrived, the distinguished Italian mathematician Vito Volterra gave a lecture (sponsored in part by the Italian Ministry of Public Instruction) on notions of space, time, and mass in modern physics, ideas that were "already beginning to come into vogue." The material was "more or less new for nearly the entire audience" (Dassen, 1924, 71). The published talk (Volterra, 1910), depending heavily on Hermann Minkowski's writings, argued for a "profound revolution in modern critical thought (224)." Volterra invoked the notion of revolution for Maxwell (227) and Minkowski (230); Einstein is mentioned only as a contributor on the road from Lorentz to Minkowski. Dassen tells us (71) that Volterra's text was translated by Santiago E. Barabino. I have found no other echoes in Argentina of Volterra's talk.

[3] (Crelinsten, 2006), for much on the eclipse expeditions to detect the Einstein Effect, 51–61 on the failure of using existing photographic plates from the Intra-Mercurial search. (Rieznik, 2011; Perrine, 1916, 1927). It is clear that a handful of the world's astronomers systematically followed Einstein's early calculations on light-deflection by the sun.

diplomatic career in Europe under a progressive Argentine politician; stolid Perrine found the means to prosecute the line of observational astronomy he had mastered, in an isolated research environment with a staff from the US, France, and Germany. Both Laub and Perrine, unlike Dassen and Coutaret, separated from wife and family. But one circumstance is common to the four intellectuals: None trained research students. The following pages phrase this counter-disciplinary omission in Dassen and Coutaret's institutional setting.

8.2 Claro C. Dassen and Emilio Coutaret as Counter-Disciplinary Mathematicians

In the first part of the twentieth century, Argentina engaged the idealism of Modernity not with reverence for specialization, but rather with Counter-Disciplinarity. Artists and scientists were expected to write convincing, discursive prose on a wide range of topics and to hold forth constructively as public intellectuals. They embraced material progress at the same time that they held fast to older European aesthetic conventions.[4] Dassen and Coutaret shared this outlook, contributing to the visual arts, architecture, philosophy, history, and creative literature, as well as geometry and other areas of mathematics.

Dassen and Coutaret situated themselves at an accepted academic interface of art and science—geometrical drawing—that is today largely subsumed in computer-assisted design. This professional niche they shared with instructors who stood before young Picasso and Einstein. In the following pages, I examine their teaching activity with a view toward understanding how they conveyed a central part of Picasso's and Einstein's revolutionary innovations, Non-Euclidean geometry. I also identify a common, practical sensibility among the younger, academically trained elite in Greater Buenos Aires who resisted embracing Cubism and Relativity.

Dassen and Coutaret both extended nineteenth-century Argentine academic traditions into the first third of the twentieth century: They engaged mathematics as a teacher without generating a school of acolytes. The author of no striking intellectual innovation, each professor participated fully in the drive to make Greater Buenos Aires a world metropolis. The uneven adoption of their ideas in Non-Euclidean geometry informs our understanding of Argentina's reluctance to embrace both Picasso and Einstein. But by supporting new and reformed institutions, the two mathematicians are in the background of a flowering of mathematics and

[4] Dassen's mathematician colleague Camilo Meyer (1854–1918) fits this model from the time of his training in Nancy, France, where he took a doctorate in law before obtaining a licence in exact sciences. In Argentina by 1908, he began to publish furiously (much of his writing was high popularization) on mathematics (Hilbert's geometry), philosophy, psychology, and theoretical physics. He studied the work of Einstein and in 1914 he lectured at the University of Buenos Aires on the electron theory of Hendrik Antoon Lorentz (Dassen, 1924, 68–69); H. M. Levylier called him an "encyclopedist above all" (Damianovich & Levylier, 1918, 62).

mathematical physics beginning in the late 1930s that assimilated and impacted scientific thought in the Northern Hemisphere and that underlay a current of geometrical art beginning in the 1940s.[5]

Modern mathematicians sometimes restrict their disciplinary identification to the glory and the agony of posing and proving new theorems, that is, to demonstrating the incontrovertible truth of propositions about numbers, shapes, or immaterial relationships. Many mathematicians, however, have been and are conveyors of proofs, reviewers of truth for growing minds, and appliers of the results for practical or aesthetic ends. The church of mathematics is broad, extending over number theorists, designers of mortality tables, and students of ornamentation. Probably the largest group of mathematicians are teachers who write curriculums and textbooks, decanting old wine into new bottles. By these criteria, Dassen and Coutaret were undoubtedly mathematicians.

The problem with a commitment to proofs is cadging time to think about them. Into Postmodernity we have asked all erudite teachers to discover new things. In some places professors are paid to devote part of their day to this activity. Until the inception of the labor-union contract, with a living-wage in exchange for a full-time commitment to one institution, that was not the norm. Rather, professional academics who were not independently wealthy angled to accumulate a diverse collection of course assignments and administrative appointments. In practice, this meant less time for seeing their new ideas through press. The system in nineteenth-century France gave rise to *cumul*—the accumulation of salaried assignments at a number of institutions. By 1900, it was an established, counter-disciplinary feature of academic life in Argentina.

The payment of university professors in early twentieth-century Argentina shows how cumul favored entrepreneurial polymaths. In universities, a faculty executive council, consisting of senior professors, received an annual sum for pedagogical needs, including salaries and laboratories of various kinds. The council allocated the funds on the basis of proposals from directors of schools and institutes, and it considered nominations for teachers of individual courses. By tradition, senior professors taught several courses for two or three times the salary of a young specialist who had not yet defended a doctorate. In effect, the council members controlled their own salaries and the appointment of their protégés. Well into middle age, Dassen and Coutaret took on outside work as opportunities presented themselves in government agencies and private enterprises.

Serving diverse masters encourages transferring ideas from one specialty to another, but it is different from the mutual exchange expected of interdisciplinary activity. The counter-disciplinary imperative in Argentina bred superficiality, even though both Dassen and Coutaret remained loyal to the disciplines of their

[5] Notably in the solid-state physics of Leopoldo Máximo Falicov (1933–1995), the Concreto-Invención/Perceptivism of artist Raúl Lozza (1911–2008), the Spatialism of artist Lucio Fontana (1899–1968), and the Kineticism of artist Julio Le Parc (b. 1928). My latest thoughts about art and science across the twentieth century in Argentina, in particular the turn to "interiorization" during the 1940s, in (Pyenson, 2023).

formative years. Both were slow to embrace new notions.[6] Possibly never having met, they qualify as an example of Plutarch's parallel lives in the early twentieth-century Age of Argentine material excess.

8.3 Dassen at Buenos Aires and Coutaret at La Plata

Claro C. Dassen took a diploma in civil engineering at the University of Buenos Aires before in 1901 that university conferred on him an early Argentine doctorate in mathematics for an epistemological dissertation that examined the infinitesimal, mathematically and philosophically (Dassen, 1901).[7] He taught mathematics at the Colegio Nacional in Buenos Aires, a secondary school of which he was a distinguished graduate, and he became professor of geometry at the University of Buenos Aires. He was also employed by Public Works in Buenos Aires for 20 years (1897–1917), where he pioneered use of new building materials. His fortune secured, he designed his residence in the Buenos Aires quarter of Monserrat in collaboration with one of Argentina's most prominent architects, his university colleague Alejandro Christophersen. Completed in 1914, the house is today known as the Palacio Dassen. As a young mathematician Dassen was a committed internationalist who corresponded with Louis Couturat, the French mathematical logician, advocate of the universal language Ido (a modification of Esperanto), and close friend of Bertrand Russell's who died just as the First World War began (Dassen, 1939; Babini, 1942; Rossell, 1943). Over the last half of his life, Dassen was a tireless promotor of the Sociedad Científica Argentina, a counter-disciplinary scientific association resembling omnibus national homologues that flourished in Europe and North America during the nineteenth-century.

Dassen did not paint or sculpt, but a colleague of his at the University of Buenos Aires did, Ernesto de la Cárcova. The latter artist was close to the intellectuals who in 1876 had founded the Sociedad Estímulo de Bellas Artes (SEBA), the Society for the Promotion of Fine Arts, which also sponsored an art school. De la Cárcova assumed direction of the school in 1878. Following its inception at the apex of the art and architecture community in Buenos Aires, SEBA and its school were nationalized in 1905 and given the title Academia Nacional de Bellas Artes y Escuela de Artes Decorativas e Industriales. The school awarded a teaching diploma. De la Cárcova, a social-realist painter, taught drawing in the University of Buenos Aires

[6] (Larsen, 2017) presents Argentine museum curators of this period as operating in a small intellectual community sharing common feelings and European norms. Some of them were counter-disciplinary thinkers who embraced the values of the local bourgeoisie, as did figures I discuss here. Larsen's focus is directed away from the practical task of natural-resource inventory: surveying land and biota.

[7] It was not the first doctorate at the University of Buenos Aires in mathematical-physical sciences, and previously the National University of Córdoba conferred doctorates *ad eundem gradum* on its science professors.

engineering faculty (FCEN, 20 December 1905). He and Dassen would have known each other.

Pertinent information about the founding plan of studies of the new art academy's school is available in the minutes of the academic council of the Faculty of Exact, Physical and Natural Sciences at the University of Buenos Aires. Late in 1905, the council allowed graduates of the art school to be exempt from taking a number of university courses figuring in the first two years of the plan of studies for the degree of architectural engineer. These courses ranged from the humanities to the sciences of nature and to mathematics. In particular, the academy courses on plane geometry and solid geometry (*geometría del espacio*) satisfied the architectural plan (FCEN, 20 December 1905). I shall return to the topic in the following pages, but I want to signal two points here. The first point is that these courses are the very titles of two geometry books intended for the higher secondary schools and published by Dassen in 1904 and 1905, and the books refer to descriptive and projective geometry. The second point is that descriptive and projective geometry figure explicitly in the plan of studies of the elite colegios nacionales at this time.

Like Dassen, Emilio Coutaret trained in architecture and engineering, but at the Ecole des Beaux-Arts in Le Havre in his native France. As a young man he emigrated to Argentina to work in the railways and then, around 1910, collaborated in designing the Cathedral of La Plata and refiguring the Jockey Club of Buenos Aires. Each imposing structure was a focus for the local élite. Different from Dassen, Coutaret became a private art teacher who distinguished himself in free-hand drawing and Impressionist canvases. Coutaret taught geometry and drawing at the National University of La Plata for 30 years, following its inauguration in 1906.[8]

8.4 Art and Design in the Work of Dassen and Coutaret

At the outset I shall address art and design in the work of Dassen and Coutaret. First to Dassen's Buenos Aires residence. Because the townhouse is a Belle Epoque fantasy atypical in Christophersen's oeuvre of Second-Empire designs, we are not unreasonable to attribute much of it to Dassen himself, especially given that the top half of the structure is an astronomical observatory, one of Dassen's passions. It is situated on an urban lot 7.7 m wide not far from the National Congress. For decades from the time of its completion, it was one of the tallest constructions in the city, rising to 60 m, and one of the first to use the new technique of reinforced concrete. The street level was reserved as a high-end commercial rental. The two floors above it were high-ceiling living quarters for the Dassen family with an enormous balcony overlooking the street. Stairs with 365 steps led to the observatory under a cupola.

[8] For Coutaret's early artistic activity in La Plata, (Rueda, 2019). A survey of private art schools circa 1900 in La Plata in (Nessi, 1982, 1–5), an entry by Nessi.

The interior walls were paneled in wood with Masonic symbols and silk fabric, supporting bookcases for Dassen's collection of some 1200 rare imprints.[9]

Dassen was a sometime architect, but architecture was Coutaret's profession. Coutaret enters the present story because he taught geometry to both engineers and artists at the National University of La Plata. Nevertheless, throughout his life, Coutaret also kept a brush in his hand, painting portraits, landscapes, and natural-history murals. He followed the academic styles of Europe, from Beaux-Arts to the explosion of color in Impressionism, a sensibility also favored by his colleague Miguel Rosso in the School of Drawing. Rosso explicitly challenged the spontaneity favored by another colleague, the masterful Impressionist Martín Malharro. Rosso was a classicist who, in the opinion of his younger colleague José María Rey, seemed like he stepped out of the European Renaissance, whereas Malharro embraced the effervescence of late nineteenth-century Paris.[10]

Even-tempered Coutaret remained on good terms with both Malharro and Rosso, whose lives he memorialized (Coutaret, 1910, 1911). In part this was because both Malharro and Rosso, who died weeks apart in 1911, worked firmly in the context of traditional European aesthetics, a sensibility Courtaret shared, rejecting the allure of both autochthonous design and radical early twentieth-century artistic movements. Coutaret's classroom with its collection of plaster models was initially in the Museum of Natural History, a repository for mounted paleontological specimens. He used his painterly art for murals bringing the bones back to life.

Dassen and Coutaret each landed a university position at a time of educational renewal following the consolidation of the Argentine state in basically the form it is known today: From 1880, the city of Buenos Aires became an autonomous federal capital of a land mass that extended from the Bolivian border to Tierra del Fuego. The capital was and is, along with its suburbs, the intellectual and economic center of the country. Notwithstanding resistance from priests and lawyers, its schools and universities scrambled to provide thousands of practical modernizers with the means to build a civilization that imitated European cities in architecture, transportation, and public health. Dassen and Coutaret's common task was to apply their skill and knowledge in building an austral Paris.

---

[9] (Bedrossian, 2021), as the residence appears under its present owners. Comments from Dassen family members appear in https://www.fervorxbuenosaires.com/actores/. Dassen's rare books are in the library of the Sociedad Científica Argentina in Buenos Aires: chrome-extension://efaidnbmnnnibpcajpcglclefindmkaj/https://www.bn.gov.ar/resources/conferences/pdfs/presentacionCapenayCrea.pdf. Dassen's wall-treatments of silk, wooden paneling, and stucco were those favored by the *porteño* bourgeoisie, for example the mansion of Matías Errazúriz and Josefina de Alvear, completed in 1916 and now the National Museum of Decorative Arts.

[10] (Andruchow & Castillo, 2019), citing (Rey, 1938). Rey was a director of the School of Drawing (Kraselsky & Andruchow, 2021).

8.5 Instruction at the Austral Universities

At the beginning of the twentieth century, Argentina had two major universities funded by the national government. The older of the two, at Córdoba, had shaken off its Jesuit heritage by replacing its theological faculty with a medical one and by instituting a program in engineering. In the 1880s it made a decision to separate research from teaching by hosting Argentina's first national academy of sciences and populating it with luminaries from the United States and Germany with the goal of providing an inventory of land (natural history and geology) and sky (astronomy and meteorology). The academy fellows squabbled over money while producing large and unreadable surveys without training acolytes. Moreover, the academy and the university wrestled awkwardly with a major intellectual turn early in the twentieth century—the eclipse of the field sciences by the laboratory sciences, and the replacement of the illustrated natural-history tome by the mathematical elaboration of particular laboratory or observatory findings in timely journal articles (Pyenson, 2021, 400–439).

The University of Buenos Aires was a creature of independence from Spain. Some of its professors were notable in the world of science (for example, Alexander von Humboldt's companion Aimé Bonpland and the Italian exile Ottaviano-Fabrizio Mossotti), but it is an open question if they inspired a tradition of learning (Penchaszadeh & de Asúa, 2010). When Dassen obtained a position there, his most notable disciplinary colleague was Jorge Duclout, an older civil-engineering alumnus of the Zurich Polytechnic.[11] Duclout was instrumental in bringing Einstein to Argentina. In the 1890s he had been an early advocate in Argentina of Maxwell's electrodynamics, and an adept at geometry.[12] Duclout's and Dassen's intellectual sociability transpired not at their university but rather in the more hospitable halls of the Sociedad Científica Argentina, with the *Anales* of which they published some of their work. The Sociedad had been founded in 1872 through the encouragement of President Domingo Faustino Sarmiento. The democratic Sociedad functioned as a scientific center in Buenos Aires until the definitive inauguration in 1935 of the elitist Academia Nacional de Ciencias Exactas, Físicas y Naturales from an appendage authorized in 1876 by the reorganization of the University of Buenos Aires.

At the time that Dassen took up his university teaching assignments, Emilio Coutaret was working in La Plata, a city founded in 1882 to serve as the capital of the Province of Buenos Aires when the city of Buenos Aires declared its administrative autonomy in the confederation. In 1897 the province instituted a university in its own capital. The fledgling center of learning foundered until, in 1905, it was

[11] (Ortiz, 1995), for Duclout and background at the University of Buenos Aires.

[12] (Duclout, 1892), an annotated translation of a lecture by Ludwig Boltzmann on Maxwell, which Duclout followed by a long explanation of Lagrange's equations of motion as they were used by Maxwell. Duclout's admiration for Maxwell and Boltzmann complements Einstein's. (Dassen, 1924, 69) for Duclout's unsuccessful petition in 1917 to teach a course at the University of Buenos Aires on higher geometry and history of mathematics.

nationalized under a plan to create a new university dedicated to the advancement of science. In particular, the university absorbed two existing research enterprises in the city's central park—the astronomical observatory (construction beginning 1883) and the natural-history museum (founded 1888) (Podgorny & Lopes, 2008; Rieznik, 2011).

Coutaret came on board as a founding professor of drawing, eventually director of the School of Drawing emerging from the Department of Cartography, quartered in the natural-history museum captained by the naturalist Francisco (known as Perito) Moreno. The cartographical operation was supervised initially by Swiss engineer Enrique A. S. Delachaux, who may have studied at the Neuchâtel Academy and who honed his talents at the Sorbonne, the Paris Natural History Museum, and with the geographer Louis Vivien de Saint-Martin before emigrating to Argentina. There he worked with Moreno to delineate the border between Argentina and Chile. Coutaret's new school belonged to an Instituto de Artes y Oficios y Artes Gráficas (the name suggests art as a part of the trades) in the Faculty of Natural and Anthropological Sciences. It had two laboratories, so to speak. First a press, which for years functioned as the printing office of the province. Second was a vague relationship to the founding collection of La Plata's fine-arts museum, envisaged as an Academia de Dibujo y Bellas Artes under the Instituto de Ciencias Geográficas y Físicas and curating paintings donated to the province in 1877 by Juan Benito Sosa.[13] Delachaux, whose counter-disciplinary notion was to educate generalists in the image of the Renaissance wonder Giovanni Pico della Mirandola rather than the modern technical specialist, died in 1908. A testament to Coutaret's vision is that his school survived Delachaux's death and shortly thereafter the death of Malharro and Rosso.

The School of Drawing, then, joined an academy of fine arts with a school of technical drawing, a common cohabitation in Europe.[14] In Coutaret's initial mission statement, the school was to train art teachers, on the one hand, and commercial designers, on the other hand. His pedagogical plan had three parts: drawing, modeling, and coloring. Coutaret had the first year copying geometrical solids as a way of representing natural forms; in the second year came modeling, including chiaroscuro with "notions of the theory of shading." Colors came in the third year. The engineer Coutaret was responsible for teaching geometry to artists and designers as well as engineers and architects. For more than a decade Coutaret and his handful of artist colleagues taught in the basement of the museum, where students were surrounded by anthropological and biological artifacts (notably human skeletal remains

[13] For more than a 100 years, the collection of the provincial art museum has been on the move. At present it is gathered in a former theatre under renovation, and its works are inaccessible to the general public (Ruvituso, 2023). (Nessi, 1982, 246–248), for a history by one of the museum's directors, Jorge López Anaya. I am grateful to the present director, Federico Ruvituso, for helpful discussions.

[14] Variations in nineteenth-century Spain are found in Picasso's art schools in A Coruña and Barcelona. (Pyenson, 2021, 157–183).

and fossils). As befitted a modern enterprise, Coutaret's school attracted a large number of women students and a handful of women instructors.[15]

8.6 Descriptive and Projective Geometry

Descriptive and projective geometry were at the center of mathematics education for European artists and engineers in the latter part of the nineteenth century. The argument was made from general pedagogy—that geometry presented immature minds with a model of rational thought. The point was emphasized by one of Coutaret's colleagues for pupils in primary schools, and Claro C. Dassen's volume on solid geometry emphasizes it in the preface.[16] The argument, however, is unconvincing. Syllogisms, the stuff of medieval logic, are surely as rational as geometry, and any competent artist or designer would not need intricate lines and curves to draw a credible image of the world. What geometry offered was a dispassionate way of evaluating students. It is hazardous to use taste as a criterion in higher learning, condemning a student for a painting that is not beautiful enough or does not represent reality well enough or evokes no feeling in a viewer, but one can certainly set geometrical problems that have objectively correct answers.

Dassen focused his early epistolary attention on a two-volume pedagogical work devoted to plane and solid geometry, appearing in 1904 and 1905 (a second edition was published in 1914 and 1916). On the title page of each volume, he was clear that he intended the book for use in the colegios, whose graduates could be expected to enroll in universities or other institutions of higher learning. To this end, he provided a detailed key in volume two for identifying which pages treated topics specifically mentioned in the published course plans of the colegios nacionales, those taking precedence among the Argentine secondary schools. Dassen's second volume would surely have been used only in the fifth, optional year at the colegios nacionales, for in reality, it was nothing less than a compendium of descriptive and projective geometry of the kind studied by Einstein at the Zurich Polytechnic.[17]

His idea, Dassen emphasizes in the preface to the first volume, was to promote a vital reorganization of secondary-school mathematics based on the principles of rational geometry. He accordingly took the liberty of rearranging Euclid's order of presentation to cluster common themes and proofs. Furthermore, we read in the preface to the second volume on solid geometry, since none of his proofs introduced

[15] The appointment of Elvira Vicentini as instructor of technical drawing: (LPMHN, IMD-13, Carta 25, Coutaret to Lafone Quevedo, 17 March 1914). I am grateful to Máximo Farro, Director of the Archives, for providing copies of Coutaret's correspondence. For the history of the School of Drawing (Rey, 1938). Unless otherwise noted, Rey's long and detailed manuscript is my source. Rey is clear, for example, that early on the School offered evening classes for working students.

[16] For an extended discussion with attention to primary schools by a professor of pedagogy at La Plata, (Mercante, 1911).

[17] (Pyenson, 2021, 220–239), for Einstein's mathematics.

in plane geometry appeal to Euclid's parallel postulate, they may be taken over to solid geometry and spherical geometry holus-bolus. This, Dassen tells us, is the Principle of Duality, which he had introduced in his dissertation without name when he emphasized the generality of theorems in projective geometry.[18]

In mathematics, the Principle of Duality, whose discovery in various forms was contested by Jean-Victor Poncelet, Julius Plücker, and Joseph Diaz Gergonne in the nineteenth century, became central to projective geometry. It asserts that in geometrical systems, theorems hold true if suitably-chosen "dual" elements (for instance, lines and points) are interchanged. In some cases, a difficult proof in one system is relatively easy to prove in its dual.[19] Placing geometrical dualities in a larger structure—making them less surprising—was a part of the middle twentieth-century program of the influential mathematical collective known as Bourbaki, for whom the disentangling of structural homologues began with imagination: For the collective, "Structures and axiomatics were deeply linked with *analogies* and *intuitions*" (Heinzmann & Petitot, 2020, 203).[20]

Just as projective geometry found a firm place in nineteenth-century mathematics, so duality found a place more widely in twentieth-century culture. It is a herald of twentieth-century Complementarity in quantum physics, which is present in the wave-particle duality for sub-atomic entities like electrons, in the equivalence of wave mechanics and matrix mechanics for quantum phenomena, and the Uncertainty Relations governing conjugate variables. Duality is also a feature of dialectics, in both its Hegelian and its Marxist elaboration. It is broadly suggestive of dualism, condemned by philosophers both materialist and idealist (Lovejoy, 1930). But it fits well with the notion of a pre-established harmony between mathematics and the real world, which suffused Europe (Pyenson, 1990, 205–231). Dualism has nothing to do with Positivism, sometimes alleged to be a main current of philosophy in Argentina at this time.[21]

---

[18] (Dassen, 1904, 1905, duality on xii; Dassen, 1901, 39 for projective geometry).

[19] (Gray, 2008, 39–55) for the early nineteenth-century development of Non-Euclidean geometry from the duals of projective geometry in two dimensions. In an essay that begins with Edwin Abbott's fantasy about multi-dimensional worlds, *Flatland* (1884), which until recently had no Spanish translation, and with a title that suggests both duality and fields of force in physics, Gabriela Siracusano refers to a general familiarity with Non-Euclidean Geometry in Argentina and Spain in the 1930s and 1940s. (Siracusano, 1998, 185–188).

[20] On the early notions about structure among the Bourbaki, see (Beaulieu, 1989). "Beaulieu's dissertation is the most extensive description of the origin and the first 10 years of activity of the group. It is the sourcebook of all biographically oriented studies on Bourbaki" (Heinzmann & Petitot, 2020, 188–189).

[21] (Ortiz, 2011; Gangui & Ortiz, 2011, 201–218). Argentina's first president elected under manhood suffrage in 1916, Hipólito de Irigoyen, claimed inspiration from early nineteenth-century German Idealist Karl Krause. (Roig, 1969, Stoetzer, 1998, 362–387). The popular Neo-Idealist Alejandro Korn dominated philosophy at La Plata in the first three decades of the twentieth century and was central in university reform after the First World War. (Stoetzer, 1998, 319–320; Dragowski, 2019). Also: (Montserrat, 1993, 70–82) signaling Korn and the Idealist Coriolano Alberini.

In the second volume of his geometry textbook, Dassen deals extensively with descriptive and projective geometry. He includes a table enduring for many pages where theorems of plane geometry are given in their dual forms for solid geometry and spherical geometry. His illustrations of duality were just the ones taught to art and engineering students in Europe. Had it been used in the secondary schools or universities, Dassen's book could have formed a base of recognition for the intersecting lines and planes of Cubism and the curved space of general relativity. It is, after all, a short jump from the designation of solid geometry as *geometría del espacio* in three dimensions to the notion of *espacio-tiempo* or space-time in four dimensions, which Hermann Minkowski illustrated with conic sections. Einstein eventually embraced space-time on his way to general relativity.

8.7 Classroom Instruction in Geometry

Picasso and Einstein did not triumph in Argentina at the same time that they triumphed in Europe. We can better understand the austral reception by looking at what transpired in the classrooms of Greater Buenos Aires.

A useful starting point is Coutaret's most illustrious pupil Emilio Pettoruti, seen by art historians as the first Argentine Cubist painter. Beginning in 1913, Pettoruti studied extensively in Europe and came to know many figures in the avant-garde. He returned to Argentina, and his convictions and contacts led to his being named director of the Museo Provincial de Bellas Artes at La Plata in 1927. Pettoruti was a pupil of Coutaret's before his first trip overseas. In an intriguing passage in his autobiography, Pettoruti recalls studying geometrical perspective in the city of his birth, La Plata, about 1912–1913:

> I walked about the museum halls copying, here and there, stones, minerals and birds, developing a collection worthy of an ornithologist. I also often copied Michelangelo's Moses, and its magnificent plaster reproduction. A personal friend of my father's, Coutaret became particularly interested in my studies, and many of the problems related to perspective, so complex they had to be solved [the Spanish text reads *debían ser soslayados*, or sidestepped] in class, he explained to me at his home. I studied with him for a year, and believe I did not disappoint him as a student.[22]

The context for this recollection is found in a curriculum proposal of Coutaret's from around 1908 which he circulated to his superior Samuel A. Lafone Quevedo, anthropologist and linguist who followed Perito Moreno as director of the natural-history museum. There Coutaret affirmed the importance of perspectival drawing, but he emphasized: "A simple examination of the drawing program in the primary schools shows us the supreme importance of hand-drawing over geometrical drawing, however reduced to elementary notions." At La Plata, an equal amount of time was given to the two approaches. "The careers of drawing teacher and technical

[22] (Pettoruti, 2006, 8). Compare (Kaplan, 2014), which overlooks Coutaret's geometrical instruction.

drawing, without being opposed to each other, require distinct aptitudes and study." His proposed three-year course of study in fact gave little time to geometry. A committee of four professors vetted Coutaret's proposal and recommended separating the artistic and the technical sides of drawing. They considered reducing the time of study for a diploma from three years to two years. For future teachers in the secondary schools, material in the history of art and aesthetics could be combined in one course of two hours weekly. Philosophy and theory could be eliminated because "pedagogy and methodology" were the responsibility of another faculty. Ultimately the committee of four recommended six hours of geometry per week for first-year students. For the diploma in technical drawing, there were four hours of "geometrical drawing" in years two and three, and eight hours in year four.

Coutaret's project at La Plata was finally approved by the dean of engineering on 12 May 1910, with an addendum on 14 June 1910 requiring two more hours in "perspective" during the second year and in "geometrical drawing" during the third year.[23] Though the course manual is unknown, we surmise that the geometry lectures did not involve much of the descriptive and projective geometry that was familiar to artists and engineers over the previous generation at a few hundred European institutions of higher learning. Evidence is found in one surprising regulation adopted in 1910: Students could be admitted to the School of Drawing having completed the fifth grade ("5. grade") of primary schools upon passing a test by the faculty of pedagogy. In any case, enrollment was modest. In 1913, Coutaret reported only twelve students in the course on geometrical drawing applied to surveying. He continued that the first-year and second-year classes, separated from engineering, had been canceled, and "instruction is quite deficient."[24]

Regulations for instruction in geometry seem to have varied across the secondary schools, art schools, and universities in Greater Buenos Aires. The minutes of the academic council of the Faculty of Exact Sciences in the University of Buenos Aires are clear that many of its degree programs, from the certificate to the doctorate, required geometry. In both 1902 and 1913, the doctorate in ciencias físico-matemáticas specified four hours per week of projective and descriptive geometry in the second year; five hours per week of applied descriptive geometry in the third year; and three hours per week of higher geometry in the fourth year.[25] The same requirement for the second and third year was in the 1902 plan for civil engineering and the 1903 and 1906 plans for mechanical engineering (FCEN, 27 August 1902, 18 September 1903, and 26 July 1906). According to a plan of 1902, architects, in

[23] LPMHN, IMD-4, Carta 66, Coutaret to Samuel A. Lafone Quevedo, 31 March 1910, with other correspondence and notes about Coutaret's proposal circa 1908.

[24] LPMHN, IMD-13, Carta 17, Coutaret to Lafone Quevedo, 20 November, 1913. Presumably primary (grade) school ended in the fifth grade for children about eleven years old. See the photograph of a plane-geometry class for the fourth grade in the primary school attached to the Pedagogical Faculty of the university, where children seem to be about ten years old: (Universidad Nacional de La Plata, *Album,* 1909), an unpaginated description of the university, with a large collection of photographs.

[25] FCEN, 27 August 1902; plan proposed 10 May 1913 and adopted 31 May 1913.

their second year, took four hours of descriptive geometry, and also in their second year, surveyors took four hours of descriptive and projective geometry (FCEN, 27 August 1902). To accommodate working students, courses in the first two years of architectural studies were to be scheduled from 5 p.m. to 7 p.m. whenever possible, instead of from 10 a.m. to 12 p.m. (FCEN, 20 December 1905). The minutes of the academic council suggest that on average fifty students registered for these courses, if indeed they were given regularly.

Another indication of unrealized ambitions in geometrical instruction comes from the International American Scientific Congress held in Buenos Aires in 1910 under the auspices of the Sociedad Científica Argentina to commemorate a century of Argentine political independence. It featured a symposium on mathematical sciences chaired by Benjamín Sal, former dean of the Faculty of Physical Sciences at La Plata and then president of the provincial Department of Engineers. Sal set seven topics for discussion. The first three were vectorial notation, notational systems in projective and descriptive geometry, and a proposal to join plane and solid geometry in secondary-school instruction, making use as much as possible of the Principle of Duality (Barabino, 1910, 22). We have seen the same use of duality in the textbooks from 1904–1905 written by Claro C. Dassen. In fact, Dassen addressed the 1910 congress on this topic, and a version of the talk appeared in the *Revista del Centro de Estudiantes de Ingeniería de Buenos Aires* (Rossell Soler, 1943, 31). The Colegio Nacional of Buenos Aires mentioned descriptive and projective geometry in published course plans around 1900, and it adopted duality in its course plan for 1910; then duality dropped from sight, as mathematical instruction became progressively more elementary into the twentieth century.[26]

Notwithstanding his promotion of geometry, by his own account Dassen explicitly resisted teaching it on an advanced level. In his history of mathematics in Argentina, published in 1924 for the golden jubilee of the Sociedad Científica Argentina, Dassen reflected on the preceding 24 years, when he was responsible for mathematics on the doctoral level at the University of Buenos Aires. Dassen expressed his counter-disciplinary views in the third person:

> In his opinion, the country should not, today, promote the doctorate in mathematics, a career currently without a future here and incapable of securing the well-being of those who cultivate it and their family, if they have one. For this reason, he has not organized the creation of anemic or artificial [*vida forzada*] courses, nor has he sought students. On the contrary, he has always spoken clearly in the above sense to those who have consulted him (Dassen, 1924, 68).

Dassen's position has the merit of denying a proposition used by countless theologians for the existence of God, *quid melius quam nihil*, but his attitude is unworthy of higher learning, which, if it does not end inevitably in the tragedy of the novel, *Jude the obscure*, nevertheless seldom leads directly to a life of ease. In Dassen's lifetime, the land of the research doctorate, Germany, was filled with underemployed doctors, a number of whom emigrated to Argentina. Students intrigued by

[26] According to the annual plans surviving in the library of the Colegio Nacional de Buenos Aires.

Dassen's publications would have been discouraged by the master himself from taking Non-Euclidean geometry seriously.[27]

A final indication follows from the penultimate paragraph of the preface to Dassen's volume on solid geometry. He points to the important rôle of mathematical models for stimulating young minds. Every secondary school, he writes, should have a cabinet of geometrical models. Given that Dassen expressed admiration for mathematicians at Göttingen, especially David Hilbert and Felix Klein, and that he was familiar with the proceedings of European mathematics symposia and congresses before the First World War, it is likely that he had in mind Klein's commercial enterprise in designing and circulating plaster models of mathematical surfaces. The models were visible to students in glass cabinets at many European and North American institutions of higher learning. Einstein would have been familiar with some of them at the Zurich Polytechnic as a student and then as a professor there.[28] I have seen no firm evidence that these models were shown at Argentine universities.

On average, an upper bound of fifty students per year studied university geometry at La Plata between 1906 and 1914, and an equal number at Buenos Aires between 1900 and 1914. Under the generous assumption of no repeating students from one year to the next, we have 1100 students in total who might have heard about descriptive and projective geometry in class. Europe and North America had a hundred times as many comparable institutions of higher learning, many of them with a higher enrollment.[29] We are not unreasonable, again, to guess that on the eve of the First World War in the global North on both sides of the Atlantic and in Japan, there were at least 100,000 engineers, teachers, and entrepreneurs who in their youth suffered through calculations in advanced geometry. In maturity, some of them sat on university councils that authorized courses and hired instructors. In Argentina, comparable people could have given serious thought to Cubism and Relativity when the topics occasionally appeared in their daily newspaper or periodicals. They certainly had the means to acquire Cubist paintings and promote Relativity.[30] We see that Greater Buenos Aires did not have enough of these university alumni to constitute an intellectual community valuing such patronage.

[27] The distinguished geometer Luis Antoni Santaló Sors cites Dassen's remark and comments that mathematics in the years 1900–1917 made no progress at the University of Buenos Aires because the faculty did not guide students through expanding areas of research, which led to declining enrollment and discouragement—in effect, that superficiality is an uninvited companion in the counter-disciplinary life (Santaló, 1970, 264–265).

[28] (Pyenson, 2021, 214–216, 239–246) for the European context of plaster casts and Einstein's familiarity with them.

[29] (Forman et al., 1975, 16) for 27,650 as an estimate of the number of students in physics lecture courses at institutions of higher learning in Europe and North America in 1900. Mathematics, requiring no expensive laboratory equipment, would have been taught more widely.

[30] (Echagüe, 1912), a virulent indictment of "Cubist" art in the Paris Autumn Salon of 1912, on 140 emphasizing the harmony and color of the garland-and-basket theme (executed by unnamed Marie Laurencin in the unnamed "Cubist House") there; 141 the watchwords of decorators: "The turn to

8.8 High Modernity and Counter-Disciplinary Mathematics

As Dassen's and Coutaret's architectural undertakings suggest, Argentine intellectual patronage was anchored in the landed and commercial class, who mistrusted ideas that advocated for a major change in taste, indeed for anything proposed by non-traditional elites. The aura of social prestige that surrounded iconoclastic European intellectuals did not extend to thinkers in Argentina, for whom it was difficult to earn a living by innovating in styles and abstract ideas. The two geometers used nineteenth-century mathematics in the service of Neo-Classical forms to build a city with Second-Empire and even Gothic residences and public buildings topped by prominent finials to delight passing pedestrians. In their time, modernist architectural structures were few.[31]

Modernity sought to transform humanity by revising living conditions. The result, seen today from Jakarta to São Paulo to Shanghai, is a severe and cold landscape of modern materials, interspersed with shanties of the dispossessed. This gargantuan concentration of people is surely not what the Neo-Idealist planners imagined around 1900. In it we see engineering practicality separated from artistic and scientific abstraction, a distinction that made possible the spectacular twentieth-century industrial rise of Japan, China, and India.[32] The world experienced Cubism and Relativity as cosmetic beauty marks, prominent in one socio-economic climate and dimly known in another.

The separation of Technology from Science in High Modernity allows for the cohabitation of impractical fields of study like Latin rhetoric and cosmology with eminently practical structural engineering and aromatic chemistry. It is a pairing achieved after generations of discussion in Europe but one embraced with little second-guessing in much of the Americas and Asia. The pairing has the advantage of allowing technical innovation to cross over the rigid specialist barriers that are a feature of nineteenth-century and twentieth-century industrial societies. In addition to producing Nobel laureates in biochemistry (a new field in the twentieth century), Argentina welcomed other thinkers who crossed disciplinary lines. When Europe was overrun by fascism, Argentina offered refuge to the International Academy of the History of Science, at the time an interdisciplinary gathering; Aldo Mieli and José Babini were prominent in this rescue mission. In any case, balancing pure

Antiquity" in "simple forms [*órdenes*}, beautiful and evident [*francas*] harmonies." Echagüe is analyzed in (Wechsler, 1998, 138–139) and (Wechsler, 2011).

[31] In addition to the well-known Edificio Kavanagh (1934), there are private residences inspired by Le Corbusier: the Casa Taller Forner-Bigatti in San Telmo, Buenos Aires (1937) by Alejo Martínez (Štech, 2020) and the Casa de Victoria Ocampo in Palermo Chico, Buenos Aires, by Alejandro Bustillo, contracted 1928 (Guerra, 2022).

[32] That is to say, technoscience did not appear there before it became widespread in North-Atlantic Postmodernity (Forman, 2007). I contend that while particular discoveries and innovations were displayed using European norms, Modernity and Postmodernity had a special timing and took a distinct form in Argentina (Pyenson, 2021, 461–575).

learning and practical expediency in Argentina was unlike the compromise achieved in Europe and North America. It was part of a cultural environment that artists and writers struggled to understand in the decades following the Centenary of 1910, even as they labored to forge a unique national style.[33]

It was and is hard for even subtle minds to sense the character of the time in which they live. Did Coutaret and Dassen in their retirement understand their rôle as a polymath in an epoch of specialization? In the end, they focused on whimsy, both fictional and non-fictional. Dassen's last work is a long monograph—we could charitably call it ultra-disciplined—filled with transcriptions of writings by nineteenth-century mathematicians in Argentina and lists of students (Dassen, 1939–1941). In many hundreds of pages, we see intellectuals at a crucial time in Argentine history. Similar documentary undertakings occupied historians a hundred years ago, and they continue in letterpress editions of the papers and correspondence of significant thinkers, Einstein included.[34]

More accessible is Coutaret's 1929 novel, The *Malvinas restored*, recounting how an enterprising naturalist-mariner used a weapon for incapacitating motors and firearms—"a discovery of a new source of energy able to give to the nation that has use of it, the most absolute supremacy on sea and land." The discovery harnessed magnetic fields and was used to resolve a nationalist cause-célèbre peacefully. The plot follows the search by an Argentine naturalist for an uncharted, volcanic island in the South Atlantic where the telemagneto device is eventually found and activated (Coutaret, 1929, 169). The international crew of Coutaret's naturalist-mariner somewhat resembles the crew pursuing Moby Dick in Herman Melville's novel. Coutaret's novel has elements of Jules Verne, H. G. Wells, Arthur Conan Doyle, and likely closer to home, the long history of harvesting marine fauna in the South Atlantic. Perhaps, too, there is an echo of the Argentine naturalist Eduardo Ladislao Holmberg's *Viaje maravilloso del señor Nic-Nac [al planeta Marte]* (1876) as well as the Patagonian expedition in 1922 by Holmberg's successor as Buenos Aires zoo director, Clemente Onelli, to find a living plesiosaur.[35]

Dassen's and Coutaret's swan song, each one a window looking out on Science in its time and place, can serve in advanced history seminars today. Dassen's documents are useful in the way that medieval historians still profit from the *Monumenta Germaniae Historia*; Coutaret's novel recounts the ethics and prejudices of Argentine intellectuals on the verge of twentieth-century political upheavals.[36] Their contemporaries Jacobo Laub and Charles Dillon Perrine, pioneers of Relativity

---

[33] For extensive debates and controversies early in the twentieth century about creating a national style in Argentina, (Muñoz, 1998).

[34] For an example of one institution created to foster this interest in the United States, (Gondos, 1981). Inventory-and-commentary was at the center of scholarship in history of science during the 1920s and 1930s (Pyenson, 1995).

[35] For Holmberg, (Ortiz, 2005; Haywood, 2012). On whaling and guano harvesting, (García, 2021).

[36] (Berkhofer, 2023), for masterful use of the *Monumenta Germaniae Historica.*

introduced at the beginning of the present text, produced nothing comparable. Which publications of today's scientists shall be read a hundred years hence, and how?

8.9 Envoi

The limitations of Dassen's and Coutaret's counter-disciplinary activity are usefully contrasted to the vision of Juan José Saldaña, an international organizer of the stature of historians of science George Sarton and Otto Neugebauer. Don Juan José continued the tradition of international scientific projects such as the nineteenth-century astronomical *Carte du ciel* and diverse academic cartels, eventually UNESCO, with which he was closely connected. As a buccinator, he called together scholars from a score of countries, who came to recognize common themes in their research and life. Impediments abound in this kind of endeavor, notably placing students in permanent positions where they can elaborate their specialized training. Don Juan José overcame obstacles with an uncommonly sympathetic personality that followed an unwavering intellectual compass. The present volume testifies to the vigor of his undertaking.

Acknowledgments I thank historian Ailin Sigal, Universidad de Buenos Aires, and historian of science Susana V. García, Museo de La Plata, for valuable help and discussion.

The author declares no conflicts of interest.

References[37]

Andruchow, M., & Castillo, P. (2019). La Escuela de Dibujo en el Museo de La Plata: La controversia Malharro-Rosso en la enseñanza. *Arte e Investigación, 15*, e019. https://doi.org/10.24215/24691488e019

Babini, J. (1942). Claro Cornelio Dassen. *Archeion, 24*(1), 97–98.

Baldasarre, M. I. (2020). Between Buenos Aires and Europe: Cosmopolitanism, pensionnaires, and arts education in late 19th century Argentina. In O. E. Vázquez (Ed.), *Academies and schools of art in Latin America* (pp. 17–31). Routledge.

Barabino, S. E. (Ed.). (1910). Congreso Científico Internacional Americano. *Anales de la Sociedad Científica Argentina, 69*(July).

Barrachina, R. O. (2003). *Una perspectiva weberiana para la biografía de Jakob Laub, primer colaborador de Einstein.* Maestría en Filosofía e Historia de la Ciencia, Facultad de Humanidades, Universidad Nacional del Comahue, Arg.

Beaulieu, L. (1989). *Bourbaki: Une histoire du groupe de mathématiciens français et de ses travaux (1934–1944).* Thèse Ph.D. en histoire et sociopolitique des sciences. Université de Montréal, Canada.

[37] Internet Material Accessed in February 2024.

Bedrossian, P. R. (2021). *El Palacio Dassen, el edificio de la Asociación Argentina de Actores*. https://pablobedrossian.com/2021/01/07/el-palacio-dassen-el-edificio-de-la-asociacion-argentina-de-actores-por-pablo-r-bedrossian/

Berkhofer, R. F., III. (2023). *Forgeries and historical writing in England, France, and Flanders, 900–1200*. Boydell.

Cortés Saavedra, D. F. (2016). Geometric abstraction and concrete art in South America. In S. Ross (Ed.), *Routledge encyclopedia of modernism* (Vol. 26, pp. 506–516). Taylor and Francis. https://doi.org/10.4324/9781135000356-REM541-1

Coutaret, E. (1910). El pintor Don Miguel Rosso [portrait]. *Ars, 3*, 489.

Coutaret, E. (1911). El pintor. *Ars*, Número Malharro, 3, no. 18 (October), reprinted in *Cuadros de viaje: Artistas argentinos en Europa y Estados Unidos (1880–1910)*, ed. Laura Malosetti Costa (pp. 257–263). Fondo de Cultura Económica, 2008.

Coutaret, E. (1929). *Las Malvinas restituidas (novela)*. L. J. Rosso.

Crelinsten, J. (2006). *Einstein's jury: The race to test relativity*. Princeton University Press.

Damianovich, H., & Levylier, H. M. (1918). Camilo Meyer. *Anales de la Sociedad Científica Argentina, 86*, 49–84.

Dassen, C. C. (1901). *Metafísica de los conceptos matemáticos fundamentales (espacio, tiempo,cantidad, límite) y del análysis llamado infinitesimal*. Tailhade y Rosselli.

Dassen, C. C. (1904). *Tratado elemental de geometría euclídea de acuerdo con las ideas modernas y metodos más rigurosos; vol. 1: Geometría plana, texto adaptable á los programas de los colegios nacionales y escuelas normales*. Coni Hermanos.

Dassen, C. C. (1905). *Traitado elemental de geometría euclídea, de acuerdo con las ideas modernas y metodos más rigurosos, vol. 2: Geometría del espacio; texto adaptable á los programas de los colegios nacionales y escuelas normales*. Coni Hermanos.

Dassen, C. C. (1924). *Las matemáticas en la Argentina* [Evolución de las ciencias en la República Argentina, 4]. Coni.

Dassen, C. C. (1939). Vida y obra de Luis Couturat. *Anales de la Academia National de Ciencias Exactas, Físicas y Naturales de Buenos Aires, 4*, 73–204.

Dassen, C. C. (1939/1941, 1941/1942). Vol. 1, La Facultad de Matemáticas de Buenos Aires (1874–1880) y sus antecedentes, Antecedentes varios (desde 1767); vol. 2, La Facultad desde 1874 hasta 1881, appearing in 1200 pages in the *Anales de la Academia Nacional de Ciencias Exactas, Físicas y Naturales de Buenos Aires 5* and *6*.

Dragowski, A. (2019). La reforma lejos de la reforma en la facultad de ciencias físico-matemáticas de la UNLP: Ricardo Gans y sus ambivalencias, 1918-1921. *Anales de la Asociación Física de Argentina, 29*(4), 129–132.

Duclout, J. (1892). Estudio sobre las hipótesis mecánicas que sirven de base a la teoría electromagnética de la luz Maxwell. *Anales de la Sociedad Cientifica Argentina, 33*, 73–96.

Echagüe, J. P. (1912). Los cubistas y el arte nuevo. *Nosotros, 9*(October), 137–141.

FCEN. Universidad de Buenos Aires, Bibliotéca Luis Leloir, Facultad de Ciencias Exactas, Fïsicas y Naturales. Libro de Actas de las Sesiones de la Facultad.

Fernández Staccio, E. (2002). *La matemática en la Argentina entre las guerras mundiales*. Appearing as Informe técnico interno, no. 77. Instituto de Matematica de Bahia Blanca, Universidad Nacional del Sur, Arg.

Fernández Stacco, E. L. (2011). *200 años de la matemática en la Argentina*. Instituto de Matemática de Bahía Blanca.

Forman, P. (2007). The primacy of science in modernity, of technology in postmodernity, and of ideology in the history of technology. *History and Technology, 23*, 1–152.

Forman, P. (2012). On the historical forms of knowledge production and curation: Modernity entailed disciplinarity, postmodernity entails antidisciplinarity. *Osiris, 27*, 56–97.

Forman, P., Heilbron, J. L., & Weart, S. (1975). Physics circa 1900: Personnel, funding, and productivity of the academic establishments. *Historical Studies in the Physical Sciences, 5*, 1–185.

Gangui, A., & Ortiz, E. L. (2011). Anti-positivismo, ciencias theóricas y relatividad en Argentina de la década de 1920. *Revista Brasileira de Historia da Ciencia, 4*, 201–218.

Gangui, A., & Ortiz, E. L. (2015). La relatividad en la cultura de la Argentina de inicios del siglo veinte. In P. Lorenzano & F. T. Molina (Eds.), *Filosofía e historia de la ciencia en el Cono Sur* (pp. 279–287). Editorial UNQ.

García, S. V. (Ed.). (2021). *En el mar austral: La historia natural y la explotación de la fauna marina en el Atlántico Sur*. Prohistoria.

Gondos, V., Jr. (1981). *Franklin Jameson and the birth of the National Archives, 1900–1926*. University of Pennsylvania Press.

Gray, J. (2008). *Plato's ghost: The modernist transformation of mathematics*. Princeton University Press.

Guerra, G. (2022). Precisiones sobre la casa Victoria Ocampa en Bario Parque. *Revue de Recherches sur Le Courbusier, 5*, 68–83.

Haywood Ferreire, R. (2012). El viaje a Marte en la imaginación, ayer y hoy. *Revista Iberoamericana, 78*, 25–38.

Heinzmann, G., & Petitot, J. (2020). The functional role of structure in Bourbaki. In E. H. Reck & G. Schiemer (Eds.), *The pre-history of mathematical structuralism* (pp. 187–214). Oxford University Press.

Herrera, M. J. (2014). *Cien años de arte argentino*. Biblos.

Kaplan, L. A. (2014). Crossing the Atlantic: Emilio Pettoruti's Italian immersion. *Artl@s Bulletin, 3*(2), 1.

Kraselsky, R., & Andruchow, M. (2021). Dibujo y enseñanza: Malharro-Rosso; aspectos del debate en el Escuela de Dibujo de La Plata. In B. Gustavino et al. (Eds.), *La constitución de las disciplinas artísticas* (pp. 45–57). Universidad Nacional de La Plata, Facultad de Artes.

LPMHN. La Plata, Museo de Historia Natural, Archives.

Larsen, C. R. (2017). "Noble and delicate sentiments": Museum natural scientists and an emotional community in Argentina, 1862-1920. *Historical Studies in the Natural Sciences, 47*, 42–75.

Lovejoy, A. O. (1930). *The revolt against dualism: An inquiry concerning the existence of ideas*. Open Court.

Mehrtens, H. (2004). Mathematical models. In S. de Chadarevian & N. Hopwood (Eds.), *Models: The third dimension of science* (pp. 276–306). Stanford University Press.

Mercante, V. (1911). Enseñanza de la geometría. *Archivos de Pedagogía y Ciencias Afinas* [Facultad de Ciencias Jurídicas y Sociales, La Plata] 9(25), 4–34.

Montserrat, M. (1993). La presencía evolucionista en el postivismo argentino. In M. Montserrat, *Ciencia, historia y sociedad en la Argentina del siglo XIX* (pp. 70–82). Centro Editor de América Latina.

Muñoz, M. A. (1998). Un campo para el arte argentino: Modernidad artistica y nacionalismo en torno al Centenario. In D. B. Wechsler (Ed.), *Desde la otra vereda: Momentos en el debate por un arte moderno en la Argentina (1880-1960)* (pp. 43–82). Ediciones del Jilguero.

Nessi, Á. O. (Ed.). (1982). *Diccionario temático de la artes en La Plata*. Instituto de Historia del Arte Argentino y Americano.

Ortiz, E. L. (1995). A convergence of interests: Einstein's visit to Argentina in 1925. *Ibero-Amerikanisches Archiv, 21*, 67–126.

Ortiz, E. L. (2005). On the transition from realism to the fantastic in the Argentine literature of the 1870s: Holmberg and the Cordoba six. In E. Fishburn & E. L. Ortiz (Eds.), *Science and the creative imagination in Latin America* (pp. 59–85). Institute for the Study of the Americas.

Ortiz, E. L. (2011). Julio Rey Pastor, su posición en la escuela matemática Argentina. *Revista de la Unión Matemática Argentina, 52*(1), 149–194.

Ortiz, E. L., & Rubinstein, H. (2009). La física en la Argentina en los dos primeros tercios del siglo veinte: Algunos condicionantes exteriors a su desarrollo. *Revista Brasileira de História de Ciência, 2*(1), 40–81.

Penchaszadeh, P. E., & de Asúa, M. (2010). *El deslumbramiento: Aimé Bonpland y Alexander von Humboldt en Sudamérica*. Museo Argentino de Ciencias Naturales.

Perrine, C. D. (1916). The total solar eclipse of February 3, 1916. *Publications of the Astronomical Society of the Pacific, 28*, 247–252.

Perrine, C. D. (1927). Contribution to the history of attempts to test the theory of relativity by means of astronomical observations. *Astronomische Nachrichten, 219*(17), 281–284.

Pettoruti, E. (2006). *A painter before the mirror.* Trans. Cristina Soares Gache. Fundación Pettoruti; Spanish edition: Solar, 1968.

Podgorny, I., & Lopes, M. M. (2008). *El desierto en una vitrina: Museos e historia natural en la Argentina, 1810–1890*. Limusa.

Pohl, C., & Hadorn, G. H. (2010). Td-net—The Swiss academies of arts and sciences' forum for transdisciplinary research. In R. Frodeman, J. T. Klein, & C. Mitcham (Eds.), *Oxford handbook of interdisciplinarity* (pp. 434–436). Oxford University Press.

Pyenson, L. (1988). Silver horizon: A note on the later years of the physicist-diplomat Jakob Laub. *Jahrbuch für Geschichte von Staat, Wirtschaft und Gesellschaft Lateinamerikas (Cologne), 25*, 757–766.

Pyenson, L. (1990). *El joven Einstein*. Trans. Rosa Alvarez Ulloa. Alianza Editorial.

Pyenson, L. (1995). Inventory as a route to understanding: Sarton, Neugebauer, and/sources. *History of Science, 33*, 253–282.

Pyenson, L. (2021). *The shock of recognition: Motifs of modern art and science*. Brill.

Pyenson, L. (2023). Un renascimiento cosmográfico en la Argentina posmoderna. In M. Rieznik & C. Lois (Eds.), *Técnica e estética de la imágenes: Elementos para pensar lo visual en las práticas cientïficas y artistícas* (pp. 258–308). Equinoctialis.

Rey, J. M. (1938). *Historia de la Escuela de Dibujo (Anexa a la Escuela de Bellas Artes) 1905–1938*. [15 November 1938]. Archivo José María Rey, caja 2, Archivo Histórico Universidad Nacional de La Plata.

Rieznik, M. (2011). *Los cielos del sur: Los observatorios astronómicos de Córdoba y de La Plata 1870–1920*. Prohistoria.

Roig, A. A. (1969). *Los krausistas argentinos*. José M. Cajica, Jr.

Rossell Soler, P. A. (1943). Claro Cornelio Dassen. *Anales de la Sociedad Científica Argentina, 135*, 21–37.

Rueda, M. d. l. Á. d. (2019). Emilio Coutaret, José María Rey y el Circulo Ars: La formación del campo artistico en ciudad de La Plata en la primera década del siglo XX. *Revista de Instituciones, Ideas y Mercados, 68*(October), 89–103.

Ruvituso, F. L. (2023). La colección fundacional en su historia: Benito Sosa y el sueño de un museo de arte público. In J. Conde (Ed.), *Catálogo razonado: Colección fundacional Museo Pettoruti* (pp. 16–33). Ediciones Bonaerenses.

Santaló, L. A. (1970). La matemática en la Facultad de Ciencias Exactas y Naturales de la Universidad de Buenos Aires en el period 1865-1930. Primer Congreso Argentino de Historia de las Ciencias. *Boletín de la Academia Nacional de las Ciencias [Córdoba], 48*, 255–273.

Schubring, G. (2002). A framework for comparing transmission processes of mathematics to the Americas. *Revista Brasileira de História da Matemática, 2*(3), 45–63.

Siracusano, G. (1998). Punto y linea sobre el "campo". In D. B. Wechsler (Ed.), *Desde la otra vereda: Momentos en el debate por un arte moderno en la Argentina (1880–1960)* (pp. 179–213). Ediciones del Jilguero.

Štech, A. (2020). Buenos Aires' Casa Taller Forner-Bigatti. *Design Miami,* 13 October. https://forner-bigatti.com.ar/fundacion-fb/

Stoetzer, O. C. (1998). *Karl Christian Friedrich Krause and his influence in the Hispanic world.* Böhlau Verlag.

Universidad Nacional de La Plata. (1909). *Album, República Argentina*. La Plata.

Volterra, V. (1910). Espacio, tiempo i masa según las ideas modernas. *Anales de la Sociedad Científica Argentina, 70*, 223–243.

Wechsler, D. (1998). Nuevas miradas, nuevas estrategias, nuevas contraseñas. In D. B. Wechsler (Ed.), *Desde la otra vereda: Momentos en el debate por un arte moderno en la Argentina (1880–1960)* (pp. 119–155). Ediciones del Jilguero.

Wechsler, D. B. (2011). Cosmopolitanism, Cubism and new art: Latin American itineraries. Trans. Antonio Bautista-Triguero. *Art in Translation 3*(1), 69–85.

Lewis Pyenson is Emeritus Professor of History, Western Michigan University. For nearly a half-century, he taught history of science, also serving as graduate dean for 10 years under four presidents and six provosts at two universities. For two years, he was a Killam Research Fellow of the Canada Council for the Arts. He held the Sarton Chair at Ghent University in Belgium, and he was Suntory Lecturer in Japan. He is a Life Member in the Academy of the Arts and Humanities of the Royal Society of Canada. He is now extending the ideas in his book, *The shock of recognition: Motifs of modern art and science* (Leiden: Brill, 2021).

Chapter 9
From Latin American to the New Native American Identity: The Importance of History of Science in the Construction of the Who We Are

Regis Cabral

Abstract This provocative essay is inspired by the professional practices of Juan Jose Saldaña. It argues that the making of a new identity for the region, a Native American identity that transcends and incorporates its Latin traditions, is interlinked with the rise of the knowledge field, sciencediplomacy. As such, very much in light with the ideas and ideals of Juan Jose Saldaña, a needed opportunity emerges for the history of sciences, that of informing and training the professionals in this strategic field for nations and cultures.

Keywords Science and diplomacy · Innovation · Knowledge based industry · Dependency · International relations · Latin American identity · Sciencediplomacy

9.1 Introduction

This is a provocative essay, illuminated by my interactions with Juan Jose Saldaña and other historians of science, in Latin America and beyond. Saldaña and his writings often inspired me in the search for the identity of the people that make the region known as Latin America. Our identity, as the identity of most peoples on our planet, is the result of interactions with other cultures. These cultural relations have evolved into diplomatic relations. As knowledge and science revealed their importance and relevance, we entered into a period of science and diplomacy. A new age has emerged, that of sciencediplomacy, where science and its methods are critical in informing decision makers in matters of strategic relevance. The practice of history of science, of history of technology and of history of innovation have emerged as central to educate and train the professionals in this new field. Part of this training is

R. Cabral (✉)
FEPRO – Funding for European Projects, Piteå, Sweden
e-mail: director@fepro.eu

M. de la P. Ramos-Lara, L. C. Arboleda (eds.), *History of Science in Latin America: The Construction of an Intellectual Field (20th century)*,
https://doi.org/10.1007/978-3-031-75536-1_9

the recovery of our identity. This process reveals that Latin America as a concept is fading away. The continent, without rejecting its Latin dimensions, is more than that. A new Native America is emerging, more than the sum of its parts, very much in light of the ideals promoted by Juan Jose Saldaña.

9.2 Knowledge Stolen: Identities Stolen

Who are we, Latin Americans? Latins because many of us have as a primary language, language with roots in Rome, Lacio, claiming to be inheritors of the Roman Empire? Americans because of a map designed and written by Americo Vespucio? Not that these are unique impositions and constructions. Quebecquians are also Latin. Vikings with helmet horns never existed except as constructions of nineteenth century nationalists but which became part of the global folklore.

This identity denial is also a denial of humanity, a justification of exploitation, slavery and unfair transfer of resources, including human capital. Nevertheless, it is a denial promoted by economic and political elites, not from outside but from Latin America. Their dream is to become European, have blond hair and be called white in their identity papers. It is a constant re-statement of how Europe "discovered" the Americas, even though there is no question about the fact that the continent on the other side of the Atlantic existed before the European voyages. It is also a denial of the achievements of non European navigators, from Kon Tiki to the growing evidence of pre—1500 Chinese presence. It also has absurd dimensions, such as the claims that whatever has been achieved by the pre-Columbian cultures, from pyramids to quipus to astronomical advances had been the result of" wise" extra-terrestrial visitors.

It is also a denial of a history, pre—and post—European, of achievements. There are some solid examples of these advances. Today over 20% of agricultural relevant species have emerged from Latin American cultures. These are the cases of tomatoes, beans, maize, potatoes, peppers, chocolate, tobacco, pineapples, peanuts, blueberries, squash, and more. One is not even sure about what Europeans really ate before the navigations. Try to imagine Italian cuisine without tomatoes, or a European life without potato chips, chocolate bars or pizzas without tomato paste.

Much of these developments were deeply linked to urban environments in pre-Columbian Americas. Modern archeology, based on aerospace technologies show that the Americas were a mosaic of networked cities, some with advanced transport communications. Today it is known that even the Amazonian basin had city conglomerates. Knowledge centers, mathematics, and even written production was part of life for many of these cultures. Tenochtitlan, Mexico City, was not the only case. Many of these centers had disappeared before the European invasions. Environmental crisis and the collapse of socio-political systems have much to teach us about our own mismanagements. That learning has to be reconstructed, as Europeans were systematic in their process of cultural destruction. Churches were built over temples and centers of learning, languages and cultural practices were forbidden, Mayan texts were destroyed. Education and the advancement of knowledge was hampered by religious control. From the "discoveries" in the 15 hundreds to independences around the 18 hundreds, Latin America was under the influence of the Catholic Inquisition. Promoting the printed word could result in penalties. Promoting free thinking, an enquiring mind, could result in death. Natural Philosophy was bypassed.

This process did not stop when Latin America became politically independent. The political independence was colored by elites trying to maintain the status quo. It was discolored by the economic dependency to colonial powers. The poverty imposed on Haiti by France is a sad illustration of this. So it is the "opening" of the Brazilian ports to friendly nations, resulting in an impossible situation for any Brazilian innovator. It was an open door to poverty. The opportunity for Brazilians existed. Since Napoleon troops had invaded Portugal, the seat of the Imperial government was in the Americas, Rio de Janeiro, and not Europe. It is proper to say that Brazilian independence was an independence in reverse. It occurred because the Portuguese demanded the return of the government to Lissabon.

Some recent cases of denial are known. José Bonifácio de Andrada e Silva contributions to mineralogy, including the large collection of Scandinavian minerals now in a Brazilian Museum, are for the most unknown. It is still difficult for Latin American scientists to receive support for nominations to the Nobel Prize. Oswaldo Cruz, Leite Lopes, just to name two, were opposed by their own Latin American counterparts. Santos Dumont who invented the first airplane to lift by its own craft (the Wright brothers Kitty Hawk needed a rail to reach liftoff) is presented sometimes as a French airman. Other of his innovations, such as the wrist watch, despite Cartier's efforts, are rarely mentioned. Even when he is mentioned or given priority, other Brazilian airmen are forgotten, such as Dimitre de Lavaud (naturalized Brazilian) who flew on 7 de Janeiro de 1910 in an airplane of his design and with 100% Brazilian components. In another special situation, the Brazilian contribution to prove Einstein General Relativity Theory, the eclipse observations at Sobral, is forgotten even by Brazilians.

Obviously this is not so for specialists in the history of these fields. But the telling of history has been hijacked by a mentality of dependency. The low level of education in the continent, despite the great words of promise by leaders and politicians, is marcant. As such the teaching of history is also not there and receives low priority.

To recover our past, we need historians of science and innovation. They are our voices from the past. But we must keep these voices alive. Their names email glide away. For instance, how much is discussed today about Ubiratan D'Ambrosio and his contributions to ethnomathematics? Or the works of Mario Bunge? To keep the contributions of Juan Jose Saldaña alive goes beyond the historiography of Latin American history of science. We need role models who motivated the construction of a constantly denied identity.

9.3 Interactions of Science and Diplomacy

Latin America, as the name implies, does not exist in a vacuum. Its global position was created in relation to external powers. Diplomatic services, with their ministers, special envoyés and attachés, always have had intelligence functions, including of technological, industrial and production capacities. For a nation, to evaluate the potentials of another nation, as ally or as competition, as market or as producer of raw materials, may be central to hegemony and even survival. Natural philosophers, and scientists later, because of their facility in obtaining "pass par tout" have never been far away from spy functions. But they could be more dangerous for the powers controlling the places they visited, as minds inquiring into nature, also questioned social and power relations. The number of natural philosophers that were forbidden by the inquisition goes beyond Copernicus, Galileo, Kepler and Newton. Natural philosophy and later science and their promoters, and thus members of scientific academies whenever traveling abroad were perceived as either messengers of new ideas or as foreign observers. José Bonifácio de Andrada e Silva as well as Francisco Miranda are important cases, with European experience that included visits to, among other countries, Sweden.

We tend to think of ourselves as a knowledge society. Human society, and maybe this the dominant aspect of humanity, for good or for worse, has always been knowledge based, memory protectors, even if reconstructed or selective, and incorporating teaching. Knowledge, science, information and innovation, while with different words, have always been there. But it has, since the outset of the scientific revolution, been gradually systematized. The method of knowledge production has been professionalized. The search for knowledge has become more than a part of production and industry. It has become an industry. It was already so in the late nineteenth century, with the chemical industry, vaccines and agricultural colleges. Nobel Prizes were, and are, awarded for useful knowledge. Nations seriously wanting to develop and modernize, sent their elites to study abroad. This was the case of Japan. But not in the case of Brazil, where progress became a word in the flag while the majority of the population remained without access to education and to schools.

As time passed, it was clear that a change was needed. The first world war was a scientific and innovation war, chemical weapons, airplanes and dreadnoughts. It was already beyond doubt that nations and cultures that invested in knowledge production had an advantage. But it was not clear to all, as scientists were still considered "not practical". To be a scientist was no valid excuse to save one from the trenchs.

This changed definitely by the second world war. Science became a weapon in itself, and as such so was education from grade school to research level. Without education, there is no scientific production, there is no defense, no glories in the fields of victory. Rockets, jet planes, and more than anything, the atomic bomb demonstrated the birth of a world dominated by systematic knowledge production.

The post war conflicts and efforts towards development highlighted this in many ways. The letter S in UNESCO is for science. But this means also that nations aiming at dominating others, gave priority to controlling education and science. Scientific applications were to be imported and not produced locally. This model became unsustainable, as production facilities and products required competent staff. But neither was the opposite model, of totally indigenous knowledge demonstrated to be a way out of dependency and poverty.

Science and science based products became critical to the relations between cultures. For Latin America the 1950s and 1960s were decades where attempts at fair trade and development, including "compensações específicas", that is when unfair prices were to be compensated by access to scientific knowledge of industrial relevance were met by the United States with suspicion. It was science and liberation, as Leite Lopes so clearly argumented for. The Brazilian nuclear programme of the 1950s, resulting in institutions like the National Research Council, CNPq, the Brazilian Center for Research in Physics, CBPF, suffered an USA intervention. Peron's Argentina nuclear programme, restricted both by the ideological biases of the regime and the difficulties of interactions with the international scientific community, collapsed on its own. In both cases, there was a clear understanding that science was key to a just position in the global community. In a cold war world, this was not welcomed and Latin America was to remain in the backyard. The major powers had instruments for this. In the case of Latin America there were US trained military elites, backed by national oligarchies.

One should be careful with generalizations. The Latin American universe is coloured by diversity. This is even so with the military. While the military regimens in cold war Latin America were known for damaging the educational system, purely out of ideological blindness, we should not forget that some of the great institutions of the region were founded by the military. Two outstanding examples are the Brazilian National Research Council, CNPq, and one of the region's greatest innovation motors, the airplane manufacturer EMBRAER.

The understanding that without science and innovation there is no freedom and independence was further reinforced by the developments of the late phases of the cold war and the years that followed: the space age, the internet, the rise of a knowledge based China as a superpower, and an environmental crisis that now can only be addressed by innovative ways of thinking. And there was more. The size of the Latin American markets offered opportunities that could only be harvested by investment in technology, which in its turn required qualified technical personnel.

Still, when I wrote my dissertation in 1986, and for some years beyond, science and innovation were two fields interacting with diplomacy. They were part of the diplomatic play, but separate, often elements to be traded or offered in deals, arrangements and treaties. It was science and diplomacy, not sciencediplomacy, where one can not exist without the other.

9.4 The Age of Sciencediplomacy

The new millennium signaled dangers and opportunities. Since 2001, humanity has had a permanent residence in space. New wars, new epidemics, the threat of global warming and the rise of AI as a major factor in international relations (and the future of humanity, as we know) have placed humanity at a life and death crossroads.

International cooperation, and survival, requires scientific inputs in quantity and quality. Strategic partnerships must be forged within and beyond the country level. The world is at war in Gaza, in Ukraine and other places, but the International Space Station still runs with the efforts of astronauts of all nations. Except China, which runs its own station. It is in the cards that competition and cooperation are the lifeline of space science. As much as space science is part of sciencediplomacy.

There is a precedent to this, the nuclear non proliferation treaties, when the scientific community paved the way to international treaties. The same is true of global warming treaties, despite their fragility. Traditional alliances and procedures can be transcended by the scientific community because of the language of science transcends religious and cultural boundaries. This is the strength of sciencediplomacy. Science is now an inseparable component of foreign policy. It is not science in diplomacy only, not science for diplomacy only, but sciencediplomacy, at the service of national and global interests. Foreign policy now must integrate the advice and evidence provided by the scientific community, as well as the criticisms. Otherwise nations and cultures could be at loss.

It is a knowledge field in the making. There is still on how it should be called. A search will give the readers results (which also vary depending on language and culture) such as science diplomacy (with the space in between), diplomacy science, science for diplomacy, science in diplomacy, science and diplomacy. It is definitely not the science of diplomacy. The latter has existed since science assumed a dominant status in society, and there was the science of everything, science of arts, science of literature, etc. It had to do with applying a method of inquiry to each field, a method nearly copied from exact sciences, particularly physics. Sciencediplomacy is about the integration of advice and knowledge.

Today, the history of sciences, of technology and of innovation has also achieved a new status. While the strength of the Latin American history of science has been to reconstruct an identity, now it has a further motivation. This is to provide tools, evidence, facts and learning to the professionals working in sciencediplomacy. In addition to recovering the past, we are participating in the making of the future.

9.5 Conclusion: From Latin America to the New Native America

As I think of this, I can not but feel that Juan Jose Saldaña would be pleased. For he planted the seeds of this development. The creation of the Latin American History of Science Society, for which Saldaña was a central person, is a testimony to that. But, in my mind, his great message is that our history, and future, did not start in 1500. Just picture the front cover of the journal Quipu. The rediscovery of what we are transcends European culture. No one is rejecting the greatness of the European contributions. Much of the European achievements resulted from the interactions with beyond Europe cultures and nature, including Latin America. A global society is larger than the sum of its native cultures. It is about time we assume the rightful place in this global society. Which is only possible that, while Latins, we are more than that. We are natives to a continent still in the making, a sum of all the contributions that make us great—a New Native America. I think Juan Jose Saldaña would like that.

Bibliography

Beltran, E. (1984). La Historia de la Ciencia en América Latina. *Quipu, Revista Latinoamericana de Historia de las Ciencias y la Tecnologia, 1*(1), 7–23.

Cabral, R. (1987a). The Mexican reactions to the Hiroshima and Nagasaki tragedies of 1945. *Quipu, Revista Latino Americana de História da Ciência e da Tecnologia., 4*, 81–118.

Cabral, R. (1987b). Science and development. In J. Heilbron (Ed.), *The Oxford companion to the history of science* (pp. 205–207). Oxford University Press.

Cabral, R. (1988). Ameaças Atômicas Norte Americanas Contra a America Latina. *Ciência e Cultura, 40*, 656–658.

Cabral, R. (1989a). Máscaras que Não Servem: Os Limites Ideológicos ao Desenvolvimento Científico e Tecnológico. In *D'Ambrosio U Anais do Segundo Congresso Latino Americano de História da Ciência e Tecnologia* (pp. 457–469). Nova Stella.

Cabral, R. (1989b). A Bomba Atômica na América Latina: Primeiras Reações à Hiroshima e Nagasaki, 1945. In *D'Ambrosio U Anais do Segundo Congresso Latino Americano de História da Ciência e Tecnologia* (pp. 449–456). Nova Stella.

Cabral, R. (1992). Latin America, Spain and Portugal: Scientific networks and exchanges with Sweden. *Science Studies, 5*, 97–107.

Cabral, R. (1993). The financing of biotechnology in developing countries. A Brazilian case in point. *Technology in Society, 15*, 311–325.

Cabral, R. (1994). Biotechnology, wheat production, and the Brazilian company for agricultural and livestock research (EMBRAPA). *Science and Public Policy, 21*, 147–156.

Cabral, R. (1998). The Cabral-Dahab Science Park management paradigm: An introduction. *International Journal of Technology Management, 16*, 721–725.

Cabral, R. (2003a). Latin America, science. In J. Heilbron (Ed.), *The Oxford companion to the history of science* (pp. 448–452). Oxford University Press.

Cabral, R. (2003b). Cesar Lattes and Jose Leite Lopes. In J. Heilbron (Ed.), *The Oxford companion to the history of science* (pp. 542–543). Oxford University Press.

Cabral, R., & Dahab, S. (1993). Science parks in developing countries: The case of BIORIO in Brazil. In *Biotechnology review no. 1: The management and economic potential of biotechnology* (Vol. 1, pp. 165–178).
Cabral, R., et al. (1989). O Estudo Histórico da Ciência e Tecnologia nas Agências das Nações Unidas: Esquema Teórico. In *D'Ambrosio U Anais do Segundo Congresso Latino Americano de História da Ciência e Tecnologia* (pp. 475–490). Nova Stella.
Diplo Cientifica. Rede de Diplomacia Cientifica en LatinoAmerica y Caribe http://linkedin.com/in/diplo-científica-7883b71b5.
European Union External Action. (2022). *What is sciencediplomacy?* European Commission. https://www.eeas.europa.eu/eeas/what-science-diplomacy_en
Ferrari, M. H. M. (1997). *As Ciências em Portugal e no Brasil (1772–1822): O Texto Conflituoso da Química*. EDUC.
Freire, O., Jr. (1999). *David Bohm e a Controvérsia dos Quanta*. CLE.
Leite Lopes, J. (1969). *Ciência e Libertação*. Editora Paz e Terra.
Moreira, I. d. C., & Videira, A. A. P. (Eds.). (1995). *Einstein e o Brasil*. Editora UFRJ.

Regis Cabral is known for the "**Cabral-Dahab Science Park Management Paradigm**", used to study, plan and evaluate science parks, areas of innovation, business incubators and similar organizations.

He is a physicist and historian of science and technology, a specialist in the interactions between the production of knowledge and international relations. In addition, he is an evaluator of business plans, funding applications, as well as of science parks, business incubators and accelerators, and areas of innovation.

Dr. Cabral is recognized internationally as an expert evaluator of higher education programmes, of international organizations such as UNESCO, UNCTAD, for the European Commission, of international aid and development projects. He is a member of the Reading (Reviewer) Committee for the Washington (US Department of State) White House Mandela Fellowship for Young African Leaders, and evaluator of candidates to University of California, Berkeley, Awards.

He is also a referee and reviewer for high level projects, journals and conferences, having received awards from AOM, Academy of Management for contributions to Organisation Management and Theory and Entrepreneurship, as well as for IASP, International Association of Science Parks and Areas of Innovation.

With over 300 articles, books, reports, debate entries and presentations published, Dr. Cabral has been a Professor, Lecturer and speaker, including at the University of Lund, Sweden.

He has been elected to political positions of trust in the City of Piteå, Sweden, for instance at the Committee for Employment and Adult Education and, at the moment of writing is a Judge (lay) at the Administrative Courts of Sweden and at the Immigration Courts of Sweden.

Dr. Cabral is currently Director of FEPRO—Funding for European Projects involved in the preparation, funding, including crowdfunding, management and coordination of projects, evaluations, start ups, SMEs. In this capacity he is a member of the **Early Warning European Mentor Academy**, EWEMA, assisting start ups, SMEs, family business and other organisations in moments of crisis, so that they can survive, become sustainable and resilient.

Currently Dr. Cabral's research focus is on the historical relations between inflation and innovation.

Chapter 10
Science and Cultural Diversity in Latin American Countries

Olival Freire Junior and Michel Paty

Abstract Juan José Saldaña González played an important role in fostering the studies on the history of science and cultural diversity by leading the creation of the International Association for Science Cultural Diversity (IASCUD) in the early twenty-first century in Mexico City. Since then, the relevance of studies on the relation among science and culture and cultural diversity and the approach in terms of cultural history and history of knowledge has been increasing. They are instrumental for the future development of history of science and technology as a scholarly field. This has been particularly important in Latin America, where there is at least a confluence of three distinct cultural trends: the European conquerors, the native people who lived here since immemorial times, and the African people who were enslaved and brought to America. In addition, later, there were waves of migrations from other European and Asian countries. In addition, the current development of science requires interaction among scientists and the knowledge of traditional communities.

Keywords Cultural diversity · Latin America · History of science · IASCUD-IUHPST

10.1 Introduction

When the XXI International Congress of History of Science was gathered in Mexico City in 2001, Juan José Saldaña González, who was the organizer of that meeting, led the creation of a new scholarly body, the International Association for Science

O. Freire Junior (✉)
Instituto de Física, Universidade Federal da Bahia (UFBa), Salvador, Brazil

Conselho Nacional de Desenvolvimento Científico e Tecnológico – CNPq, Brasilia, Brazil

M. Paty
Centre National de la Recherche Scientifique (CNRS), Université Paris Diderot, Paris, France

M. de la P. Ramos-Lara, L. C. Arboleda (eds.), *History of Science in Latin America: The Construction of an Intellectual Field (20th century)*,
https://doi.org/10.1007/978-3-031-75536-1_10

Cultural Diversity (IASCUD) of the International Union of History and Philosophy of Science and Technology (IUHPST). This new organization was created during the Symposium on Science and cultural Diversity at that Congress. It was preceded by a meeting in the city of Pachuca, Mexico, the year before the congress, during the VII Mexican Congress of the History of Science and Technology. Among the attendees at these meetings there were Michel Paty, Claude Debru, Luis Carlos Arboleda, Pablo Mariconda, Paulus Gerdes, Ubiratan D'Ambrosio, Irfan Habib, Dhruv Raina, Roy MacLeod, Antonio Lafuente, Teresa Rabiela, and Wesley Stevens, among others (Saldaña, 2003).[1] However, what was the novelty in joining the history of science and other cultural traditions than Europe? After all, studies on the history of science in areas outside of Europe were not new, as testified by studies of Joseph Needham on science in China, Otto Neugebauer on mathematics and astronomy in the clay tablets from the ancient Babylonians, and Roshid Rashed on mathematics in the Arabic world, who, incidentally, was one of the keynote speakers in the congress in Mexico City. The creation of IASCUD, however, was intended to foster studies beyond these previous area studies. For instance, Michel Paty [incidentally one of the authors of this paper] argued for the need for both history and philosophy, understood as rationality, in a larger meaning, for understanding science developments, exemplifying that one approach calling for concepts such "scientific style", "scientific program", or "scientific tradition" could contribute to dissolving the intractable issue of the context of discovery and context of justification in the scientific changes (Paty, 2001, p. 171).

Since then, the relevance of studies on the relation among science and culture and cultural diversity and the approach in terms of cultural history and history of knowledge has been increasing. They are instrumental for the future development of history of science and technology as a scholarly field. This has been particularly important in Latin America, where there is at least a confluence of three distinct cultural trends: the European conquerors, the native people who lived here since immemorial times, and the African people who were enslaved and brought to America. In addition, later, there were waves of migrations from other European and Asian countries. In addition, the current development of science requires interaction among scientists and the knowledge of traditional communities. This chapter gives a brief overview of the IASCUD and its intellectual challenges. Then, it identifies key subjects for research on the history of science and cultural diversity in countries from the Latin American area.

10.2 IASCUD's Rationale and Intellectual Challenges

Since its creation, the International Association for Science Cultural Diversity (https://iuhpst.org/pages/inter-division-commissions/iascud.php) has been ongoing. Throughout these activities, it has attempted to go beyond those already traditional

[1] For the proceedings of the 2001 meeting in Pachuca, see *Cuadernos de Quipu*, 5, 2001.

works on the history of science and technologies in Europe and America. The challenge is surpassing activities circled the so-called area studies. As declared in its Mission Statement, "diversity is conceived around national, ethnic, linguistic and religious identities. These identities are perceived as homogenous and essentially distinct from each other." The history of science and technology is not immune to these trends as they "also begin with similar assumptions, and at times these researches contribute to the shaping of such identities. 'Area studies' still constitute a way of organizing research in social sciences and humanities." Still, according to its Mission Statement, "IASCUD sets itself the task of promoting a critical analysis of such trends in the field of the history of knowledge. It diagnoses that the divide of history of knowledge based on the areas studied is detrimental to a fruitful deployment of the field. IASCUD thereby aims at bringing together all those who are convinced that a global approach to the history of knowledge provides the right framework for a fully theoretical approach to science and technology. It sets itself the task of building bridges among various groups that develop the history of knowledge, wishing thereby to promote a new understanding of what may be conceived as 'cultural diversity' in relation to science."

As we notice, IASCUD uses "history of science" and "history of knowledge" as interchangeable terms. However, some authors argue differently. Sven Dupré and Geert Somsen (2019, p. 186) acknowledge that many authors present the history of knowledge as a new field but one that "is often presented as a mere expansion of the history of science". However, they have a greater ambition: "The re-definition of the historiographical domain of the history of knowledge urges us to ask new questions about the boundaries, hierarchies, and mutual constitution of different types of knowledge as well as the role and assessment of failure and ignorance in making knowledge." Lorraine Daston (2017, 132) goes in the same direction. While warning history of knowledge is "an emergent field (not yet a discipline)", she claims the need to connect such a wider field with the history of science because the latter deals with issues that go beyond science itself, issues such as (Daston, 2017, 131):

> We are the discipline charged with historicizing just such questions about the how, what, and wherefore of knowledge: What counts as knowledge in a given epoch and culture and why? How do classifications and hierarchies among kinds of knowledge (say, theology and mathematics in thirteenth-century Paris, or Confucian ethics and silk manufacture in Ming dynasty China, or the sciences and the humanities right now) reshuffle? How do new ways of knowing—the controlled experiment, constructing a genealogy in historical philology or evolutionary biology, error analysis—emerge, and how are they braided together with older methods into a stronger rope of inference.

We think this wider perspective on the relationship between science and knowledge, and their histories, is more akin to Saldaña and IASCUD initial perspective because it creates more room for the scholarly study of science in the area of Latin America. Still, Saldaña both looked for understanding science in Latin America as something more complex than histories of mere receptions and had in high consideration all the knowledge of the native people who lived there before the arrival of the conquerors.

10.3 Latin American Contexts

The perspective of studying the history of science and technology, paying attention to the contextual cultural diversity has, at least a double implication when Latin American countries are concerned. As pointed out in the Introduction of this paper, Latin and Northern American regions have been the crucible of diverse cultural trends. While capitalism and its associated Western science and technology developed earlier in the northern area, diverse cultural wealth is everywhere. In addition, archeological investigations are bringing to the fore more information about so far unknown civilization, such as the huge reminiscences from towns in the Amazonian forest, in the Upano Valley in Equator, dating from 2500 years before current times.

Saldaña was very aware of the first implication of such a perspective for the history of science in Latin America. The American colonies were not a cultural vacuum when European science arrived there. The model of diffusion, preached by Basala, did not cope with this. One decade before the Mexico Congress, Antonio Lafuente and Juan José Saldaña (1987) had already emphasized the role of this context of reception in the volume they organized, bringing together classical from the historiography of science translated into Spanish and studies about theory and case studies of science developments in developing countries. The issue has been a hot topic in the history of science in developing countries. In addition to the criticisms of the idea of center and periphery in the diffusion of Western science, more promising and fruitful perspectives have appeared. This is the case of Kapil Raj's proposal that scientific knowledge's features should not be understood as intrinsic features of this kind of cultural product; instead, they should be investigated in the processes of cultural circulation and interactions (Raj, 2007). If interactions and circulation are so central to write the history of science, they always happen in concrete, and diverse, cultural contexts.

The other, and possibly stronger, implication of the approach suggested by Saldaña is that taking together science and cultural diversity implies taking history of science and technology as part of a broader endeavor, namely history of knowledge. Taking the pair science and cultural diversity may be a powerful antidote against the Eurocentrism still influential in the history of science and shed light on forms of knowledge produced the native people who lived in America before the conquerors arrived there. Let us illustrate this with a few cases.

The historian Gerardo Aldana y Villalobos (2022) renovated the studies on Mayan astronomy by taking distance from the previous trend to compare this astronomy with European astronomy, particularly in which concerns accuracy. Maya's achievements in this domain were so impressive that such a quest for comparisons had a long survival. Instead, Aldana y Villalobos looked for the meaning of such astronomical activities for the own Mayan society. Doing this, he not only was able to unearth the accuracy of the Mayan Venus tables and identify a female Mayan astronomer but also locate such feats in the cultural Mayan context. More precisely, mastering historical and anthropological records, in addition to astronomical ones, he was able to identify that the Mayans mastered the Venus table around the same time that the fascination by the feathered serpent transited from other Mesoamerican societies to the Mayans. Aldana y Villalobos (2022) then suggests that the Mayans identified Venus appearances with the serpent (Freire Junior, 2023).

The case of Mayan astronomy easily fits in the field of history of science, while previously better situated as area studies. In other cases, for instance, the cultural developments of the native people in Brazil, harshly fits in the domain history of science. In such cases, the perspective of history of knowledge allows better to explore the relations among knowledge and cultural diversity. Still keeping the case of the indigenous people who lived where nowadays is the country Brazil, their ability to domesticate wild varieties of plants and roots, for instance manioc, may be fitted in a larger chronological pallet of a history of agronomy and botany, or even environmental history. Here, we are not dealing with relics from the past, as plants are part of most Brazilians' current diet. In the same vein, it would be impossible to make sense of the Portuguese language spoken in Brazil without considering how much the formation of this language was influenced by the interactions among the Portuguese colonizers and the native indigenous people (Freire Junior, 2020; Baniwa, 2022).

The Brazilian composer Gilberto Gil realized this connection between science and the wider cultural diversity. In the most recent difficult times of Brazilian history he wrote a manifesto titled "Shining of science and culture are going to take us from the darkness" (Gil, 2022). This manifesto is a nice illustration of the perspective on science and cultural diversity envisioned by Saldaña, a quarter of century ago. Gil says[2]:

> Imagining the future for Brazil, and from Brazil, is to promote the warp between the most advanced sciences and popular knowledge, between the sensitivity of the people of the forests and that of the quilombos, between the methods of social scientists and the wisdom of the peripheries, between biomedical science and the knowledge that emerges from encounters on the asphalt, on the ground and in the forest.
>
> [...]
>
> Our indigenous, riverside, quilombola and other forest people traditions demonstrate in practice the power of popular knowledge in premeditating and complementing, in the fabric of their lives, the discoveries of science that in recent years have shown the calamity of the climate emergency. In the activism of indigenous and peripheral youth today, the great intelligence of the Brazilian people meets the most advanced science and the urgent global climate policy.

We conclude by arguing that the perspective of science and cultural diversity may be a powerful intellectual resource for the current civilizational challenges. A recent

[2] In Portuguese: [Imaginar o futuro para o Brasil, e a partir do Brasil, é promover a urdidura entre as ciências mais avançadas e os saberes populares, entre a sensibilidade dos povos das florestas e a dos quilombos, entre os métodos dos cientistas sociais e a sabedoria das periferias, entre a ciência biomédica e o conhecimento que brota dos encontros no asfalto, na terra e na mata]. [...] [Nossas tradições indígenas, ribeirinhas, quilombolas e dos demais povos da floresta demonstram na prática a potência dos saberes populares em premeditar e complementar, no tecido de suas vidas, as descobertas das ciências que nos últimos anos mostraram a calamidade da emergência climática. No ativismo de indígenas e jovens periféricos hoje, a grande inteligência do povo brasileiro se encontra com a ciência mais avançada e com a urgente política climática global.]

science and cultural diversity example may illustrate this inspiration potential. The historian Federica Gigante discovered that a medieval astrolabe (around 1060) stored at the Fondazione Museo Miniscalchi-Erizzo, in Verona, had scripts in Arabic and Hebrew languages. Looking at the data and inscriptions, she suggested that such an instrument testified to the existence of a scientific collaboration among Muslim, Jewish, and Christian scholars across current Spain, Italy, and North Africa, as it had been built in Andalucia. Inshallah, this peaceful coexistence between different people and countries may inspire us nowadays, particularly because of these difficult times when peace seems to be an elusive dream (Jones, 2024; Gigante, 2024).

References

Aldana y Villalobos, G. (2022). *Calculating brilliance – An intellectual history of Mayan astronomy at Chich'en Itza*. University of Arizona Press.

Baniwa, G. (2022). As contribuições dos povos indígenas para o desenvolvimento da ciência no Brasil: os povos originários colaboram de diversas formas com a sociedade brasileira desde a chegada dos portugueses até os dias de hoje. *Ciência e Cultura, 74*(3), 1–6.

Daston, L. (2017). The history of science and the history of knowledge. *Know: A Journal on the Formation of Knowledge, 1*(1), 131–153.

Dupré, S., & Somsen, G. (2019). The history of knowledge and the future of knowledge societies. *Berichte zur Wissenschaftsgeschichte, 42*, 186–199.

Freire Junior, O. (2020). History of science and technology in 20th-Century Brazil. In: Guillermo Palacios; Lauren (Robin) Derby; Brenda Elsey; Martin Nesvig. (Orgs.), *Oxford research encyclopedia of Latin American history*. Oxford University Press.

Freire Junior, O. (2023). Understanding the Mayans on their own terms. *Physics Today, 76*, 47–48.

Gigante, F. (2024). The seventeenth-century collection of Ludovico Moscardo. *Nuncius, 39*, 163–192.

Gil, G. (2022, June 25). Brilho da ciência e da cultura vai nos tirar da escuridão. *Folha de São Paulo*.

Jones, S. (2024, March 5) 'Extraordinary': Islamic and Jewish science merge in 11th-century astrolabe. *The Guardian*.

Lafuente, A., & Saldaña, J. J. (Eds.). (1987). *Historia de las ciencias*. CSIC.

Paty, M. (2001). A Letter from Paris. In Juan José Saldaña (Ed.), *Science and cultural diversity – Filling a gap in the history of science* [Cuadernos de quipu, 5], pp. 171–177. Sociedad Latinoamericana de Historia de las ciencias y de la Tecnologia.

Raj, K. (2007). *Relocating modern science: Circulation and the construction of knowledge in South Asia and Europe, 1650–1900* (p. 2007). Palgrave Macmillan.

Saldaña, J. J (Ed.) (2003). *Science and cultural diversity, proceedings of the XXIst international congress of history of science*, Volume I, Plenary Lectures, Sociedad Mexicana de Historia de la Ciencia y la Tecnología & Universidad Nacional Autónoma de México.

Olival Freire Junior is Professor of physics and history of science, Universidade Federal da Bahia, Brazil. He published, among other articles and books, Quantum Dissidents—Rebuilding the Foundations of Quantum Mechanics 1950 1990., Springer, 2015; David Bohm—A Life Dedicated to Understanding the Quantum World, Springer, 2019; and was the editor of The Oxford Handbook of the History of Quantum Interpretations, OUP, 2022.

Michel Paty is Emeritus Research Director, CNRS—Université Paris Diderot, France. He published, among other articles and books, La matière dérobée. Paris: Archives contemporaines, 1988; L'analyse critique des sciences, ou le tétraedre épistémologique (sciences, philosophie, épistémologie, histoire des sciences). Paris: L'Harmattan, 1990; and Einstein philosophe. La physique comme pratique philosophique. Paris: PUF, 1993.

Chapter 11
Networks of Latin American Historians of Medicine and Public Health and Their International Visibility

Ana María Carrillo

Abstract The chapter discusses how, after a group of academics—whose main leader was Professor Juan José Saldaña—convened historians of science in Latin America and the Caribbean to organize themselves and disseminate the results of their research internationally, and thus contribute to a non-ethnocentric history of science, historians of medicine and public health have been involved in this process. It analyzes the influence that Latin American and international congresses on the history of science and technology had on the creation of a network of historians of medicine and public health in Latin America and the Caribbean, and how this has contributed to their international visibility.

Keywords History · Public health · Medicine · Networks · Latin America

11.1 Introduction

When, in 1982, a group of seven Latin Americans called the First Latin American Meeting of Science Historians, there were scientists and historians in the region already involved in rescuing the scientific and technological past of their countries. What made a difference was that this group not only proposed to investigate the scientific, technical and cultural heritage, but also to create chairs, institutes, societies and publications, to work together, and to communicate their results worldwide; in other words, to institutionalize and professionalize the history of science in Latin America and the Caribbean, and to seek the support of governments and the collaboration of international organizations in order to do so.

A. M. Carrillo (✉)
Departamento de Salud Pública – Facultad de Medicina, Universidad Nacional Autónoma de México, Mexico City, Mexico
e-mail: farga@unam.mx

M. de la P. Ramos-Lara, L. C. Arboleda (eds.), *History of Science in Latin America: The Construction of an Intellectual Field (20th century)*,
https://doi.org/10.1007/978-3-031-75536-1_11

Historians of science answered the call to the first meeting, and were involved in the creation of the Latin American Society for the History of Sciences and Technology that same year; they enthusiastically attended six congresses convened by the Society (Havana, 1985; São Paulo, 1988; Mexico City, 1992; Cali, 1994; Rio de Janeiro, 1998, and Buenos Aires, 2004), and published the results of their research in *Quipu, Revista Latinoamericana de Historia de las Ciencias y la Tecnología*, since its foundation in 1984.

Not only that; increasingly, delegations from countries in the region began to systematically participate in the International Congress of History of Science –the world's premier meeting for history of science and technology–, to the extent that just under 20 years after the first call for the regional organization of historians of science in Latin America and the Caribbean, Mexico was able to host this important world meeting of scholars.

Another difference with traditional historiography was that, in studying the scientific past, the Latin American historian of science criticized the historiographical model that theoretically denied science in the region and affirmed the European character of its origin (Saldaña, 1986, 1987). They set out to recovery a true worldwide history of science, technology, and medicine. They pointed out that classical Greece was not the only source of rational thought and showed that registry of science had been able to sustain that in the Chinese, Hindu, Islamic and Amerindian civilizations original scientific knowledge, that arose totally independent from the West, had taken place. But in the countries of the Americas, Asia, and Africa, an abortion of non-western cognitive tradition had taken place as a result of colonization and the imposition of western science, and this western ethnocentrism was reflected in the major accounts of the history of science and technology (Saldaña, 2001b: 124).

With the aim of overcoming oblivion in mind, the constitutive meeting of the International Association for Science and Cultural Diversity (IASCUD)—as an inter-Division Commission of the International Union of History and Philosophy of Science—as well as an international symposium on the theme of science and cultural diversity took place in 2000. The initiative to convoke this meeting was intimately entailed to the decision of the organizer of the XXI International Congress of History of Science, to adopt science and cultural diversity as the congress theme. The meeting of the IASCUD attracted a community of historians from Argentina, Canada, Colombia, France, India, Mexico and Mozambique and several other colleagues from Australia, Brazil, South Korea, Spain, and the US sent contributions (Saldaña, 2001a).

At the meeting, Habib S. Irfan, from India, argued that the project of modern science itself unfolds with imperialism, and agreed with Amartya Sen on the history of cultural hegemony that imperial powers established over dominated societies, as well as with Joseph Needham in his proposal for an ecumenical history of science that would recognize the different contributions of civilizations and cultures to the growth of modern science (Habib, 2001). Less than a decade after the First Latin American Meeting of Science Historians, there was already interest in the international literature in the scientific past of American, Asian, and African countries (Lafuente & Saldaña, 1987).

All these efforts had a common denominator: the philosopher and historian of science Juan José Saldaña was the main driving force behind them. That is why, the First Latin American Meeting of Science Historians was held in Puebla, Mexico and presided by Saldaña. He was the first president of the Latin American Society for the History of Sciences and Technology and the director of *Quipu, Revista Latinoamericana de Historia de las Ciencias y la Tecnología*. The intellectual architect and the secretary general of the International Association for Science and Cultural Diversity—which is still active today—was also Juan José Saldaña, and he chaired the First International Symposium on the theme in Pachuca, Mexico, in 2000. As the secretary general of the International Union of History and Philosophy of Science, he presided the XXI International Congress of History of Science, that gathered in Mexico at the beginning of the twenty-first century. In his country, he revived the abandoned Mexican Society for the History of Science and Technology, which he chaired, and created and ran a seminary for over forty years: the Research and dissertation seminar in the history of science and technology, which was fruitful for the development of the discipline. This was a seminal program for researchers, and his work as a trainer of human resources was outstanding.[1]

Saldaña's works and the theses of his disciples dealt with what in different moments and circumstances of Latin American history was understood as science and technique. They maintained the need to include the local context as an explanatory category of what occurred in a defined socio-historical circumstance; that is, geographical accidents, individuals and groups of individuals, institutions and ideologies, scientific concepts, and theories, interacting with each other in a defined social and cultural context (Saldaña, 1996b: 7). They emphasized the structuring role played by the State with respect to science since the foundation of the national State, as well as the contextual factors responsible for the peculiarities of indigenous science: it was a new history of science, fundamentally understood as social history.

The Executive Committee, appointed at the First Latin American Meeting of Historians of Science included a medical historian: The Cuban José López Sánchez, and among the researchers trying to recover the memory of the scientific past in the region there were several other medical historians. This was not surprising since the history of medicine is perhaps the oldest of specialized histories (Peset, 1987) and as early as the eighteenth and nineteenth centuries a mutual hybridization between history and epidemiology (the science of epidemics) took place, with the first pragmatically used as a source of practical knowledge and the second contributing to the creation of the history of medicine as a specialized discipline (H Haeser and A Carreras Panchón cited by Perdiguero et al., 2001: 668).

[1] I was very fortunate to be a member of his seminar throughout my graduate studies. I wrote, for my teacher, an In memoriam (Carrillo, 2022); also, a text about the seminar (Carrillo, in press).

11.2 Antecedents of the History of Medicine in the Region

In the Latin American and Caribbean region there were chairs of medical history in medical schools, and in general attempts to recover scientific developments since the 1800s. In Mexico in the last third of the nineteenth century, the pharmacist and physician Francisco A. Flores pioneered an attempt to historicize Mexican medicine and public health, from Ancient Mexico to the time in which he wrote. As Gonzalo Aguirre Beltrán pointed out in his 'Advertencia' or 'Notice' to the second volume of the facsimile edition of his book, Flores' work was part of the nineteenth-century preoccupation with formulating grandiose schemes to explain the progress made by society, institutions, technology, medicine, and science in general (Aguirre Beltrán, 1992).

Influenced by Comtian philosophy and his law of the three stages, Flores divided the scientific history of Mexico into periods; and referred to the medicine of Ancient Mexico as primitive or theological; to that of Colonial Mexico as metaphysical, unlike the medicine of his time, which he considered positive or scientific. However, on the basis of his research he surrendered to the facts. He recognized the value of the physicians of the colonial university and spoke of the past glories of the Indian peoples. 'Let us discover with respect when pronouncing the venerable names of the last representatives of the Aztec medical faculty', he proposed (Flores, 1886–1888: 379), which made his teacher Porfirio Parra write in the prologue of the work: 'We feel less enthusiasm than the author for the degree of civilization reached by the Aztec culture' (Parra, 1886–1888: XXVII).

Also in Peru, interest in the history of medicine began in the nineteenth century, and in the first half of the twentieth century there was already a society dedicated to it—although it had an ephemeral life—and also publications and chairs. Marcos Cueto analyzes these origins, as well as the concerns of Carlos Paz Soldán and Juan B. Lastres, two Peruvian medical historians of the 1930's and 1940's, through their correspondence with the most important medical historian of the time, Henry Sigerist, and shows that they were concerned with both research and teaching of the history of medicine, and with developing an international network of researchers (Cueto, 2022).

In their work on the history of science, both Eli de Gortari and Elias Trabulse had analyzed aspects of the history of medicine, and by the late 1980's, historians of medicine were publishing regularly in *Quipu*, and were writing books with conceptual frameworks and methodology appropriate to economically underdeveloped countries, which were examples of non-self-marginalizing science. Jaime Vilchis, Xavier Lozoya and Eduardo Estrella for example, revalued the Mesoamerican and Andean medical sciences (Vilchis, 1988; Estrella, 1996; Lozoya 1996).

Lozoya pointed out In his studies on indigenous knowledge of flora and fauna, that for five centuries Latin American men and women had lived and played a leading role in the conflict that meant preserving one culture under the permanent tutelage and influence of another, which had been imposed and was the owner of power, An alternative version to that of western science was yet to be written; it was to be a version that did not establish differences of value over the knowledge acquired by any culture, whether its own or that of others (Lozoya, 1996: 119–120).

For his part, in his work on excellence in the biomedical sciences in the twentieth century, Marcos Cueto pointed out that for a long time the existence of a scientific past in the Latin American region had been denied, which had inspired the first works on scientific modernization that proposed a linear model of the development of science. But critical studies were reacting to these models. His work, in particular, attempted to show that several Latin American researchers managed to work at the international frontier of knowledge, without being alien to local social and cultural influences. He also suggested revising the term periphery because it minimized the leading role played by Latin American scientists in the construction of their own history (Cueto, 1986).

11.3 Hispalc (History of Public Health in Latin America and the Caribbean) Network

In 1995, the Pan American Health Organization launched an initiative to support the field of public health history research in the region. Signed by Anne-Emanuelle Birn —in her capacity as resident of PAHO's Health Research and Technology Development Program—. PAHO initiative had several lines of work, including a network of researchers, a bibliography of public health history, a research grant, and a syllabus to support teaching in public health history. It was hoped that the Network would facilitate communication and collaborations among researchers from different countries and serve as a bulletin of information about activities, new developments, and concerns in the field. It was also intended to allow the exchange of topics, research and teaching methods, new trends in the fields of history and history of science, and the creation of a space for intellectual debates among users. The creation of a bibliography of public health history arose from the concern that although there were many works on public health history in the region, there was little knowledge of these works from country to country, because they were published in journals and books with little circulation and was intended to give more visibility to studies in the region (Birn, 1994–1995, 2023).

Marcos Cueto—who is currently the secretary general of the International Union of History and Philosophy of Science—recalls that Birn—at PAHO—and himself—who was in Lima–, organized with Alberto Pellegrini of PAHO a contest on health reforms in history, which gave three research grants of a respectable amount for one year. At the same time, they began to organize Hispalc. Also, a reference base of historical works on Latin American health that they had been gathering at PAHO was passed on to the Casa de Oswaldo Cruz at that time directed by Paulo Gadelha. Apparently, the link still exists on the Casa's website (Cueto, 2024).

In addition to supporting three research projects, PAHO also sponsored a panel on at the IV Latin American Congress on the History of Sciences and Technology, which took place in Cali, Colombia, in 1995. After that, the plan to build a Latin America and the Caribbean network of medicine and public health historians continue and there were informal talks about it in parallel to the V Latin American Congress on the History of Sciences and Technology, which took place in Rio de Janeiro in 1998.

The constitution of the network was also discussed at the meeting 'Epidemics, Hygiene and Public Health in Latin America, Eighteenth, Nineteenth and Twentieth Centuries, organized by Emilio Quevedo in Bogotá in 1999, the 'International Network for The History of Public Health', held in Almuñecar, Spain, that same year, and the 'Congress of the Latin American Studies Association', in Washington, D.C., in 2001.

Finally, the network took shape at the 'XXI International Congress of History of Science', held in Mexico City at the beginning of the twenty-first century, within the framework of the symposium Science, Health, and Power in Latin America and the Caribbean, which lasted two days and had a large number of participants. Thirty-two papers were presented by researchers from Argentina, Brazil, Colombia, the United States, Mexico, Peru, and Venezuela, and some Europeans, such as the Spanish Esteban Rodríguez Ocaña and the Italian Bernadino Fantini. The symposium resulted in a dossier in a special issue of the journal *História, Ciência, Saúde. Manguinhos*.

In the introduction to the issue—entirely dedicated to the history of health and medicine in the region–, the guest editors argued that as an integral part of social history, the history of health was a vital space for understanding dimensions that had not yet been studied in the countries of the region, such as the interactions between health processes and social, cultural, economic and political contexts; continuity and change in the living conditions of urban and rural populations; the relations between the state and different social groups, and the relations between health and the processes of State nation building (Hochman et al., 2002).

They also stated that the studies revealed the presence of characteristics of each local context and suggested the existence of general trends that span the region as a whole. Two aspects that were common to the papers presented at the symposium were the richness of health ideas, practices, and programs, as well as their complex interaction with social, cultural, and political processes. They were also a sign of the growth of the field of the history of health, disease, and medicine in America Latina, remarkable both from the point of view of the number of papers presented in scientific meetings, books, book chapters and journal articles, as well as the growth in quality, the thematic, methodological, and conceptual diversity, and the regional scope of this production (Hochman et al., 2002).

The issue was also the result of the efforts of a group of professionals who, dedicated to research and the generation of knowledge and information in the fields of social history of health, medicine, and the biomedical sciences, sought to constitute a network for the history of public health in Latin America and the Caribbean, which was named Red Hispalc (Red de Historia de la Salud Pública en América Latina y el Caribe).

At that time in the world there was a great vitality of health history. Concepts and hypotheses arising from developments in the broad field of social sciences had been instrumental in promoting a refinement of history's own methods and approaches in regard to public health (Rosemberg and Porter cited by Perdiguero et al., 2001), and historians and public health specialists learned from one another. Something similar happened in the Latin America and Caribe: The histories of health, disease, and medicine had grown as fields of study within the region. This growth had not only been quantitative but also qualitative, reflecting a thematic, methodological, and conceptual diversity.

For José Luis Peset, the historian had gone from considering the patient's bed as the center of medical activity to being interested in the public health specialist's office or the hygienist's campaigns; and he suggested beginning to study the patient and his or her sensation in a triple aspect: physiological, psychological, and social. (Peset, 1987: 94).

With help from some students Cueto organized a list of contacts and sent out news on the internet, Nisia Trinidade Lima and Gilberto Hochman also supported the list at different times, but Hispalc was not really funded by anyone. The Hispalc-L (History of Public Health List in Latin America and the Caribbean) was a discussion list that provided information about Internet resources: virtual libraries for the history of medicine and public health, catalogues, on-line archives services, manuscript collections, databases, exhibitions, and other digital resources such as images. It also informed about fellowships and grants, and invitations to publish in books and monographic issues of journals. It also announced recent publications and new. But its main objective was to exchange experiences, techniques, and results of work, as well as to promote comparative studies (Hochman et al., 2002).

The activities of the members of the Network were numerous. At the IV Meeting of Philosophy and History of Science of the Southern Cone', which met in Buenos Aires in 2004, it organized the symposium Science and Health in Latin America, Nineteenth and Twentieth centuries. And another in 2009, History of Health and Disease in Latin America and the Caribbean (Eighteenth to Twentieth Centuries), for the 'LIII, Congress of Americanists', held in Mexico City.

Hispalc exchanged information, among others, with Conecta, a newsletter that compiles and shares news on academic and informative activity in the History of Science, Medicine, and Technology, founded in 1995. For ten years, the Hispalc Network was in operation. Brazilian public health historian Gilberto Hochman says: 'I have a generous version of its demise: the growth of publications, meetings, postgraduate courses and lines of research, as well as the institutionalization of the history of health in Latin America over the last 20 years was also—of course not only—the result of Hispalc-L (everyone's) collective effort, and it became so broad and diverse that it was difficult to have a single node. And the rise of the Internet and new forms of scientific communication made obsolete practices that made sense in 2001 (Hochman, 2023).

11.4 Publications, Comparative Studies, and International Visibility of Historians of Medicine and Public Health from Latin America and the Caribbean

The subject of disease began to gain an important place, first in Latin American and then in international historiography. According to Diego Armus, it was an attempt to renew the traditional history of medicine, in which he identified three directions: the new history of medicine, the history of public health and the sociocultural history of disease. Unlike the traditional history of medicine, which was attentive only to progress, the new history of medicine tended to highlight the uncertain developments of medical knowledge, dialogued with the history of science, and discussed the social, cultural, and political context of medical developments. The history of public health, on the other hand, emphasized the political dimension, looked at power, politics, the state, and the medical profession. While the sociocultural history of disease barely dialogued with the history of biomedicine, and concentrated on the sociodemographic dimensions of diseases, and on the processes of professionalization and medicalization (Armus, 2005).

Although historians of medicine and public health in Latin America and the Caribbean continued to publish in *Quipu*, a privileged space to do so was *História, Ciências, Saúde—Manguinhos*, a journal launched in 1994, that would soon consolidate its position as the most important medical history journal in the Latin American and Caribbean region. With an uninterrupted periodicity, it publishes in open access research articles and notes, documents and images of historical value, debates, interviews, and book reviews relevant to the history of science and health, the study of scientific dissemination and the analysis of historical and cultural heritage.

On the other hand, the monitoring of some of the main journals on the history of medicine in Spanish and English (the Spanish journals *DYNAMIS Acta Hispanica ad Medicinae Scientiarumque Historiam Illustrandam*; *Asclepio. Revista de Historia de la Medicina y de la Ciencia*, and *Frenia. Revista de Historia de la Psiquiatría*; the English journal *Social History of Medicine, The Journal of the Society for the Social History of Medicine*, and the American *Bulletin of the History of Medicine*), before and after key moments for Latin American and Caribbean historians –such as the creation of the Latin American Society for the History of Science and Technology or the establishment of the Hispalc Network– show first a growing interest on the part of historians from outside Latin America and the Caribbean to historicize the scientific development of the region, as well as to review books published by Latin American or Caribbean historians; and, on the other hand, the growing participation of historians of medicine and public health from the region.

From the meeting of historians of medicine at the Latin American Congresses on the History of Sciences and Technology, projects emerged to carry out comparative studies of the history of medicine and public health. At the initiative of Colombian colleagues –at the head of which was Emilio Quevedo– the Network of Researchers for Comparative-Historical Studies of Medicine and Public Health in Latin America was formed, comprising researchers from Argentina, Brazil, Costa Rica, Cuba, Ecuador, and Mexico. Mainly due to the difficulties in obtaining a budget for such a large-scale undertaking, the group of researchers from Colombia has been carrying out comparative research with one or another Latin American country.

Also, the Interinstitutional Network Diseases, Epidemics and Contagions. History of Health and Disease in Argentina and Mexico, worked in 2012 with the participation of a Mexican university and several Argentinean universities. Its intention was to form research teams organized through inter-institutional networks, for the critical debate of problems linked to historiography, and the programming and dissemination of results to the rest of the academic community.

11.5 Conclusion

The history of science has antecedents in Latin America and the Caribbean that can be traced at least as far back as the nineteenth century, but since the early 1980s they have been developing vigorously. In this renaissance, a key role was played by a small group of historians of science—led mainly by Professor Juan José Saldaña—who set out to professionalize and institutionalize the history of science in the region, and created the Latin American Society for the History of Sciences and Technology and the International Association for Science and Cultural Diversity, organized congresses and other academic meetings, published original works in books and in *Quipu, Revista Latinoamericana de Historia de las Ciencias y la Tecnología*. They questioned the Eurocentrism that had limited the study of world science, and proposed not only new topics, but also new methodological perspectives and conceptual frameworks appropriate to the realities of the countries of the region.

Works on the history of medicine and public health are not unrelated to those on the history of science in general, and historians of medicine and public health were part of this project from the beginning. At the society's congresses they established relationships and created the Hispalc network, to increase exchanges and promote joint and comparative work among historians of the Latin American and Caribbean region, without renouncing exchanges with other geographical areas. After a decade, the network disappeared, but not the collaborative work, or the growth in the historiography of disease, medicine, care and public health in LatinAmerica and the Caribbean, which has an increasing international visibility.

Acknowledgments I thank María de la Paz Ramos Lara for her careful reading and helpful suggestions.

References

Aguirre Beltrán, G. (1992). Advertencia. In *Historia de la medicina en México, desde la época de los indios hasta la presente (edición facsimilar)*, Flores y Troncoso, Francisco de Asís. 4 vols.: II, Mexico: IMSS.

Armus, D. (2005). Legados y tendencias en la historiografía sobre la enfermedad en América Latina moderna. In Minayo, Maria Cecilia de Souza, Coimbra JR, Carlos EA orgs. 183–205. *Críticas e atuantes: ciências sociais e humanas em saúde na América Latina* [online]. Rio de Janeiro: FIOCRUZ.

Birn, A.-E. (1994–1995). Iniciativa de la Organización Panamericana de la Salud en historia de salud pública, *História, Ciência, Saúde. Manguinhos, 1*(2), 143–144.

Birn, A.-E. (2023). Personal communication.

Carrillo, A. M. (2022). In memoriam: Juan José Saldaña (24/II/1944–31/X/2022). Página web (Blog) de *História, Ciência, Saúde. Manguinhos*. https://revistahcsm.coc.fiocruz.br/.../in-memoriam-juan.../

Carrillo A. M. (In Press). Juan José Saldaña y su Seminario de Investigación y Tesis en Historia de la Ciencia y la Tecnología. In M. P. Ramos-Lara, and L. C. Arboleda (Eds.), *La cátedra que profesionalizó la Historia de la Ciencia en México. Contribución de la FFyL-UNAM a América Latina*. UNAM.

Cueto, M. (1986). La excelencia en las ciencias biomédicas del siglo XX. In J. J. Saldaña (Ed.), *Historia social de las ciencias en América Latina* (pp. 481–492). UNAM.

Cueto, M. (2022). La historia de la medicina, Henry E. Sigerist y los intermediarios médicos en el Perú de mediados del siglo XX. *Historia (Santiago), 55*(1), 329–350.

Cueto, M. (2024). Personal communication.

Estrella, E. (1996). Medicina autóctona precolombina. In J. J. Saldaña (Ed.), *Historia social de las ciencias en América Latina* (pp. 43–70). UNAM.

Flores, F. A. (1886–1888). *Historia de la medicina en México, desde la época de los indios hasta la presente*, 3 vols: vol I. Secretaría de Fomento.

Habib, S. I. (2001). Science cultural diversity in a post-colonial context. In J. J. Saldaña (Ed.), *Science and cultural diversity. Filling a gap in the history of science* (pp. 145–149). SLHCT.

Hochman, G. (2023). Personal communication.

Hochman, G., Cueto, M., Carrillo, A. M., de Romo, R., & Cecilia, A. (2002). Editor's note. *História, Ciência, Saúde. Manguinhos, 9*(suplemento), 8–10.

Lafuente, A., & Saldaña, J. J. (1987). Introducción. In A. Lafuente & J. J. Saldaña (Eds.), *Nuevas tendencias en historia de las ciencias* (pp. 1–4). CSIC.

Lozoya, X. (1996). Historia natural y herbolaria medicinal en el siglo XVI en América. In J. J. Saldaña (Ed.), *Historia social de las ciencias en América Latina* (pp. 119–149). UNAM.

Parra, P. (1886–1888). Prólogo. In *Historia de la medicina en México, desde la época de los indios hasta la presente*, Francisco Flores (Vol. I, pp. IX–XXXI). Secretaría de Fomento.

Perdiguero, E., Bernabeu, J., Huertas, R., & Rodríguez-Ocaña, E. (2001). History of health, avaluable tool in public health. *Journal of Epidemiology and Community Health, 55*, 667–673.

Peset, J. L. (1987). Historia del cuerpo, historia de la mente. In *Nuevas tendencias en historia de las ciencias* (pp. 87–94). CSIC.

Saldaña, J. J. (1986). Marcos conceptuales de la historia de las ciencias en Latinoamérica: positivismo y economicismo. In J. J. Saldaña (Ed.), *El perfil de la ciencia en América* (pp. 57–80). SLHCT.

Saldaña, J. J. (1987). Introduction. In J. J. Saldaña (Ed.), *Cross cultural diffusion of science: Latin America* (pp. 5–8). SLHCT.

Saldaña, J. J. (1996a). *Historia social de las ciencias en América Latina*. UNAM.

Saldaña, J. J. (1996b). Introducción. Teatro científico americano. Geografía y cultura en la historiografía latinoamericana de la ciencia. In J. J. Saldaña (Ed.), *Historia de las ciencias en América Latina* (pp. 7–41). UNAM.

Saldaña, J. J. (2001a). Foreword. In J. J. Saldaña (Ed.), *Science and cultural diversity. Filling a gap in the history of science* (pp. 7–8). SLHCT.
Saldaña, J. J. (2001b). Western and non-western science: History and perspectives, science, and cultural diversity. In J. J. Saldaña (Ed.), *Science and cultural diversity. Filling a gap in the history of science* (pp. 123–143). SLHCT.
Vilchis-R, J. A. (1988). Medicina novohispana del siglo XVI y la materia médica indígena. Hacia una caracterización de su ideología. *Quipu. Revista Latinoamericana de Historia de las Ciencias y la Tecnología, 5*(1), 19–48.

Ana María Carrillo has a PhD in History from the National Autonomous University of Mexico (UNAM). She works at the Department of Public Health of the Faculty of Medicine, UNAM, where she is a full professor and coordinates the International Seminar on the History of Medicine and Public Health. She has published nearly 80 works on the history of epidemics, public health, and the health professions in Modern and Contemporary Mexico. She is a member of the National System of Researchers, Level II.

Chapter 12
The History of Science in Latin America and the International Agenda: History, Reflection on Our Themes, Local Scientific Practices and Perspectives

Maria Rachel Fróes da Fonseca

Abstract In the context of the conceptual and thematic renewal of Latin American historiography of the sciences in the last decades of the twentieth century, several points stood out. These included the recognition of scientific practice in Latin America, understanding the social significance of scientific practice, and the development of studies aimed at analyzing the processes of transmission and reception of scientific knowledge. In this agenda, the idea of a social protagonism of science in Latin America was evident, as it acquired a social and political connotation. The intention is to reflect, in theoretical and methodological terms, on the historical construction of the specificities of the formation of scientific practice in colonial Latin America, especially from the perspectives and trajectories of enlightened men such as mexican José Antonio Alzate and brazilian Manuel Arruda da Câmara. In the scientific practice developed by these figures, we seek to perceive the characterization of the elements of science, their concept of science, and the way scientific knowledge is applied to address the problems experienced by that society.

Keywords History of sciences-Latin America · José Antonio Alzate · Manuel Arruda da Câmara

12.1 Introduction

For a long time, historiography produced papers on the scientific experience in Latin America that, in reality, were nothing more than extensive chronologies and celebratory accounts, completely disregarding the specificities of that formation of scientific practice.

M. R. F. da Fonseca (✉)
Casa de Oswaldo Cruz, Fundação Oswaldo Cruz, Rio de Janeiro, Brazil
e-mail: rachel.froes@fiocruz.br

M. de la P. Ramos-Lara, L. C. Arboleda (eds.), *History of Science in Latin America: The Construction of an Intellectual Field (20th century)*,
https://doi.org/10.1007/978-3-031-75536-1_12

It is important to note that until the 1970s, the studies produced denied the possibility of the existence of a scientific tradition in this region of the globe. Latin American societies were understood solely as recipients of scientific models produced in the Old World. Studies focused on the scientific production of European centers and their process of diffusion to the so-called peripheral regions.

George Basalla's conception of science, which is very present in the historiography of science in Latin America, proposed a model (three stages) on how Western science had penetrated non-European regions (Basalla, 1967). It conceived science as a univocal and universal body of knowledge that did not take into account the historical asymmetry of the scientific system, as well as its polarized nature localized in a small number of countries. In reality, it was a reproduction of ethnocentric vision and cultural colonialism. Countries like those in Latin America, which had not directly participated in the Scientific Revolution, were regarded in a marginal manner, with no contributions worthy of elaboration in studies concerning them.

Exponents in Latin American historiography, such as Antonio Lafuente and José Sala Catala, were already emphasizing the inadequacy, or even the antiquity, of Basalla's proposal (Lafuente & Catala, 1989) in the 1980s. It was considered that such an approach completely ignored the entire scientific history specific to these regions. Science historian Alberto Saladino García pointed out the importance of questioning the Eurocentric perspective present in Basalla's approach, and the challenge that arose in breaking away from these epistemological and methodological issues:

> La crítica eurocéntrica implícita en la historiografía representa un reto que han asumido nuestros historiadores, por eso es conveniente identificar, denominar, sistematizar, sustanciar y describir las metodologías de la historia de la ciencia puestas en práctica en América Latina. (….).[1] (García, 2015)

The concept of dependency applied to science, which characterized some countries as retrogade in relation to the so-called centers of science production, proved to be rigid, hindering the understanding of Latin American realities. Therefore, it is believed to be important to understand the history of science outside the confines of a strictly progressive line necessarily linked by advancements and discoveries.

The historian of science Juan José Saldaña, especially, pointed out that in the historiographical production of the last decades of the twentieth century, we can perceive a growing awareness that the history of sciences presented epistemological questions, especially due to its geographic and cultural specificities, which should not be disregarded. It was important "pensar nuestra ciencia",[2] the science of Latin

[1] Translation made by the author: "The Eurocentric criticism implicit in historiography represents a challenge that our historians have taken on, which is why it is convenient to identify, name, systematize, substantiate and describe the methodologies of the history of science put into practice in Latin America".

[2] Translation made by the author: "think about our science".

America, as a history of science with reference to its own identity and culture, "entendida como ciencia en su contexto"[3] (Saldaña, 1996).

"Doing science" is comprehended here as inseparable from the economic, social, and political conditions characteristic of the context in which scientists are operating. As emphasized by Mexican historian Roberto Moreno de Los Arcos, it was important to construct a history of science that did not appear isolated from its social context and that highlighted both moments of success and failure (Moreno y De Los Arcos, 1994, p. 10).

Therefore, the aim is to think our science by considering the existence of scientific production in regions such as Latin America. Science ceased to be understood as an autonomous entity governed by internal laws of rationality and began to be understood as a social activity subject to the context in which it was produced. Brazilian historian Maria Amélia M. Dantes characterized the history of science that was sought to be made as:

> escrever a história da ciência dos países latino-americanos deixa de ser simplesmente buscar sua contribuição para uma suposta ciência universal, mas procurar registrar, nos vários momentos de suas histórias, traços da existência da atividade científica: ação de indivíduos, existência de espaços institucionais, existência de trabalhos científicos, etc.[4] (Dantes, 1992, p. 2)

So as to advance in this direction, science historians have resorted to different theoretical and methodological approaches and abandoned the perspective that had been previously attributed to them, being seen only as systematizers and reproducers of memory, in order to become builders of this history themselves (García, 2015).

According to Juan José Saldaña, studies have been showing the development of a historiographical perspective of science in Latin America, grounded in social history, which not only sought to restore the value of scientific activity produced in the region but also to highlight the existence and unique form of scientific practice in Latin America. (Saldaña 1989) It was essential to seek the foundations of a new history of sciences for Latin America.

Research in the field of the history of sciences, especially since the last decades of the twentieth century, has asserted the need to abandon some conceptions that had been hitherto prevalent in traditional approaches. This includes the triumphant view of science expressed in the emphasis on "great" events and figures, as well as the idea of linear evolution in the trajectory of institutions and in the production of scientific knowledge.

This revision was guided by the elimination of these conceptions and the adoption of an approach that comprehended science from a different perspective, namely, the social history of sciences. All this analytical effort was part of the line of social history of science, understanding the interaction of various societal instances in the

---

[3] Translation made by the author: "understood as science in its context".

[4] Translation made by the author: "writing the history of science in Latin American countries is no longer simply seeking their contribution to a supposed universal science, but seeking to record, at various moments in their histories, traces of the existence of scientific activity: the action of individuals, the existence of institutional spaces, existence of scientific works ".

production of knowledge. From this perspective, it was believed that one should not adopt a decontextualized view of scientific activity, nor focus solely on the internal dynamics of this process, that is, on the study of the conceptual and logical structure of scientific knowledge.

This process of historiographical transformation and renewal was accompanied by fundamental actions for the scientific community of science historians. Along these lines, we recall the holding of the Primera Reunión Latinoamericana de Historiadores de las Ciencias (First Latin American Meeting of Historians of Sciences) in 1982, in Puebla (Fefer, 2009). In addition to the theoretical and methodological transformations implemented in the 1980s, as previously mentioned, significant changes occurred in the organization and professionalization of activities in the field of history of sciences in Latin America. This included the creation of courses, academic programs, study groups, and societies such as the Sociedad Latinoamericana de Historia de la Ciencia y la Tecnología (Latin American Society for the History of Science and Technology) in Mexico, in 1982, of which Juan José Saldaña was its first president.

Rafael Fefer, in his article on the genealogy of Latin American historiography of science, reconstructs the emergence of Latin American historiography of sciences and the rise of Latin America as a research object for international historiography. In this process, not only studies on the institutionalization of sciences in the region emerged, but also reflections on conceptual themes came about. Fefer, in his analysis, brings into the conversation the work of Juan José Saldaña, who stated that the community of Latin American historians of sciences, in the 1980s, had promoted not only the organization of this field but also the professionalization of the discourse of science in this region (Fefer, 2009).

In this agenda of historiographical renewal, Marcos Cueto emphasized that certain themes and concepts stood out, such as reception, local adaptation, and institutional history. These concepts aligned with researchers' efforts to challenge the view of passive reception of European science and to highlight previously disregarded local aspects and dynamics. The peruvian historian also highlighted the growth in the number of historians of sciences in Latin America and their diversity of academic backgrounds, with an emphasis on training in history (Cueto, 2020).

In this scenario, some works were quickly introduced due to their analytical perspectives, such as Antonello Gerbi's work on the history of the debate about the nature of the New World. This analysis corroborated the entire debate on recognizing the specificities of nature in the American continent and strongly questioned the perspective of inferiority attributed to this nature in comparison to European nature.

Thus, a conceptual and terminological modernization was configured, adopting a new language that allowed for naming processes and scientific activities that had previously been overlooked by science historians.

To understand the history of sciences being produced in Latin America, it is important to have a specific methodology for a specific object of study, as highlighted by Juan José Saldaña (Saldaña, 1986). Latin American science historians have certainly contributed to recognizing the existence of a unique scientific culture in their countries, as well as enabling the incorporation of archives and documentary sources into the history of sciences (Saldaña, 1986, p. 65–66). Changes have occurred regarding the use of archives and the expansion of sources, especially primary ones, corresponding to the perspective of understanding the history of sciences through the construction of scientific knowledge, and the discourses, institutions, and scientific practices.

Thus, progressively, the scientific culture in Latin America has been highlighted, emphasizing the importance of studies that seek to understand the history of sciences in Latin America and perceive, in a differentiated way, the relationship between so-called universal science and regionally developed scientific practices. It was important to seek to understand the dynamics specific to each of the societies in question. There was a turn towards empirical investigation to demonstrate the richness of scientific activity in these regions, thereby providing a reassessment of the history of science itself.

In this context of emergence and transformation of Latin American historiography of sciences, recognition of scientific practice in Latin America, understanding the social significance of scientific practice, and the development of studies directed towards analyzing the processes of transmission and reception of scientific knowledge stood out (Quevedo, 2000).

Other perspectives of analysis, in a more recent context, have impacted this historiography, such as the idea of the various aspects resulting from the circulation of scientific knowledge, aiming to transcend the boundaries of traditionally defined political borders.

According to Fefer, this historiography of sciences produced in Latin America still presents some gaps and points to be further explored. These include expanding the thematic and chronological scope in research on the institutionalization of sciences in Latin America, analyzing scientific disciplines, the specificities of different localities beyond politically central regions, science education, the formation of audiences, and scientific trajectories (Fefer, 2009).

In his work, Juan José Saldaña emphasized the cultural and social role of scientific institutions, through their political and social repercussions. In this sense, the institutionalization of science has not been, and is not, a uniform or homogeneous process in all countries and historical contexts. Its development depends on the peculiarities of social organization, culture, history, and the conditioning factors of the physical environment in each country.

12.2 Social Protagonism of Science in Latin America

In the more recent historiographical production on science in Latin America, we perceive an assertion of the need for a broader understanding of the history of the production and dissemination of scientific knowledge. It is no longer limited to investigating scientific and technical heritage but rather perceived as a process in which a set of factors act and articulate with one another.

The notion of scientific tradition has been being introduced, allowing for the understanding of moments and terrains that were previously overlooked, especially those preceding the implementation of scientific institutions and policies during the colonial period. Evidence of the construction of this "national" or patriotic science has been present since the eighteenth century, when scholars gathered to cultivate science, giving it a social protagonism.

The concept of "national" science has emerged as one of the most prevalent themes in Latin American historiography of science, abolishing the analysis from a decontextualized perspective. Evidence of the construction of this "national" or patriotic science has been present since the eighteenth century when scholars gathered to cultivate science, giving it social protagonism.

The idea of scientific and cultural void, the notion that everything came from outside, has been clearly questioned. Regarding scientific practices in the colonial context, it is important to affirm that it was not a simple diffusion into empty territory. The characterization of the colonial condition has played an important role in Brazilian historiography since the nineteenth century. Historiography has highlighted the role of local groups, traveling naturalists, acclimatization gardens, cabinets of natural history, cataloging and acclimatization activities, scientists and institutions, professional schools, museums, and scientific associations in the practices and development of sciences in the colonial imperial context of Brazil.

The process of diffusion and reception of scientific knowledge has come to be conceived as a historical object rooted in social interests. The issue of the transmission and diffusion of science has also been deserving of studies and has presented significant results, aiming to understand the forms of localization and reception of scientific theories and practices. The understanding of science as culture, as a cultural product, has been guiding studies on the cultural role played by science in the countries of Latin America.

This analytical effort is part of the social history of science, understanding the interaction of various societal instances in the production of knowledge. It is essential not to adopt a decontextualized view of scientific activity or to focus solely on the internal dynamics of this process. Thus, through the analysis of scientists' trajectories, their production, their conception of science, and their "homeland," we believe it is possible not only to confer this contextualized dimension but also to demonstrate the existence of this interaction between culture and ideology.

In terms of a specific hypothesis, we believe in the idea of a social protagonism of science in Latin America. I intend to highlight the process through which the construction and affirmation of scientific practice (scientific production, scientists' trajectories, scientific and/or literary associations, journals) clearly contributed to the awareness of the American being. In this sense, science acquired a connotation of social and political extent.

Starting from these reflections on the perspective and analytical category that have been guiding and characterizing Latin American historiography, we can take a closer look at the history of sciences in this region. Going back to the scenario of Spanish and Portuguese colonization in the late eighteenth century, we can find the production of a discourse and scientific practice fundamentally characterized by the pursuit of knowledge of its reality, its fauna, and flora, aiming at two main, non-exclusive but complementary goals. The first of these, inserted in a utilitarian conception of science, consisted of seeking recognition of its own resources, potentially applicable for the benefit of those societies. In this way, self-sufficiency could be demonstrated, and subsequently achieved.

Alongside this aspiration, and no less relevant, the aim was to contest, through knowledge and the affirmation of the potentialities of American nature, the degrading vision that European thinkers, notably Georges-Louis Leclerc, Comte de Buffon (1707–1788), and Cornelius De Pauw (1739–1799), had projected onto the American continent. There was a struggle to counteract the forces that sought to maintain and justify the subjugation of those societies in the name of a supposed incapacity, immaturity, and inferiority.

The discourses of scientists like José Antonio Alzate in Mexico and Manuel Arruda da Câmara in Brazil, whether explicitly or implicitly, expressed new ideological assumptions. Their memoirs and essays pointed towards the pursuit of public prosperity, the aggrandizement of the "homeland," through the proposition of new techniques, new crops, and new medicines, entirely directed and suitable for that reality. The sense of "homeland" was clearly distinct from its current meaning, however, it represented their connection to the territory in which they lived. We believe this sense of belonging represented the embryo of an awareness of the American being. In this regard, we emphasize the correlation between scientific practice and the ongoing process of emancipation, which, at a later moment, would lead to another goal, political independence itself.

The history of scientific production and practice developed in colonial spaces, such as New Spain and Brazil around the turn of the 18th/19th centuries, allows us to take a closer look at this perspective of understanding one's own reality and the capability of its inhabitants, as expressions of the process of constructing a sentiment that sought to affirm the peculiarity of America in relation to Europe. It was a process of becoming aware of being from America. There was a need for mental emancipation. The expressed thought was one that served a particular action and, therefore, a thought that encompassed different issues: political, economic, social, and cultural.

We understand the history of the construction of knowledge and scientific practice in the context of colonial Latin America as central to many of the themes and issues in Latin American historiography of the sciences, to which we have referred earlier. Issues such as the recognition of scientific activity produced in the region, the existence and form of scientific practice in Latin America, and the awareness that the history of sciences presents epistemological questions, especially due to its geographical and cultural specificities, are themes of the history of scientific production and practice in the colonial context of the region. It is in this sense that we propose to take a closer look at the trajectories and production of enlightened men, such as the Mexican José Antonio Alzate and the Brazilian Manuel Arruda da Câmara.

12.3 The Building of American Nature Knowledge

Men of science, such as Mexican José Antonio Alzate (1738–1799) and Brazilian Manuel Arruda da Câmara (1752-Vila de Pombal, Paraíba/Brazil; 1810—Pernambuco/Brazil), who dedicated themselves to studying and adopting new scientific knowledge to promote understanding of the flora, fauna, and physical characteristics of the lands they inhabited, as well as to recover their historical past, simultaneously faced all those ideas that, due to an alleged intellectual, physical, and moral incapacity, contributed to the perpetuation of the colonial condition of their peoples.

In the understanding of these enlightened men, the nature of America was different, peculiar, and unique, but not inferior to that of other parts of the world, as European thought sought to suggest. Its peculiarity should be recognized and catered to in its needs, and therefore, decisions and definitions made by those overseas were not relevant.

They sought recognition of their differentiated existence, their intellectual capacity, their autonomy in order to benefit from their resources and strengths, but not political rupture. The process of autonomy began, therefore, in the intellectual and cultural fields, before projecting itself politically. Political independence in the first half of the nineteenth century resulted from a series of movements of different natures, making any univocal interpretation a partial one.

In formulating this knowledge, the impact of theories of inferiority and immaturity attributed to American nature in the second half of the eighteenth century by European thinkers such as Buffon and Cornelius De Pauw was decisive. These theories were based on the prevailing trend of the time, namely, the assertion of a close organic connection of living beings with nature in a deterministic way. Hispanic-americans and brazilians enlightened men sought to respond to these assertions, to these "calumnies" as they called them, by stimulating the production of knowledge that would refute those derogatory views of the american continent. It was important, therefore, to emphasize more than ever the potentialities and virtues of that nature. The studies should not only present the physical descriptions of those lands but also their economic and human potential.

Alzate, a scholar, humanist, and scientist of broad expertise (physics, astronomy, chemistry, mathematics, and natural sciences), as well as a polemicist, was the editor of important journals such as "Asuntos varios sobre ciencia y artes" (Various Matters on Science and Arts), "Observaciones sobre la física, historia natural y artes utiles" (Observations on Physics, Natural History, and Useful Arts), "Diario Literario de México" (Literary Daily of Mexico), and "Gaceta de Literatura" (Literary Gazette).

In the development of scientific knowledge about Brazilian nature, the trajectory of the botanist Manuel Arruda da Câmara stands out. Arruda da Câmara, who held dregrees in philosophy, mathematics, and medicine (University of Coimbra), was the founder of the Areópago de Itambé, a liberal and masonic society, and a member of the Literary Society of Rio de Janeiro.

Their reflections on scientific knowledge, along with those of other scholars, were evident in scientific and/or literary associations and societies conducive to scientific debate. Associations such as the Scientific Academy of Rio de Janeiro and its successor, the Literary Society of Rio de Janeiro, favored scientific investigation of nature, dissemination of knowledge, and its use for the benefit of the population. Many of the memoirs and studies elaborated by these enlightened men subsequently found broader dissemination channels, such as the newspaper "O Patriota. Jornal Literário, Político, Mercantil do Rio de Janeiro" (The Patriot. Literary, Political and Commercial Newspaper of Rio de Janeiro), considered the first Brazilian newspaper specifically dedicated to the dissemination of sciences and literature, regarded as a catalyst for progress. "O Patriota," founded in Rio de Janeiro by Manuel Ferreira de Araújo Guimarães (1777–1838), circulated from January 1813 to December 1814, and was dedicated to disseminating so-called useful knowledge: Literature, Mineralogy, Topography, History, National and Foreign Politics, Medicine, Statistics, Commerce, Botany, Eloquence, Geography, Agriculture, Chemistry, Hydrography, Arts, Hydraulic Navigation, Mathematics, Philosophical Grammar, and Published Works.

Over the course of affirming new scientific ideas, during the awakening of concern for nature, hispanic-americans and brazilians enlightened men vehemently expressed, as a mission, their interest in serving the "homeland" and outlined their ideas about the "nation," which were beginning to be "imagined."

The brazilian lands presented peculiarities that could not be overlooked, as it was precisely these specificities that distinguished them and endowed them with a rich existence. The knowledge of Brazilian nature promoted the enrichment of the country, through the results arising from the cultivation of crops, which not only promised great profits and the appreciation of lands and ports but also meant job opportunities for many.

In the writings of these enlightened men, expressions such as "*nuestra nación*," "*nuestra América*," "*patriotas americanos*," "*bem comum da pátria*," "*brasileiros*," "*patrícios*,"[5] were frequent, outlining the configuration of a sentiment that those lands, New Spain and Brazil, were not only territories that differed from Spain and Portugal but also communities that were taking steps towards a "homeland." The issues addressed by these enlightened men originated, according to their own words, in the "honor que se debe a la Patria y a la Nación".[6]

The affirmation of natural resources and their specificities, cultural values, and the empowerment of Brazilians and Mexicans contributed to the crystallization of an indigenous culture and the formation of a Mexican identity and a Brazilian identity.

The ideas, attitudes, and behaviors of these scientists expressed an awareness, albeit diffuse, and a premonition of the new social and political entity that was being

[5] Translation made by the author: "our nation", "our América", "americans patriots", "common good of the homeland", "brazilians", "patricians".

[6] Translation made by the author: "honor due to the country and the nation".

gestated. They expressed a new Latin American mentality. This movement of renewal in scientific practice evidenced not only the concern for the adoption of modern science but also the demand for and recognition of a distinctly American science. The pursuit of this recognition was grounded in the controversies sparked by European thinkers such as Buffon (1707–1788) and Cornelius De Pauw (1739–1799), whose approaches to the American nature were profoundly degrading towards the climate, animals, and the American man.

Such a vision transcended the field of scientific controversies, assuming representation in the political and social realm by serving as a legitimizing tool for the subjugation of the colonies. In response to these approaches, numerous hispanic-americans scientists and thinkers reacted emphatically, reaffirming, in turn, the positive aspects of the american environment.

Having established the context of New Spanish Enlightenment, it's important to analyze the main aspects in the formation of scientific practice in this scenario: the adoption of European scientific models, the conditions for establishing an organization specific to scientific activity, and the relationship between this activity and other forms of social practice. It is within this perspective that we intend to perceive the relationship between science and the aspirations for the emancipation of the colonies.

Rafael Feder highlights the emergence of a scientific-literary and technical tradition before the birth of the Mexican "homeland," represented by restless enlightened men such as Alzate:

> los letrados de los años coloniales eran inquietos hombres que colaboraron y participaron en las novedades del conocimiento moderno. Así que nuestros queridos Sigüenza y Góngora y Sor Juana no fueron estrellas novas que brillaron con luz propia en la oscuridad colonial. Por el contrario) la época virreinal tuvo claroscuros em los que participaron grupos de literatos) teólogos, técnicos cirujanos, médicos botánicos, matemáticos, astrólogos, mineros, albéitares ingenieros militares, exploradores, burócratas, letrados. editores, abogados, arquitectos, artistas, frailes, monjes, actores y universitarios. Todos estos personajes son los demiurgos de la incipiente comunidad científica y del espíritu. humanista que había antes de que se expidiera el acta de nacimiento de este país: inician nuestra tradición científica-literaria, técnica y artística.[7] (Feder, 2009, p. 276)

The hispanic-americans and brazilians, like José Antonio Alzate and Manuel Arruda da Câmara, were equipped with a new science that allowed them to study the American nature through direct experience, without the need for intermediaries.

[7] Translation made by the author: "the literates of the colonial years were restless men who collaborated and participated in the developments of modern knowledge. So our beloved Sigüenza and Góngora and Sor Juana were not new stars that shone with their own light in the colonial darkness. On the contrary, the viceregal era had chiaroscuros in which groups of literati participated, theologians, surgical technicians, botanical doctors, mathematicians, astrologers, miners, albéitares military engineers, explorers, bureaucrats, lawyers. Editors, lawyers, architects, artists, friars, monks, actors and university students. All these characters are the demiurges of the incipient scientific and spiritual community. Humanism that existed before the birth certificate of this country was issued: they begin our scientific-literary, technical and artistic tradition".

The absence of mediation represented the possibility of destabilizing the idea of authority that had hitherto underpinned scholastic science.

It is precisely in this environment of cultural renewal that the groundwork is laid for the germination of political emancipation, as this transformation of the cultural world implied the affirmation of the ideology of an independent scientific practice, representing a true challenge to authority in the intellectual sphere.

12.4 The Builders of American Science and Nature: José Antonio Alzate and Arruda da Câmara

José Antonio Félix Alzate y Ramírez de Cantillana was born in Ozumba, Province of Chalco, in 1737. He was the son of the Spaniard Juan Felipe de Alzate and Josefa María Ramírez Cantillana, and he was descended from the Mexican poet Sor Juana Inés de la Cruz (1648–1695). Alzate studied at the College of San Ildefonso in Mexico City, graduating with a bachelor's degree in Arts from the Royal and Pontifical University of Mexico in 1753, and in Theology in 1756. He was ordained as a priest in 1758.

He presented several projects to the local government to improve Mexico City, including studies such as the "Proyecto para desaguar la laguna de Tezcuco" (Project for Draining the Lake of Texcoco) in 1767, "Ensaio sobre a utilização da madeira nas tubulações para água na cidade" (Essay on the Use of Wood in Water Pipes in the City) in 1768, "Modelo de carro para recolhimento de lixo" (Model of a Cart for Garbage Collection) in 1790, and "Propostas para a limpeza da cidade" (Proposals for City Cleaning) in 1791. The colonial administration also commissioned Alzate for various services, such as observing the transit of Venus across the Sun's disk in 1769 and writing an essay on the cochineal insect in 1774.

Alzate participated in various scientific associations and institutions such as the Real Primitiva Sociedad Bascongada, the Real Jardín Botánico de Madrid, and the Real Academia de las Ciencias de Paris, and among his circle of acquaintances were mexican intellectuals like José Mariano Mociño, Antonio León Gama, and José Ignacio Bartolache, as well as foreigners like the american botanist Alexander Garden.

The search for a so called "criolla"[8] culture was fully realized with José Antonio Alzate, one of the best examples of "criollo" scientist, one that had limitations imposed by the colonial situation itself and who could not reach high positions in colonial administration.

Alzate's work was characterized by an emphasis on scientific knowledge, a rationalist attitude, a focus on scientific methods, the idea of dissemination, and the encyclopedic nature of the topics he addressed. In his conception of science, mastery over nature was seen as the path to achieving society's happiness through the use of scientific knowledge, especially in the natural sciences. As Alberto Saladino

---

[8] Translation made by the author: "creole".

Garcia aptly noted, Alzate "destacó el hecho de concebir a la ciencia como conocimiento que respalda el beneficio social".[9]

As a disciple of Francis Bacon (1561–1626), Alzate attributed to science the function of creating human well-being and emphasized the importance of mastering nature to meet human needs. The relationship between Alzate's conception and Baconian thought was based precisely on these fundamental aspects: the emphasis on experimental method as opposed to the Aristotelian and the utilitarian function of science. His philosophical conception centered fundamentally on human mastery over natural forces through the discovery of facts empirically and rational explanation.

Alzate proposed an understanding of Mexican reality from the perspective of new knowledge. Scientific interest was present, driven by the desire to showcase the resources of Mexican lands and to find suitable means to exploit them. This scientific vocation was assumed as a mission, namely, to bring true science to everyone so that they could develop it for their benefit. Observation and experimentation guided Alzate's efforts in pursuit of a science specific to the American land.

Useful knowledge was indeed the true knowledge, as the notion of utility defined science in Alzate's view. He understood useful science as the set of subjects that solved or led to solutions for the economic and social problems of the country, enabling their resolution. Sciences such as physics, chemistry, medicine, botany, natural history, and geography would be the great promoters of the so-called "useful arts." Alzate's conception thus presented a very clear social dimension, as it was directly related to the social and economic conditions prevailing in Mexican society at that time.

To dedicate oneself to natural history meant observing the practical utilities that could derive from this science. The New Spanish nature, declared Alzate, offered many natural wonders that could not be ignored by those seeking to understand nature and its resources. Among these wonders, he pointed out the variety and quantity of plants from which effective remedies for human ailments could be extracted, such as the plague, syphilis, and yellow fever.

Alzate firmly opposed preconceived systems, such as systematic knowledge of botany or chemistry. The existence of rules, nomenclatures, and axioms limited the understanding of nature. He argued that by observing nature, one could perceive that it could break free from the constraints, rules, and axioms that naturalists sought to impose upon them. The variety and diversity of American nature reinforced his opposition to the idea of generic concepts, which he believed did not adequately capture it.

The scientific knowledge, the driving force of his thinking, should be universalized, popularized, and disseminated in Spanish rather than Latin so that all social segments could have access. In this sense, the creation of dissemination vehicles was crucial, and for this purpose, Alzate committed himself as an editor of

[9]Translation made by the author: "highlighted the fact of conceiving science as knowledge that supports social benefit ».

specialized journals such as "Asuntos varios sobre ciencia y artes" (1772) and "Observaciones sobre la física, historia natural y artes utiles" (1787). He also edited non-specialized publications such as "Gazetas de Literatura" (1788–1795) and "Diario Literario de Mexico" (1768), where he similarly dedicated significant space to science.

He dedicated himself to the elaboration of studies on various scientific subjects understood from the utilitarian perspective of science. In 1831, several of his studies were reprinted in Puebla, and others were reproduced in the journals "La Naturaleza," the organ of the Society of Natural History, and "El Estudio," of the National Medical Institute.

The thematic breadth of his works gave him an encyclopedic profile, covering almost all sciences, such as botany, medicine, astronomy, natural history, mineralogy, zoology, and chemistry. He presented studies on animals (migration of swallows, cultivation of cochineal), astronomical (incidence of eclipses) and meteorological observations, mining, experiments with electricity, various crops (cotton, indigo, and cochineal), their characteristics, and techniques.

For Alzate, the production of scientific knowledge took place not only under the internal conditions of the dynamics of science but also under the influence of social and political factors. Thus, the mexican enlightened man pointed out the need for greater support from the country's leaders to enable the establishment of the foundations for scientific development.

The utilitarian character of science was not limited to meeting the immediate needs of the population but extended, in Alzate's conception, to the broadening of the horizons of existence itself. Science, in Alzate's view, was imbued with the idea of the "homeland," understanding the transformative potential of science and the benefits derived from it, especially economic self-sufficiency. The knowledge of the resources of the American nature and their rational utilization through scientific studies, instruments, and procedures would provide economic self-sufficiency, inevitably impacting the political situation. Science alone would not bring the country into being, but it would enable it, through the discovery of resources, for its future growth.

In Alzate's conception, useful knowledge was that which provided or directed measures that benefited society. The utilitarian nature of science guided many of his studies, where he sought to address important aspects for the development of medicine, agriculture, mining, botany, physics, geography, and astronomy.

Medical themes and matters held a privileged space in Alzate's production, with studies on diseases, medical treatments, and the medicinal properties of Mexican plants. The discussion about the peculiarities of American nature also encompassed the nosological framework of New Spain, seeking to demonstrate that there were no endemic diseases specific to the Valley of Mexico and vehemently questioning the attribution of unhealthiness conferred on physical conditions such as the climate and soil of Mexico. On the other hand, he indicated the possibility of alleviating the symptoms of some highly contagious diseases in Europe, such as syphilis, when they occurred in Mexican lands. He also addressed the issue of the causality of diseases, following the discoveries made after the invention of the microscope, and questioning the idea of the determining role of insects or small animals in the etiology of diseases. In his studies, he dealt with smallpox, the prevention and cure of the plague, scurvy, dental cavities, syphilis, yellow fever, measures and devices for the treatment of various ailments, the medicinal virtues of products and plants in curing diseases. On one hand, he analyzed the healing potential of plants, and on the other, he pointed out the ignorance or disregard existing regarding the virtues of native plant species and indigenous medical knowledge.

His words advocated the rejection of the colonial past, the power of reason, the freedom of thought, and the recognition of the potentialities of his land. The potentialities of American nature had been ignored by many naturalists, but "no deben ser ignorados por los que se dedican a saber lo que es la naturaleza y sus raras producciones"[10] (Apud. García, 1990, p. 131).

On the other hand, the development of scientific research, based on procedures of modern science, its dissemination among social sectors, and its application for the benefit of that society, would signify fundamental steps towards the greatness of the "homeland". In this way, the possibility of a distinct science, or rather, of a scientific knowledge constructed according to peculiar conditions of the American reality, was envisaged. Through the defense of culture, the construction of the "nation" was sought.

Experience and observation, in his view, were fundamental for understanding the virtues of plants because, according to the climate and soil in which they grew, plants did not present significant differences, but were distinguished by variations of local nature (heat, chemical operations).

For Alzate, mining also required new knowledge to enhance it, so that it could help eradicate the poverty in which a large part of the population lived. The knowledge of the geography of America, as Alzate referred to, was indispensable for human life, for handling matters of the greatest importance. Observations of physical (earthquakes) and astronomical (eclipses) phenomena were the subjects of several articles, always imbued with the idea of the utility of this knowledge.

The cultivation of cotton and indigo, considered of great necessity and economic interest due to their various uses, was described in detail, presenting their qualities, potentialities, and cultivation methods. The cultivation of cochineal, also of economic interest (for dyes) and with quite favorable conditions in Mexican lands, also deserved Alzate's efforts.

The recognition of natural potentialities represented one of the paths to the creation and affirmation of national consciousness. It was also important to confer upon that community a past of its own. The defense of the ancient Mexicans, of the Aztec past, was of great significance in this process, as colonial presence, with its prerogatives, had deprived them of their history. The production of literature about the Indians, their characteristics, customs, and inclinations became indispensable to put an end to superficial ideas that were distant from the truth.

[10] Translation made by the author: "They should not be ignored by those who are dedicated to knowing what nature and its rare productions are".

It particularly provoked him, the idea that botany had never developed in Mexican soil, leading him to invoke his patriotic aspirations to refute it:

> En honor de la patria y de la nación concluyo con esta refleja. Se dijo en una de las arengas que la botánica no se había cultivado en Nueva España; si esto se dice respecto al conocimiento de las virtudes de las plantas es proposición que desmiente la historia.[11] (Alzate y Ramírez, 1789, p. 68)

In his conception of science, knowledge should be elaborated and disseminated in such a way that everyone, learned or ignorant, had access to its benefits. This led him to yet another criticism of Linnaeus' work, as the structure in which it was organized hindered the understanding of true plant knowledge by using a nomenclature that did not respect the names already given by its inhabitants, especially by indigenous peoples.

The teaching and dissemination of useful knowledge, new crops, new techniques, new treatments for diseases, and new species indicated to the new-hispanic the potential of his land to produce and to free itself from foreign products.

He emphasized that the study of nature was extremely useful when pursued solely for people's instruction, and this utility expanded when directed towards the public good. There were countless examples, he proclaimed, of natural resources from his land that could be turned into profitable products and represent economic self-sufficiency. The desire to be useful to the "homeland" was expressed as a mission, transcending the boundaries of his scientific and cultural trajectory.

Similarly, we can observe the trajectories of enlightened men in portuguese colonization spaces, such as that of Manuel Arruda da Câmara, for whom the recognition of natural potentialities also represented one of the paths to the creation and affirmation of national consciousness. Manuel Câmara (1752, Vila de Pombal, Paraíba—1810, Pernambuco), who was the son of Francisco de Arruda Câmara and Maria Saraiva da Silva, began his studies in Goiânia (Pernambuco). He consecrated himself as a Carmelite religious in 1783 with the name Friar Manuel do Coração de Jesus, at the Convent of Carmo de Goiânia, and obtained Secularization Brief in 1791/1792. His academic training began with the Philosophy Course at the University of Coimbra in 1786, where the following year he also graduated in Mathematics. He later enrolled in the Medicine Course at the same university, interrupting it due to some problems that arose, as a result of his sympathetic position towards the ideas of the French Revolution. He completed his medical training at the Faculty of Medicine of Montpellier, France, obtaining his doctorate in 1791 with the thesis titled "Physiological and chemical investigations on the influence of oxygen on animal economy, mainly on heat and color in men," which, based on Antoine Lavoisier's theory, dealt with the influence of oxygen on animal metabolism.

Arruda da Câmara became renowned for his studies on various crops, especially on the cultivation of cotton plants. He wrote dissertations on their introduction to

[11] Translation made by the author: "In honor of the country and the nation, I conclude with this reflection. It was said in one of the harangues that botany had not been cultivated in New Spain; If this is said regarding the knowledge of the virtues of plants, it is a proposition that refutes history."

the country, their economic potential, the description of different species, and the most suitable conditions and methods for planting them.[12]

He sought to highlight in his studies, as did José Antonio Alzate, the inadequacy of explanatory systems produced by Europeans, especially Linnaeus' classification, which did not correctly include all Brazilian species, such as Cassia, or described them inadequately.

The issue of the peculiarities of each region was addressed through the analysis of cotton cultivation. He adopted as a starting point the observation that this plant grew naturally in two regions of the world, Asia and America, and therefore any attempts to naturalize it in other lands would be fruitless. Thus, many completely ignored the variety of cotton species and were unaware of the existence of a type specific to the region of Maranhão, detailed by Arruda da Câmara in his observations. The misunderstanding of the natural character of each plant species led to a lack of knowledge about cultivation properties and planting specifics (location, climate, and soil). The lack of accurate information about cotton cultivation, or any other crop, inevitably led to the failure of its results and, consequently, its economic profitability. In this way, he affirmed the positivity of the brazilian nature and questioned the widely proclaimed benignity and superiority of the Old Continent. He believed that it was precisely in the tropics where the most suitable regions for human presence were found, where one could benefit from the richness and variety of lands and climates.

Although he did not specifically refer to the approaches of the philosopher Cornelius De Pauw and Buffon, Arruda da Câmara demonstrated familiarity with them, as well as the need to refute them.

Just like Alzate had done, Arruda da Câmara also stated that scientific knowledge should necessarily be based on the idea of usefulness for society, for the common good referred to at the time:

> Eu me apliquei então cuidadosamente a fazer todas as observações de que era capaz, segundo as luzes, ainda que tênues, que eu tinha adquirido, para que meus patrícios tivessem alguma coisa que lhes fosse própria, e não mendigassem de livros estranhos, que são raros, as noções que necessitavam. (...) é que eu devo consagrar este pequeno trabalho, com o qual procurei contribuir, segundo a fraqueza de minhas forças, para o bem comum da

[12] Author of: "Aviso aos lavradores sobre a inutilidade da suposta fermentação de qualquer qualidade de grãos, ou pevides, para aumento da colheita" (1792); "Do fruto e raiz do mucumã utilizado pelo povo faminto do sertão de Pernambuco" (1796); "Memória sobre a cultura dos algodoeiros e sobre o método de escolher e ensacar o algodão, em que se propõem alguns planos novos para seu melhoramento" (1799); "Memória sobre o algodão em Pernambuco" (1810); "Discurso sobre a instituição de hortas nas principais Capitais do Brasil" (1810); "Memória sobre as plantas de que se pode fazer barrilha no Brasil" (1814); "Discurso sobre a utilidade da instituição de jardins nas principais províncias do Brasil, oferecido ao Príncipe Regente" (1810); "Dissertação sobre as plantas do Brasil que podem dar linhos, próprias para muitos usos da sociedade e suprir a falta de cânhamo" (1810); "Flora pernambucana ou Centúria das plantas pernambucanas"; "Tratado de Agricultura" (inédito); "Insectologia ou coleção de desenhos de insetos" (inédito); "Estampas representando assuntos de história natural, insetos, peixes e pássaros, desenhados a lápis, nanquim e aquarela", por Arruda Câmara; "Dicionário de Botânica ou Compêndio dos vegetais do Brasil, tanto indígenas como aclimados, ampliado e publicado por Joaquim de Almeida Pinto" (1873).

> Pátria (...). O bem comum é o edifício para cuja construção todos os particulares têm obrigação de trazer os materiais, conforme os seus talentos (....)[13] (Câmara, 1982, p. 109–111)

Although these concepts had their limitations, there is no doubt about their significance in the process of raising awareness within the imagined community. The ideas and reflections of the enlightened men them as men of science who sought to confront the circumstances of their time. The affirmation of natural resources and their specificities, cultural values, and the capability of the "Mexican" and the "Brazilian" contributed to the crystallization of an autochthonous culture and the formation of a Mexican identity and a Brazilian identity. Many enlightened men sought recognition of a differentiated existence, their intellectual capacity, and their autonomy to benefit from their resources and strengths, but not a political rupture. Although the thoughts and trajectories of these enlightened men did not propose political rupture, they expressed a differentiated sentiment regarding the colonial condition, a concern about their belonging to that land.

In this line of reflection, we revisit the words of Juan José Saldaña, who noted in the historiographical production of the sciences in the last decades of the twentieth century, the awareness that the history of science presented epistemological issues, especially due to its geographical and cultural specificities, which should not be disregarded. That is, to think about science and the history of science in Latin America, with reference to its own identity and culture (Saldaña, 1996).

12.4.1 Final Considerations

Starting from the analytical perspective of the social protagonism of science in Latin America, when it acquired a connotation of social and political extent, efforts were made to reconstruct and reflect on the specificities of the formation of scientific practice in Latin America. Thinkers, enlightened men and "criollos," such as the mexican José Antonio Alzate and the brazilian Manuel Arruda da Câmara, through the defense of the American nature, the knowledge of its resources, and their utilization, affirmed constitutive values for the formation of another "homeland" to which they were dedicated.

José Antonio de Alzate fostered a mexican consciousness. It was the mexican identity understood in the exaltation of nature and its people, in the affirmation of its peculiarities, and in the knowledge of oneself and one's land. Corroborating the

---

[13] Translation made by the author: "I then applied myself carefully to making all the observations I was capable of, according to the light, however faint, that I had acquired, so that my countrymen would have something of their own, and not beg for strange books, which are rare, the notions that needed. (...) is that I must consecrate this small work, with which I tried to contribute, according to the weakness of my strength, to the common good of the Fatherland (...). The common good is the building for the construction of which all individuals are obliged to bring the materials, according to their talents".

prevailing scientific pragmatism, Arruda da Câmara, equally, affirmed that scientific knowledge should necessarily be guided by the idea of usefulness for society, in order to contribute "para o bem comum da Patria".[14]

The ideas and trajectories of these enlightened men exposed a consciousness, albeit diffuse, and a sense of the new social and political entity that was being gestated. They expressed a new Latin American mentality. They sought to disseminate the so-called "useful knowledge," that is, knowledge that promoted public happiness and was fully directed and suitable for these societies. They recognized the ability of Mexicans and Brazilians for science, observation, and experience. Furthermore, knowledge of nature and its potentialities was fundamental for the affirmation of national consciousness.

The enlightened men Alzate and Câmara, whose trajectories we have sought to reconstruct, pointed out the importance of defending American nature through the knowledge of its resources and their use for the benefit of their societies. The knowledge of nature was not understood by both only as theoretical knowledge but fundamentally presented a very defined practical significance.

Their trajectories were guided by the knowledge and classification of plant species, and by the introduction and improvement of new crops, such as cochineal cultivation. In the path of affirming new scientific ideas, following their concern for nature, mexican and brazilian enlightened men vehemently expressed their interest in serving the "homeland" as a mission.

Looking at the history of the construction of scientific knowledge and practice in the context of colonial Latin America, based on the analysis of the trajectories and production of spanish-american enlightened men, such as José Antonio Alzate and Manuel Arruda da Câmara, leads us to reflect on central questions of the Latin American historiography of sciences itself, such as the recognition of the existence and form of scientific activity in Latin America, and the awareness that the history of sciences presents questions of an epistemological nature, especially as a result of its geographical and cultural specificities.

We understand that it leads us to corroborate the parameters of rupture of a methodological mimesis, and the need to "pensar nuestra ciencia", as the historian Juan José Saldaña brilliantly stated in his studies:

> En fin, que así como en otras latitudes fue fundamental para la construcción de la capacidad científica nacional "aprender de sí mismo" escribiendo para ello la historia de la ciencia, en América Latina no puede continuar la amnesia colectiva sobre nuestro pasado científico y tecnológico sin riesgo de perder el futuro.[15] (Saldaña, 1996, p. 41)

---

[14] Translation made by the author: "for the common good of the Homeland".

[15] Translation made by the author: "Anyway, just as in other latitudes it was fundamental for the construction of national scientific capacity to ´learn from oneself` by writing the history of science, in Latin America the collective amnesia about our scientific and technological past cannot continue without risk of losing the future".

References

Alzate y Ramírez, J. A. (1788). *Gaceta de Literatura de México, I*, 419.

Alzate y Ramírez, J. A. (1789). Carta al autor de esta gaceta. *Gaceta de Literatura de México, I*(16), 68.

Basalla, G. (1967). The spread of western science. A three – Stage model describes the introduction of modern science into any non – European nation. *Science, 156*, 611–622.

Carreras, S., & Zeiter, K. C. (2014). Las ciencias en la formación de las naciones americanas. Una introducción. In *Las ciencias en la formación de las naciones americanas* (pp. 9–23). Iberoamericana Vervuert.

Cueto, M., & da Silva, M. A. D. (2020). Trayectorias y Desafíos en la Historiografía de la Ciencia y de la Medicina en América Latina. *Asclepio. Revista de Historia de la Medicina y de la Ciencia, 72*(2), 1–14.

da Câmara, M. A. (1797). *Memória sobre acultura dos algodoeiros oferecida ao Ilustríssimo e Excelentíssimo Senhor D. Rodrigo de Sousa Coutinho Ministro e Secretário de Estado da Repartição da Marinha e Negócios Ultramarinos por Manuel Arruda da Câmara Doutor em Medicina pela Universidade de Montpellier, da Academia das Ciências da mesma Cidade, Correspondente da Sociedade de Agricultura de Paris e da Academia Real das Ciências de Lisboa e Naturalista empregado no serviço de S. M. Fidelíssima na Capitania de Paranambuc.*

Dantes, M. A. M. (1992). *A Ciência na América Latina – uma história a ser contada. Módulo de História da América do Curso de Atualização da Fundação para o Desenvolvimento da Educação.*

Fefer, R. G. (2009). Notas sobre la genealogía de la historiografía reciente de la ciencia latinoamericana o de cómo se inventaron historias para ser esgrimidas contra los embates del atraso. In Álvarez, I. S. (coord.). *América Latina: enfoques historiográficos* (pp. 49–64). UNAM.

García, A. S. (1999). Idea de la ciencia en José Antonio Alzate y Ramírez de Santillana. Ciencia UANL. *Revista de Divulgación Cientifica y Tecnologica de la Univ. Autonoma de Nuevo León, II*(3), 227–229.

García, A. S. (2015). *Elementos para una teoría latinoamericana sobre historia de la ciencia.* Universidad Autónoma del Estado de México, Facultad de Humanidades.

Hernández Luna, J. (selec.). (1945). *José Antonio Alzate*. Secretaria de Educación Pública.

Lafuente, A., & Catala, J. S. (1989). Ciencia colonial y roles profesionales en la América Española del siglo XVIII. *Quipu, 6*(3), 387–403.

Moreno y De Los Arcos, R. (1989). *Linneo en México. Las controversias sobre el sistema binario sexual 1788–1798*. UNAM.

Moreno y De Los Arcos, R. (1994). *Ciencia y conciencia en el siglo XVIII mexicano*. UNAM.

Quevedo, E. V. (2000). Os estudos histórico-sociais sobre as ciências e a tecnologia na América Latina e na Colômbia: avaliação e perspectivas. In *Um olhar sobre o passado: história das ciências na América Latina* (pp. 33–95). Editora da UNICAMP.

Saldaña, J. J. (1983). Hacia una crítica histórica, teórica y metodológica de la historiografía latinoamericana de las ciencias. *Ciencia, Tecnología y Desarrollo, 7*(3), 291–444.

Saldaña, J. J. (1986). Marcos conceptuales: la historia de las ciencias en Latinoamérica. Positivismo y economicismo. In *Cuadernos Quipu 1* (pp. 57–80). SLHCT.

Saldaña, J. J. (1989). La formación de la comunidad científica en México. In *Historia de la Ciencia y la tecnología: el avance de una disciplina*. Editorial Tecnológica e Costa Rica.

Saldaña, J. J. (coord.). (1996). *Historia Social de las Ciencias en América Latina*. UNAM.

Maria Rachel Fróes da Fonseca has a master's degree in History (Univ. Federal Fluminense), doctorate in Social History (Univ. de San Pablo), and held her Postdoctorate at El Colegio de México (México). She is a researcher at the Casa de Oswaldo Cruz/Fundación Oswaldo Cruz, professor of the Postgraduate Program in History of Sciences and Health, and coordinator of the

"Historical-Biographical Dictionary of Health Sciences in Brazil (1832-1970)". Author of the thesis "The only science is the country: the scientific discourse in the construction of Brazil and Mexico, 1770-1815" and the article "The scientific vulgarization in Mexico: the leyenda or short novel by José Joaquín Arriaga (1831–1896)".

Part IV
The Emergence of Local History of Science Projects, Associations, Journals, Educative and Research Programs

Chapter 13
The Development of the History of Science as a Discipline in Argentina

Carlos D. Galles and Pablo Lorenzano

The present would be little less than an enigma if one did not resort to the past to explain and understand it. Norberto Piñero and Eduardo Bidau.

Abstract Historiographical texts often prove to be tedious readings, as the discourse, if any, risks getting lost in a maze of citations. However, nowadays, we have a valuable aid in the form of the internet, which, albeit requiring caution, provides a plethora of information for each author to consult. Therefore, the reader will not find much bibliographical information in these notes, but rather, the authors trust, a guide to understanding the development of studies in the history of science in Argentina.

Keywords Historiography · History of science · Argentina

The order of the authors is strictly alphabetical and does not reflect the contributions made to the work, which were made by both authors in a truly collaborative effort.

C. D. Galles (✉)
Independent Researcher, Rosario, Argentina
e-mail: carlos.galles@gmail.com

P. Lorenzano
Center of Studies in Philosophy and History of Science, (CEFHIC)-National University of Quilmes (UNQ)/National Scientific and Technical Research Council (CONICET), Bernal, Argentina

M. de la P. Ramos-Lara, L. C. Arboleda (eds.), *History of Science in Latin America: The Construction of an Intellectual Field (20th century)*,
https://doi.org/10.1007/978-3-031-75536-1_13

13.1 Introduction

Juan María Gutiérrez served as rector of the University of Buenos Aires between 1861 and 1874. Years before, he was a fervent anti-Rosas figure and spent extensive periods in exile. Following the Battle of Caseros, he held significant positions in the government of the Argentine Confederation. He distinguished himself as a man of letters. Several of his articles were compiled into a seminal work for the study of the evolution of education in Buenos Aires from the foundation of the Colegio de San Carlos in 1783 to the creation of the University of Buenos Aires in 1821. This comprehensive text contains biographies, royal orders, appointments, various documents, and commentary. This work is recognized as the first to address the history of our higher educational institutions.[1]

In 1910, on the centenary of the May Revolution, the University of Buenos Aires published a text providing information about its constituent faculties.[2] Magnificent buildings of some faculties are showcased, the total roster of deans and professors is provided, and the study programs of different disciplines are included. "Scientific investigations" are mentioned only once to describe an activity for students of the Faculty of Philosophy and Letters, presumably involving the preparation of a sort of monograph. It contains an unsigned Historical Review by the author. In a brief note at the end, it is stated that the historical data were taken from a work by Norberto Piñero and Eduardo Bidau, published over 20 years earlier.[3] The following paragraph will be dedicated to it.

Piñero and Bidau were lawyers and had, especially the former, significant careers in the service of the State; they were also professors at the university. This text has a historical introduction where the narrated events are corroborated with documentation. The authors partially based their work on Gutiérrez's work, which they consider deserving of praise but criticize its methodology for making history based on biographies. They point out, as early as 1888, that history must undergo a methodological change and become a "social science". They take as their mentors two scholars from the French positivist school: Eugène de Roberty[4] and Arthur Bordier.[5] Despite being respectively the Secretary and the Prosecretary of the University, Piñero and Bidau's book is by no means a bureaucratic text, although copies of ordinances, decrees, notes, and regulations abound; they are present, of course, some showing the tone of the time, and some episodes lend themselves to humour within the gravity always present in the actors. The book provides a detailed index of the contents of the more than four hundred pages of the historical essay. It

[1] Gutiérrez, Juan María. *Origen y desarrollo de la Enseñanza Pública Superior en Buenos Aires*, Buenos Aires: La Cultura Argentina, 1915.

[2] *La Universidad Nacional de Buenos Aires*, Buenos Aires: Imprenta Tragant, 1910.

[3] Piñero, Norberto; Bidau, Eduardo, L. "Historia de la Universidad de Buenos Aires", *Anales de la Universidad de Buenos Aires*, Vol. I, Buenos Aires: Imprenta de Martin Biedma, 1888.

[4] De Roberty, Eugène. *La Sociologie*, Paris: Librairie Germer Ballière et cie., 1881.

[5] Bordier, Arthur. *La Vie des Sociétés*, Paris: O. Reinwald, 1887.

narrates events and offers explanations. The work also presents a list of all professors and graduates of the University of Buenos Aires from its foundation until 1888: precious material for researchers.

Nicolás Besio Moreno prepared a study titled "Historical Synopsis of the Faculty of Exact, Physical, and Natural Sciences of Buenos Aires and the Teaching of Mathematics and Physics in Argentina" (1915).

13.2 "The Series 'Evolution of the Sciences' of the Argentine Scientific Society"

In the second decade of the twentieth century, the Argentine Scientific Society, founded in 1872, decided to commemorate its fiftieth anniversary with a series of publications "aimed at making known the scientific and technical movement in the Argentine Republic" [Taken from the Foreword preceding each of the monographs]. The authors are well educated professors, practitioners of the science they expound upon, who write with the elegant and fluid prose of their century. Although not historians, they fulfill their task exemplarily. Of course, they are imbued with the positivist character of their time, and their texts reflect this. The admirable texts were published during a time of splendour for the republic and confidence in permanent progress.

These are the years that José Babini would later refer to as "the glorious decades" of national science in a section of his 1949 text on the history of Argentine science.

The data they present may be corrected, and new ones may be discovered over time by successive researchers, and, which is key in all historical description, the interpretations of the selected facts will be considered according to new approaches and points of view. History is something alive that is written and rewritten. Fortunately, it is not a dogma.

Perhaps it is opportune to quote directly Hélène Metzger when referring to the work of the historian of science:

> If we manage to give our spirit, constantly on alert, outside of all dogmatism and all skepticism, useless and sterile, a little of that active plasticity which is a condition for all fruitful research, the history of science would render as much to science itself as to the soul of the researcher liberated from an imperceptible and lazy routine, which damages the spontaneity of his judgments, a service of which the value could not be underestimated.[6]

It is worth adding to these noble words that it is also intended to illustrate the way in which events unfolded, using the information we gather and interpret as an enlightening tool to plan and correct national projects, respecting and valuing diversity. All under the hypothesis that history weighs on our shoulders and we can draw valuable lessons from it for the future, as the memory of the events that unfold tends

[6] Metzger, Hélène. *La méthode philosophique en histoire des sciences* (ed. Gad Freudenthal), Paris: Fayard, 1987, p. 38. The translation is ours.

to remain in the minds of all men and women, regardless of the fortuitous circumstance that reflecting on them may be for some their main occupation.

The SCA had set the interval of 50 years since its foundation to narrate history. Some authors adhere to this, but others, like Ramón G. Loyarte, dedicate a chapter to the study of the evolution of physics during the Colonial period; and he does so in a highly critical manner based on meagre documentary evidence. The remaining chapters are dedicated to the periods up to the foundation of the University of Buenos Aires in 1821 and the National University of La Plata in 1897. Loyarte was one of the first three graduates with a Ph.D. in Physics from La Plata. His study is largely a description of progress in the teaching of Physics in various establishments. A final chapter is dedicated to research at his alma mater, the Institute of Physics at the National University of La Plata. These research efforts, although not numerous in terms of articles, represent top-quality science published in their own journal *Contribuciones* and reproduced in articles in specialized German journals. And for a reason: the first two directors of the institute were Emil Bose and Richard Gans, both Germans and Ph.D. holders in Physics.

In his text dedicated to Astronomy, Enrique Chaudet adheres to the conditions set by the SCA and deals with the history from 1872 onwards. Benjamin Gould, the first director of the National Observatory of Córdoba and creator of the renowned Uranometria, is naturally highlighted. The progress of the La Plata Observatory is also addressed, with its long years of stagnation, despite being well endowed with observation resources.

Cornelio Dassen focuses on mathematical studies. It is a very comprehensive study that deals with the early scholars of mathematics in the country. His conclusion is that the activity "has not yet produced creative minds in the mathematical field", only demonstrating "a good and healthy inclination towards the cultivation of mathematical sciences".

The part concerning chemical sciences was entrusted to Enrique Herrero Ducloux. His writing shows a good understanding of modern atomic theories. He reviews predecessors from the university's renewal under the rectorate of Gutiérrez until describing how chemistry was taught in the early decades of the twentieth century and how it was applied, for example, at the National Department of Health. He emphasizes that scientific research in this field is not progressing, although applications are achieved in various industries. The text is complemented by what Ducloux calls a select bibliography, which includes 36 items under the title "History of Chemistry".

In this Evolution series, a total of eight works were published. In addition to those already mentioned, four others were dedicated to Botany (Cristobal M. Hicken), Meteorology (Guillermo Hoxmark), Mineralogy and Geology (Franco Pastore), Public Hygiene and Sanitary Works (Nicolás Lozano and Antonio Paitovi). Seven more were planned (but did not appear) on the following topics: Zoology; Anthropology and related branches (such as ethnography, philology, folklore, linguistics); Paleontology, Medicine, Statistics, Industry, History and the importance of the SCA. The positivist temperament of the promoters of the collection is reflected by their dedicating a volume to the topic of Sanitary Works. A volume could also have been dedicated to railways, for example, which use science but are technically, imported in this case, or to the port of Buenos Aires designed by Madero. Perhaps they were influenced by the splendour of modern techniques. For example, the building on Avenida Córdoba, which contrasts with colonial architecture, which has contained since its inauguration in 1894 the tanks for the supply of running water to the city (they have long been out of use for that purpose), aptly named the Palace of Running Water.[7] The texts were published in a time of grandeur and confidence in permanent progress.[8]

The mere dedication in the university to teaching without given rise to research had already been pointed out by Bernardo Houssay, future Nobel Prize winner, in his thesis of medicine in 1911 when he stated "We have our Faculties housed in comfortable palaces and with richly installed laboratories that improve day by day. Our professionals graduate with an excellent general preparation, and yet, a curious phenomenon, our original scientific production is very scarce".[9]

13.3 A Historian of Science, Member of the Society of Jesus

The reaction, in the best sense of the word, to such neglect of the colonial period was to come from a Jesuit of Irish descent born in 1889 a few kilometers north of Rosario, in the town of Pueblo Aguirre, now Arroyo Seco. Guillermo Furlong was a true historian, both by training and by his dedication to research. He earned his doctorate in science from Georgetown University in 1913; he had entered the novitiate of the Society of Jesus and was ordained a priest in 1924. The most important part of his teaching activity was carried out at Colegio del Salvador and at the Seminary of Villa Devoto, both institutions in Buenos Aires. He died in Buenos Aires in 1974. Furlong's written work is immense. In order to appreciate his creativity, a reading of Mario Tesler's article which deals only with his unpublished writings is highly recommendable.[10]

---

[7] https://es.wikipedia.org/wiki/Palacio_de_Aguas_Corrientes

[8] Fifty years later, the Argentine Scientific Society published another series of works on the evolution of science between 1922 and 1972. They are as follows: Mathematics (Luis A. Santaló); Physics (José F. Westerkamp); Cybernetics (Máximo Valentinuzzi and Osvaldo Skliar); Genetics (Luis B. Mazoti and Juan H. Hunziker); Meteorology, Oceanography, and Radiopropagation (José Ángel Álvarez); Botany (Ángel L. Cabrera); Astronomy (Simón Gershanik and Luis A. Milone); Geophysics and Geodesy (Otto Schneider and Rafael N. Sánchez); Chemistry (Noemí G. Abiusso); Geography (Carlos A. De Jorge); Anthropology (Argentine Center of American Ethnology); Entomology (Luis De Santis).

[9] Barrios Medina, Ariel; Paladini, Alejandro C. *Escritos y Discursos del Dr. Bernardo A. Houssay*, Buenos Aires: Eudeba, 1989, p. 12. The translation is ours.

[10] Tesler, Mario. "Los inéditos de Guillermo Furlong", *Hilario. Artes Letras Oficios*, Revista Digital N° 13: https://www.hilariobooks.com/blog-article.php?slug_es=los-ineditos-de-guillermo-furlong

In this study, only two works will be considered among Furlong's extensive production. The first[11] of these is the one in which the author vindicates the figure of Buenaventura Suárez (1679–1750), the Jesuit who, on the banks of the Paraná River, devoted part of his time to Astronomy, building his first telescopes on-site, using quartz crystals for the lenses. His crowning achievement was, as is well known, el *Lunario de un siglo*, published in Lisboa in 1748.

In this first edition, the chapter dedicated to Suárez goes from page 79 to 140. The book deals with three Jesuits born in the city of Santa Fe, like Suárez himself. Regarding his documentary sources, Furlong mentions about 20 archives ranging from Buenos Aires and Córdoba to Madrid and Simancas.

In 1969, his *Social and Cultural History of the Río de la Plata 1536–1810* was published. It consists of three volumes, richly illustrated, the third of which is entitled *The Cultural Transplant: Science*, namely the science disseminated by the Jesuits and some aspects that can be considered as the search for new knowledge. Jesuit explorers helped develop magnificent cartographic plates very close to the geographical reality.[12]

They also collected specimens to reflect the flora and fauna of the regions where their empire extended. Special mention must be made of Thomas Falkner, originally an English Protestant, whom life's vicissitudes found ill in Buenos Aires, compassionately cared for by a Jesuit who eventually won his adherence to the Society of Jesus. He later traveled vast areas of the country, gathering his knowledge in a work published in 1774 in Europe.[13]

Work on science in the colonial era has continued to this day.[14]

13.4 The Institute of History and Philosophy of Science of Santa Fe

The National University of Litoral was founded by National Law on October 17, 1919, during the presidency of Dr. Hipólito Yrigoyen. The extensive campus of the new university stretched along the Paraná River for 700 km; from the city of Corrientes, passing through the cities of Santa Fe and Paraná, to the major river port of Rosario: it had faculties in all these cities. The University's Superior Council established the Institute of History and Philosophy of Science by ordinance on July

[11] Furlong, Guillermo. *Glorias santafesinas*, Buenos Aires: Editorial Surgo, 1929.

[12] Furlong, Guillermo. *El trasplante cultural: ciencia*. Tercer tomo de *Historia social y cultural del Río de la Plata, 1536–1810*, Buenos Aires: Tipográfica Editora Argentina, 1969.

[13] Falkner, Tomás. *Descripción de la Patagonia y de las partes contiguas de la América del Sur. Con estudio preliminar de Raúl J. Mandrini*, Madrid: Taurus, 2004.

[14] The following works have recently been published on the subject. de Asúa, Miguel. *Science and Catholicism in Argentina (1750–1960)*, Berlin/Boston: Walter de Gruyter, 2022. Lértora Mendoza, Celina. *Newton en el Río de la Plata: sus teorías en la enseñanza colonial*, Buenos Aires: FEPAI, 2023 [Digital book: http://bibliotecafepai.fepai.org.ar/Libros/Filosofia/NewtonRioDeLaPlata.pdf]

29, 1939. The mission entrusted to the institute was to conduct original research, train new researchers, offer courses, and establish bibliographical resources. On the same date, Aldo Mieli was appointed as the director of the Institute. This distinguished Italian historian of science arrived equipped with his extensive knowledge, vast library, and his journal *Archeion*.

The institute and the publication of the journal came to an abrupt end following the revolution of 1943 and the subsequent intervention in universities. The journal managed to publish four volumes totaling nearly 1800 pages.[15]

A close collaborator of Mieli was the engineer and mathematician José Babini, who had already made incursions into the history of science. In 1943, shortly before being dismissed from the university, Babini managed to publish an article in *Archeion* titled "The history of science as a scientific discipline" (*Archeion* 25 (2/3/4) (1943): 101–107). There is a comprehensive publication containing all the articles and books published by José Babini[16]; there, one can see that his production on the history of science gradually manifested with Mieli's arrival in Santa Fe. Mieli's presence in the country enabled some Argentines to take a more serious interest in the History of Science, and it allowed José Babini (1897–1984) to be trained as a professional historian of science.

Another important historian of science at that time was the Hungarian Desiderio Papp (1895–1994), who emigrated to Argentina in 1942. He was appointed Professor of the History of Science at the Faculty of Pure and Applied Sciences and the Faculty of Philosophy and Letters of the National University of Tucumán in 1945.

Following the abrupt end of *Archeion*, Aldo Mieli settled in Buenos Aires and began the publication of his *Panorama general de historia de la ciencia* (*General Overview of the History of Science*).[17]

The work initially comprised five volumes, with the last seven volumes being the result of a collaboration with Desiderio Papp and José Babini.

In 1954 the editors of *Ciencia e Investigación* decided to conduct a survey among researchers from various disciplines to obtain an overview. José Babini was tasked with responding to the state of the History of Science in the country. His response is preceded by a brief presentation of the author, mentioning that he has been a corresponding member of the International Academy of the History of Sciences since 1948.[18]

Babini emphasizes that the discipline has gained recognized autonomy worldwide with the purpose of analyzing and critiquing scientific activity throughout

[15] Galles, Carlos; Ferrari, Roberto. "La etapa santafesina del 'Archeion' de Mieli", *Actas de las Segundas Jornadas de Historia del Pensamiento Científico Argentino*, Buenos Aires: FEPAI, 1984, pp. 191–199.

[16] Babini, Nicolás. *José Babini: Bio-bibliografía 1897–1984*, Buenos Aires: Editorial Dunken, 1994 [2001].

[17] Mieli, Aldo. *Panorama general de historia de la ciencia*, in 5 Volumes. The last two in collaboration with Desiderio Papp and José Babini. Published by Espasa Calpe (1945–1952).

[18] Babini, José. "La historia de la ciencia", *Ciencia e Investigación* 10 (dic. 1954): 536.

history. He highlights Mieli's work in Santa Fe and its short duration. He also notes the absence of an appropriate institution dedicated to such studies in the Republic.

It is evident that during those years the panorama was not the most favourable for the development of the discipline. Private activity, in a way, filled the university gap. In this regard, the publication of the *Panorama general de historia de la ciencia* initiated by Mieli in 1945 and continued by Desiderio Papp and José Babini is noteworthy. Babini mentions it briefly in his response to the questionnaire.

In the same year of 1954, José Babini had published his *La evolución del pensamiento científico en la Argentina* (*The Evolution of Scientific Thought in Argentina*) (Buenos Aires: Ediciones "La Fragua", 1954). Between pages 236 and 239, he provides an annotated bibliography on the history of science, especially the monographs of the "Evolución de la SCA" series. Father Guillermo Furlong S.J. is mentioned for his contributions since 1944 on the history of science "during Hispanic domination", although Furlong's writings on the subject date back much earlier.

These texts by Babini lack what has been called "bibliographic apparatus", meaning the assertions made are not supported by citations from previous publications or references to archival material.

Previously, Babini had published in 1949 a *Historia de la ciencia argentina* (*History of Argentine Science*) (Mexico: Fondo de Cultura Económica). Later, a text for the Higher Cycle of Teaching also appeared.[19]

In the early 1980s, the great Argentine master, Don José Babini, already in the twilight of his life as a scientist, humanist, and defender of democratic values, received an award for his work as a historian of the sciences. When interviewed by journalists, he pointed out that he had devoted himself to this field because he lacked opportunities in the country to pursue research in the science he truly desired to dedicate himself to: mathematics. The example is illustrative; in Latin American countries, the discipline had distinguished practitioners in the past, but their activity did not achieve the desired diffusion; the lack of professionalization among historians of science was most common among these solitary navigators who only managed to survive in their isolation thanks to their inherent intellectual capacity and enormous enthusiasm, which they fortunately managed to convey. In the last 20 years, some changes have occurred in the situation, and in a positive direction, which will be referenced towards the end of this brief exposition.

[19] Babini, José. *Las ciencias en la historia de la cultura argentina* (*Sciences in the History of Argentine Culture*), Buenos Aires: Editorial Estrada, 1952.

13.5 Studies in the History of Medicine[20]

There were few individuals who dedicated themselves to the History of Medicine. The first was Nicolás Albarellos, who in the mid-nineteenth century took charge of a chair titled Legal Medicine, Pathological Anatomy, and History of Medicine at the University of Buenos Aires. In 1916, Félix Garzón Maceda presented a medical history of the city of Córdoba.

In 1922, at the Faculty of Medical Sciences, located in Rosario and dependent on the National University of the Litoral, Ricardo Caballero established the chair of History of Medicine and Medical Doctrines. Caballero, in addition to being a professor at the Faculty, was a Senator of the Nation, which enabled him to secure a generous state subsidy with which a magnificent library of scientific classics was formed, which still exists today.[21]

In 1936, the chair and Athenaeum of History of Medicine were founded at the Faculty of Medicine of the University of Buenos Aires, with fruitful uninterrupted work until now. Over the years, various studies and journals have been published by the Athenaeum. Those that remain today are: *Revista de Bioética y Socioantropología en Medicina* (since 2005), *Médicos y Medicina en la Historia* (since 2001), *Revista de Historia de la Medicina y Epistemología Médica* (since 2009).

Distinguished medical historians have been members of the Athenaeum. It is worth mentioning Argentino Landaburu, Alfredo Kohn Loncarica, Federico Pérgola, Norma Sánchez, Juan Ramón Beltrán, Anibal Ruiz Moreno, Abel Luis Agüero, Delia Outomuro, among others, without being exhaustive.

Enrique Pedro Aznarez, a member of the Argentine Group of History of Science, was in charge of the first free Course of History of Medicine at the Faculty of Medicine of the National University of Córdoba in 1943. In 1969, he inaugurated the Museum of History of Medicine. In 1971, he established the Institute and Chair of History of Medicine.[22]

The Nobel prizes awarded to Argentine researchers, especially Bernardo A. Houssay,[23] but also Luis Federico Leloir,[24] have been the subject of numerous studies.

---

[20] Here it is presented a synthesis of a comprehensive article by Norma Isabel Sánchez: "Maneras de hacer historia de la medicina argentina (1916–1936)", *Revista de Historia de la Medicina y Epistemología Médica* 8(1) (2016): 1–40.

[21] Ivern, Andres. "Doctor Ricardo Caballero: Semblanza del profesor de Historia de la Medicina y de las Doctrinas Médicas", *Revista de Historia de Rosario*, Año 6, N° 15–16 (jul-dic 1968): 184.

[22] Aracerbi Cremades, Norma, Pedro Aznares, Enrique. "Primer profesor de Historia de la Medicina en Córdoba", *Revista de Salud Pública* XIV (2) (dic. 2010): 63–67.

[23] For instance, Barros Medina, Ariel. *Bernardo Alberto Houssay*, London: Editorial Académica Española, 2012.

[24] Lorenzano, César. *Por los caminos de Leloir*, Buenos Aires: Editorial Biblos, 1994.

13.6 Grupo Argentino de Historia de la Ciencia

In 1927, the chair of Epistemology and History of Science was established in Buenos Aires, with Albert Franceschi being its first professor. Later, the prestigious Spanish mathematician Julio Rey Pastor assumed the chair. Rey Pastor, along with the chemist and historian Humberto J. Paoli, was instrumental in 1933 in founding the Argentine Group of History of Science, a branch of the International Union of History of Science.

It is not possible in this article to present all those who over the years belonged to the Group. But one of them, due to the lack of knowledge about their work, deserves to be mentioned. Umberto Giulio Paoli was born in Lucca in 1876 and received an outstanding classical and scientific education in Europe, especially in chemistry and chemical engineering. His contact with Pierre Duhem inclined him towards the history of sciences. He arrived in Argentina in the early twentieth century and began working as the director of various chemical companies. Parallelly, he devoted himself to the history of science, especially the critical re-edition of Greco-Latin texts and the critical re-edition of texts related to Spanish America. He was a high-flying independent researcher.[25]

This issue also presents a list of those who were part of the Argentine Group of History of Science in 1960. They are: José Babini, Nicolás Besio Moreno, Juan Dalma, Venancio Delofeu, Bruno V. Ferrari Bono, Gustavo Fester, Josué Gollán, Bernardo A. Houssay, Teofilo Isnardi, Beppo Levi, Juan T. Lewis, Rodolfo Mondolfo, Desiderio Papp, Alberto Palcos, Lorenzo Parodi, Cortés Plá, Ladislao Reti, Ricardo Resta, Julio Rey Pastor, José Luis Romero, Aníbal Ruiz Moreno, Francisco Vera.

Most of these members had far surpassed their prime years. The Group presents itself as a closed academy for scholars of the history of science distinguished by their trajectories in other disciplines, with exceptions such as Babini, Papp, Reti, and Plá.

13.7 Cortés Plá: Evolution of a Science Historian

The arrival of Mieli, who brought with him an outstanding specialized library, had a catalytic effect on the interests of some Argentine professors. Soon enough, the first writings on historical-scientific topics emerged. As an example, we will refer to Cortés Plá, one of the organizers of the University Reform in Córdoba. An engineer by profession, he held an extensive career both as a teacher and as a leader at the National University of Litoral. Initially, he contributed to the History of Science by coordinating simple, short conferences dedicated to some historical event in

[25] See the article by Roberto Ferrari, "Umberto Giulio Paoli (1876–1953), químico industrial e historiador de la ciencia", *Saber y Tiempo* 4(9) (ene-jun 2000): 61–75.

particular. For instance, he presented a biographical sketch of Joseph Thomson (the discoverer of the electron), a tribute to Newton, and an article dedicated to the work of Jean Perrin.

In 1942, he published an article in *Mathematicae Notae*, which later appeared in the form of a small book entitled *Las leyes de Ohm. Ensayo de historia científica y humana* (*Ohm's Laws. An Essay on Scientific and Human History*). In this work, he conducted a detailed study of the experiments that led to the discovery of the famous law, while not forgetting to mention the academic challenges faced by the German scholar. That same year, his work *Galileo Galilei* was published by Espasa-Calpe, a publication characterized by its accessible tone of dissemination that achieved an uncommon resonance. This was followed by a book dedicated to the life and work of Newton.

In 1943 he was dismissed from the university, along with many other professors, after the coup d'etat of June 1943. This meant an interruption in his academic activities, which allowed him to devote more time to intensively studying the History of Science. Furthermore, he undertook the translation of classical works in this field. This led to the publication by Editorial Losada, in the magnificent collection *Biblioteca de Teoría e Historia de las Ciencias*, of a volume containing the principal writings of Huygens and Fresnel titled *La teoría ondulatoria de la luz* (*The Wave Theory of Light*), accompanied by numerous notes and an excellent historical introduction by the translator himself. March 1946 was an undoubtedly challenging time for Plá, as he had returned to his position as university dean for only a few months. During this complicated period, he managed to translate the principal works of Henri Becquerel (published by Espasa-Calpe) under the title *El descubrimiento de la radioactividad* (*The Discovery of Radioactivity*). The publication was preceded by a biographical study of the French scientist and a monography on the significance of the discovery of radioactivity. During those years, he also translated two books by Louis de Broglie.

The history of the determinations of the speeds of sound and light was studied in detail in a subsequent work. Cortés Plá's contribution to the History of Science would be crowned in 1950 with the publication by Editorial Guillermo Kraft of *El enigma de la luz* (*The Enigma of Light*). This text filled a gap in the international bibliography, which lacked a clear synthesis of the evolution in the Western world of conceptions about this elusive entity. The book was introduced by none other than George Sarton, considered at that time the greatest historian of science and certainly the one who contributed the most to establishing the discipline in the university sphere. Although the content of this work has today been surpassed by the abundant subsequent literature of optics historians, its conceptual clarity undoubtedly warrants a reprint.

These initial years of Cortés Plá's longest time spent away from university life (he only returned in 1955) paradoxically allowed him to develop his most enduring literary work. This ended with the publication in 1950 of *Ciencia y Sociedad* (*Science and Society*), a work that addresses many of the questions about the role of science for humanity. The reading of this article could have prevented many errors by those who, recently encouraged by relativistic trends regarding the value of theories, have fueled certain public debates.

13.8 The Intermediate Period

In 1940, the National University of La Plata published a history of the institution in two volumes totaling over 800 pages. The author was Julio S. Castiñeiras, president of the university from 1935 to 1938. The first volume is very interesting as it outlines the history of Argentine universities in the first hundred pages, including those of Buenos Aires, Córdoba, La Plata, and the more recent ones in the Littoral and Tucumán. The rest of that first volume is dedicated to the history of the university in La Plata, initially provincial and eventually national. The second volume contains a review of numerous administrative acts, councils' constitution, assemblies, various events, and also the piano version of the university's anthem, presented presidency after presidency. It is a significant piece for the scholar of the National University of La Plata, both for what it says and for what it does not say. It only presents brief references to research activities.[26]

After the closure of the institute directed, and until 1955, by Mieli, university activities in the history of science were perhaps minimal. The Spanish mathematician Julio Rey Pastor, greatly influential in the development of mathematics in the country, taught a course on the Epistemology of Engineering at the National University of La Plata.[27] This was not Rey Pastor's debut in the history of science and technology; years earlier, he had dealt with Hispanic and Lusitanian cartography. He had already published *La ciencia y la técnica en el descubrimiento de América* (Buenos Aires: Espasa-Calpe, 1942), *Historia de la matemática* (with J. Babini, Buenos Aires: Espasa-Calpe, 1951), and *La técnica en la historia de la humanidad* (with N. Drewes, Buenos Aires: Atlántida, 1957).

The National Academy of Sciences, located in Córdoba, founded by Domingo Faustino Sarmiento, organized the First Argentine Congress of the History of Science in 1969. It commemorated the centenary of the academy's foundation. The papers presented were compiled into two thick volumes totaling just over a thousand one hundred pages.[28]

In 1972, the Second Argentine Congress of the History of Science was held; chaired by Engineer Cortés Plá. According to the call, it was oriented towards "knowledge of the origin, trajectory, and evolution of Argentine scientific

[26] Castiñeiras, Julio S. *Historia de la Universidad de La Plata*, Buenos Aires: Talleres gráficos "Tomás Palumbo", 1940.

[27] Rey Pastor, Julio. *Lecciones de Epistemología de la Ingeniería*, La Plata: Asociación de profesores de la Facultad de Ciencias Fisicomatemáticas de la Universidad Nacional de Eva Perón (La Plata), 1954.

[28] *Boletín de la Academia Nacional de Ciencias*, Vol. XLVIII (First part) (1970), Vol. XLIX (Second part) (1972).

institutions, whether state or private". There were a total of 81 contributions. There was a decrease in the participation of foreign authors and delegates. Some of the attendees were: Enrique P. Aznárez, Francisco Cignoli, Juan Cuatrecasas, Horacio J. Cuccorese, Juan Dalma, Venancio Deulofeu, Virgilio G. Foglia, Telasco García Castellanos, Enrique Gaviola, Leticia Halperin Donghi, Alfredo G. Kohn Loncarica, Alfredo Lanari, Vicente Oddo, Desiderio Papp, Luis A. Santaló, Alberto A. Taquini. It was modest, carried out in a very difficult time in national history, under a de facto government.

In 1994, the Third Argentine Congress of the History of Science and Technology was held. Chaired by Arturo Otaño Sahores and Andrés Stoppani. Fifty-four papers were presented, distributed in thematic groups: Personalities in the history of science; General themes of the history of science; Topics in the history of medicine and Institutions of the history of science and technology. Some of the attendees: Miguel de Asúa, Hugo E. Biagini, Horacio H. Camacho, Laura Levi, Alfredo G. Kohn Loncarica, César Lorenzano, Marcelo Monserrat, Juan Carlos Nicolau, Gregorio Weinberg, Alcira Zarranz, and a group of young people approaching the discipline. It was promoted by the Babini Foundation. Proceedings were not published.

Adhering to the sesquicentennial of the May Revolution, the Argentine Group of the History of Science held sessions on the state of science in Argentina at the beginning of the nineteenth century.[29]

The Introduction was given by Engineer José Babini, and Alberto Palcos ("From Azara to Bonpland"), Bruno Ferrari Bono ("Malaspina's Journey and its connection with the revolutionary movement"), Anibal Ruiz Moreno ("The Professionals of the Art of Healing in the May Week"), Nicolás Besio Moreno ("Pure and Applied Science in the Revolutionary Moment") presented their papers.

This booklet also includes a list of those who were part of the Argentine Group of the History of Science in 1960. They were: José Babini, Nicolás Besio Moreno, Juan Dalma, Venancio Delofeu, Bruno V, Ferrari Bono, Gustavo Fester, Josué Gollán, Bernardo A. Houssay, Teofilo Isnardi, Beppo Levi, Juan T. Lewis, Rodolfo Mondolfo, Desiderio Papp, Alberto Palcos, Lorenzo Parodi, Cortés Plá, Ladislao Reti, Ricardo Resta, Julio Rey Pastor, José Luis Romero, Aníbal Ruiz Moreno, Francisco Vera. Most of these members had far exceeded their prime years. The Group presents itself as an academy for practitioners of the history of science, many of them distinguished by their careers in other disciplines.

It is worth rescuing the figure of Juan T. Lewis, a physiologist and professor at the Faculty of Medical Sciences of Rosario. Dismissed from his university position between 1946 and 1955, he was a researcher in a private center. For many years, he wrote editorials for the journal *Ciencia e Investigación*, deploying an incessant struggle for quality university education with attached scientific research. The journal, founded by Bernardo Houssay in 1945, is a source of high value for the historian of science, especially the issues that appeared during the first Peronism.

[29] Lecture Series, August 22 and September 5, 1960, Buenos Aires. *Ciclo de disertaciones sobre el estado de la ciencia en Argentina a comienzos del siglo XIX, Buenos Aires: Comisión Nacional Ejecutiva del 150 Aniversario de la Revolución de Mayo, 1960.*

The Argentine Scientific Society has an Institute of History of Science, with Norma Isabel Sánchez as its director. Among its activities are the organization of events, guidance for thesis students, maintaining contacts with the community, etc. It has published seven volumes of Annals with topics related to this orientation, and has the 2024 edition in press. In April 2024, it held the Second Ibero-American Congress on the History of Science and Technology, on the occasion of the 150th anniversary of the Domingo Faustino Sarmiento library (which belongs to the SCA), rich in books, maps, documents.

13.9 Foundation for the Study of Argentine and Ibero-American Thought

By the late second half of the twentieth century, those writing about the history of science held a predominantly conformist position, with few exceptions to be found. In order for the discipline to advance, there arose a necessity to facilitate the exposition of ideas and debate. It was then that an institution emerged, which persists to this day, and brought together a significant portion of historians of science from the late twentieth century.

The Foundation for the Study of Argentine and Ibero-American Thought (FEPAI) began its activities in 1981, with Celina Lértora Mendoza serving as President, a position she continues to hold. As a researcher, Celina Lértora Mendoza has dedicated her studies to philosophy and science in the colonial era, with a vast production of articles and books.

In order for the discipline to advance, there arose a necessity to facilitate the exposition of ideas and debate. While the works produced were of high quality, the position of those writing was somewhat conformist, with sporadic exceptions. Shortly thereafter, in July 1982, the First Conference on the History of Argentine Scientific Thought was organized in Buenos Aires by FEPAI, with José Babini as Honorary President. The organizing committee included Luis Santaló, Desiderio Papp, Telasco García Castellanos, and Diego Francisco Pró. This initial meeting served to bring together some of the few individuals who, at that time, devoted part of their time to the history of science. A total of 12 papers and lectures were presented.

The second conference was held in Buenos Aires in July 1984. Over 12 papers were presented. Not long before, José Babini had passed away. In his honor, a tribute was held, featuring, among others, Juan José Saldaña in his first visit to Argentina. Juan José and Celina also presented the recently launched journal *Quipu*. Among other activities, a collection of ancient scientific texts, some very rare, was presented by Roberto Ferrari. Among the variety of topics, Mario Visiconte presented a very interesting paper on police physicians during the era of Rosas, a presentation that reflects the openness of the conferences. This mention is made with reference to the deep divide among Argentine historians of different ideas about the era of Juan Manuel de Rosas. It is well known that there is a deep divide among historians when considering the era in which Juan Manuel de Rosas governed the country. The functioning of the University of Buenos Aires, for example, is often disregarded, when in reality many of those who were authorities after the Battle of Caseros in 1852, which marked the end of the Rosas regime, were educated there. It would have been surprising if Visiconte had been able to present his articles in the closed meetings of the Argentine Group for the History of Science.

The conferences, which have continued to this day, have always been open to all speakers, even without a close university affiliation, which in some regions of the world is called an "independent researcher".

Since 1982, FEPAI has also published a Bulletin of the History of Science, which has had successive directors over its long history. Readers are invited to read some of these bulletins, not all of which are available on the mentioned website. The bulletin for the first half of 1985 presents an outline of an Argentine bibliography on the history of science, which, of course, cannot be transcribed in this article, totaling 50 pages and including about five hundred publications.

The topics in these bulletins are very diverse, and as an example, an article by Ariel Barrios Medina, a history professor mainly dedicated to the work of Bernardo Houssay and his disciples, is discussed.[30] This article deals with the frustrations, and some small successes (such as finding in dust and oblivion the vessels with which insulin was first produced in the country), in a journey by the author through document centers in Buenos Aires. The extreme seriousness with which the facts are narrated leads one to wonder if there is not a deeply hidden sense of humor underlying. Barrios Medina concludes that in his opinion, historians of science made little use of archival documents. Perhaps it can be added because they have been burned too many times. Barrios Medina concludes with an attack on what he calls "carefully neutral research".

13.10 Saber y Tiempo

In 1996, Nicolás Babini founded the *Asociación biblioteca* that bears his father's name. Soon afterwards, the first issue of the magazine *Saber y Tiempo* was published. Nicolás Babini himself served as the director, with Leticia Halperín Donghi as the Editorial Secretary. The consultants for the magazine included Miguel de Asúa, Néstor T. Auza, Guillermo Boido, Horacio H. Camacho, Roberto A. Ferrari, Carlos D. Galles, Alfredo G. Kohn Loncarica, Celina A. Lértora Mendoza, Marcelo Monserrat, Luis Alberto Romero, Mario Tesler, and Gregorio Weinberg.

Nicolás Babini was an architect, politician of the developmentalist line, and author of, among other books, the following work: *La otra Argentina. La ciencia y*

[30] Barrios Medina, Ariel. "Acerca de nuestros archivos y nuestros historiadores", *Boletín Informativo Historia de la Ciencia* 4(8) (1985): 6–10.

la técnica desde 1600 hasta 1966 (*The other Argentina. Science and Technology from 1600 to 1966*) (Buenos Aires: Centro de Estudios de Historia de la Ciencia y de la Técnica José Babini, 2007).

The magazine always displayed an excellent presentation and high-quality articles. It is commonly regarded as the first Argentine magazine dedicated to the History of Science, excluding Mieli's *Archeion*, which lasted only a few years in Santa Fe and was more of an outlier in Argentine bibliography, and the concise Bulletin of Information on the History of Science of FEPAI, which still persists and contains research articles. Naturally, the magazine also included articles on the History of Technology. Over the years, the magazine, which initially had its headquarters at the Argentine Scientific Society, transitioned to being digitally published by the Center for Studies in the History of Science and Technology José Babini, affiliated with the National University of San Martín. The last issue available on the internet is from the year 2018. It would be a true loss if continuity cannot be achieved.

13.11 Brief Synthesis of the Panorama to Date

In the following, we will attempt to provide a brief overview of the situation to date. Over the past 20 years, there has been a flourishing of studies in the History of Science in the country, accompanied by a professionalization of researchers thanks to favourable policies from various university institutions and the National Council for Scientific and Technical Research (CONICET). This development builds upon the history of the discipline during the previous century, which laid the foundation for the evolution of studies. It is worth mentioning that during the VI Congress of the Latin American Society for the History of Science and Technology (LASHST), held in Buenos Aires in 2004, numerous young Argentineans initiated in the History of Science participated.[31]

Numerous faculties across the country offer courses on the History of Science. Among others, it is worth mentioning, in a list that is surely incomplete, the Faculty of Exact and Natural Sciences of the University of Buenos Aires (a chair previously held by José Babini and later by Luis Santaló), the Faculty of Exact Sciences, etc., of the National University of Rosario (where postgraduate courses in History of Technology have been taught), and the Faculty of Chemical Engineering of the National University of the Litoral (a chair initially held by José Babini). Additionally, the subject is part of the programmes of numerous teacher training colleges spread throughout the country, where students are trained as secondary school teachers. Santos B. Velazquez taught the subject of History of Science and Technology for many years at the Faculty of Chemical Engineering of the National University of Misiones. It appears that this subject has not been continued at that university.

---

[31] See the article by Carlos D. Galles and Norma I. Sánchez in this present book.

It is worth pointing out the figure of Eduardo Ortiz, eminent mathematician, Professor Emeritus at Imperial College in London, had a distinguished performance as a historian of science. He was a frequent visitor at conferences in Argentina, noted for his ideas and precision when presenting them in articles. He especially studied the development of Mathematics and Physics. In 1990 he was the winner of the J.Babini History of Science National Prize, Conicet. He was also a founding member of the LASHST. He had graduated with a doctorate from the University of Buenos Aires, and his thesis was supervised by Mischa Cotlar.[32]

The Epistemology and History of Science Symposium has been held since 1991, in the pleasant setting of the Cordoba mountains. Starting modestly, they gradually raised their level until reaching the current high standard in both specialties. The promoter of these symposia from the outset was Victor Rodriguez. These symposia led to the publication of the digital journal *Epistemología e Historia de la Ciencia*, which has been published since 2016, under the auspices of the Faculty of Philosophy and Humanities of the National University of Córdoba. In this faculty, there is a chair of history of science.[33]

Finally, it should be noted that the history of science is commonly taught as graduate courses at the young national universities created in what is called Greater Buenos Aires. These include those of San Martin, Tres de Febrero, Quilmes, and General Sarmiento. All of them have chairs in the History of Science, usually associated with courses in Scientific Policy, Sociology of Science, and Philosophy of Science.

At the National University of Tres de Febrero, César Lorenzano organized a structured two-year graduate program in Epistemology and History of Science (the master's degree program since 1999 and the doctoral program since 2000). The Academic Committee was initially composed of Antonio Castorina, Ricardo Gómez, Gregorio Klimovsky, Enrique Marí, and Félix Schuster. The curriculum includes subjects such as Semantic Views of Science, Epistemology of the History of Science, History of Argentine Science and a thesis seminar. As a culmination, students must students must submit a written master's thesis or doctoral dissertation and defend it orally. At the same university, together with the Grupo Argentino de Historia de la Ciencia, of which he is president, César Lorenzano organises the *Jornadas de Historia de la Ciencia Argentina*. Five events have been held since 2003, each of which has resulted in a publication.[34]

Among the institutes of the Argentine Scientific Society is the Institute of History of Science directed by Norma Sánchez. It promotes the dissemination of the subject and, armed with its excellent library, is a reference point for researchers. The research is published in the *Anales* of the SCA. It is also worth mentioning the figure of Lewis Pyenson, Emeritus Professor in the Department of History at the

[32] https://www.imperial.ac.uk/news/234654/emeritus-professor-mathematics-history-mathematics-eduardo/

[33] For more information, visit the website https://revistas.unc.edu.ar/index.php/afjor/about

[34] Lorenzano, César. *Historias de la ciencia argentina*, I-V, Caseros: EDUNTREF, 2003, 2005, 2007, 2011, 2023.

Western Michigan University. Since his work "Cultural Imperialism and Exact Sciences, German Expansion overseas" (1985), in wich a chapter is dedicated to the studied of Physics and Astronomy in Argentina, he has continued with a remarkable activity in the field of History of Science.

At the National University of San Martín, the Center for Studies in the History of Science and Technology José Babini (CEJB) is located within the School of Humanities. It was established in July 2002 based on the donation of 9000 volumes and the documentary background housed by the Babini Library Association, formed in 1991 from the library and personal archive of José Babini (1897–1984). The purposes of its creation, under the direction of Diego Hurtado, were to promote research in the area of history of science and technology in Argentina and Latin America, achieve its consolidation as an academic discipline, and disseminate its productions.

The *Second Meeting of Philosophy and History of Science of the Southern Cone* was held at the National University of Quilmes in 2000. During the closing ceremony, the Asociación de Filosofía e Historia de la Ciencia del Cono Sur (AFHIC, Association of Philosophy and History of Science of the Southern Cone) was founded.[35] Since 2016, the National University of Quilmes has been home to the Center for the Philosophy and History of Science (CEFHIC), directed by Pablo Lorenzano. In addition to its research lines in philosophy of science, in history of science and in integrated history and philosophy of science, the Center collaborates actively with the Graduate Program in Epistemology and History of Science at the National University of Tres de Febrero. They jointly publish the journal *Metatheoria*,[36] and since 2023 the Graduate Program in Epistemology and History of Science has been co-directed by Pablo Lorenzano, who has been a professor since its inception.

The Natural Sciences Museum of La Plata was founded in 1884. In 1999, it was decided to create a Historical Archive to house the numerous documentation held by the museum. In this archive, besides the tasks implied by its name, various lines of research are carried out in History of Sciences. Some of them adopt a perspective called "bureaucratic history of sciences". Other lines include materia medica, science education, scientific photography, and the commercialization of nature. Irina Podgorny is the director of this archive.

In 1999, the Physics Museum at the University of La Plata was inaugurated under the direction of Cecilia von Reichenbach, who held the position for more than two decades. Currently, Marcela Taylor serves as Scientific Director, and Mariana Santamaría as Executive Director. In addition to exhibiting instruments, the museum conducts various activities, including research in the history of science.[37]

[35] http://www.afhic.com/es/historia-de-afhic/

[36] For more information, visit the website https://metatheoria.unq.edu.ar/index.php/m

[37] http://museo.fisica.unlp.edu.ar/institucional. One can find the history of the first physicists from La Plata (and Argentina) on the pages of the Physics Museum. http://museo.fisica.unlp.edu.ar/investigacion/investigacion-en-historia-de-la-fisica-9781

13.12 Concluding Remarks as Follows

The authors trust the present work is useful for those who are only recently entering the world of history of science research and, also, for those who are simply occasional readers of the subject. It is our most sincere wish that you have found motivation in these lines and, at the same time, grasped the idea that the establishment of this discipline was no easy task. For those specialized in Argentine history of science, you will surely have spotted several gaps in some descriptions. We beg your pardon for this.

Acknowledgement C. G. thanks Prof. Roberto Ferrari for numerous fruitful conversations. Also, C. G. thanks his wife, Carmen J. Gallagher, for her assistance in editing the English manuscript.

Bibliographic Appendix

Several authors have collaborated to outline a historiography of the History of Science in Argentina. We will mention a few of these historians, some have been a help for the writing of this article.

- Lértora Mendoza, Celina A. "Los estudios de historia de la ciencia argentina", *Quipu* 3(1) (1986): 135–147.
- Babini, Nicolás, de Asúa, Miguel. "La historia de la ciencia en Argentina en el último cuarto de siglo", *Llull* 26(56) (2003): 731–738.
- de Asúa, Miguel. "Introducción", in de Asúa, Miguel. *La ciencia en la Argentina, Perspectivas históricas*, Buenos Aires: Centro Editor de América Latina, 1993.

In 2002, the National Academy of History published *Nueva Historia de la Nación Argentina* (Buenos Aires: Editorial Planeta). The chapters of this work dedicated to the history of science and technology cite the following bibliography:

- Lértora Mendoza, Celina A. "La enseñanza elemental y universitaria", Tomo III—*La Argentina en los siglos XVII y XVIII*, pp. 369–402.
- Monserrat, Marcelo. "La ciencia", Tomo VI—*La configuración de la República Independiente (1810–1914)*, pp. 403–427.
- Barrios Medina, Ariel. "Las ciencias biomédicas", Tomo IX—*La Argentina del Siglo XX*, pp. 501–533.
- Galles, Carlos D. "Las ciencias exactas, físicas y naturales", Tomo IX—*La Argentina del Siglo XX*, pp. 535–558.
- Buch, Tomás. "La tecnología", Tomo IX—*La Argentina del Siglo XX*, pp. 559–584.

Carlos D. Galles Graduated in physics from Universidad de Rosario. Magister in History and Philosophy of Science for the Université de Paris I (Panthéon-Sorbonne). Assisted at the seminars of professors Marie-Antoinette Tonnelat and René Taton in Paris. Ex-president of the LASHST. In charge, for a decade, of the History of Science chair at the Faculty of Exact and Natural Sciences of the University of Buenos Aires.

Pablo Lorenzano is a philosopher specialized in history and philosophy of science (HPS), in particular of biology. He is known for his philosophical (synchronic and diachronic) analyses of empirical theories by means of metatheoretical structuralism, and for his historiographical analysis of the origins of genetics. Having a Ph.D. from the Free University of Berlin, he has published the book *Geschichte und Struktur der klassischen Genetik* (Frankfurt am Main: Peter Lang, 1995), and edited *History and Philosophy of Science and Technology* (Lorenzano, Rheinberger, Ortiz & Galles (Eds.), Oxford: EOLSS, 4 Vols., 2010) and *History and Philosophy of the Life Sciences in the South Cone* (Lorenzano, Martins & Regner (eds.), London: College Publications, 2013), among others. He was Chair of the Joint Commission of the International Union of History and Philosophy of Science and Technology. He is a member of the Académie International de Philosophie des Sciences as well as of the International Academy of the History of Science. He is full professor, and director of the Center of Studies in Philosophy and History of Science (CEFHIC), at the National University of Quilmes, besides main researcher of the National Scientific and Technical Research Council (CONICET), from Argentina, and editor-in-chief of the journal *Metatheoria*.

Chapter 14
The Institutionalization of the Social History of Sciences in Brazil

Heloisa Maria Bertol Domingues

Abstract The process of institutionalization of Social History of Sciences in Brazil, which began in the 1980s, was the outcome of a twofold movement: Brazilian scientists and their meetings with Latin American scientists, led by Juan José Saldaña. The meetings and congresses helped to structure the research, disseminate the theoretical principles of the social history of the sciences and the ways of conducting social history. Brazilian scientists organized research, structured an identity that, above all, expanded their work, founding an anti-colonial historiography that was taken beyond the country's borders. Ethnocentrism was a reality. Forty years have passed and the social history of the sciences has been consolidated in the country, with important work on document preservation. Its work is now autonomous and enjoys international recognition.

Keywords Brasil · Social history of sciences · Juan José Saldaña · Institutionalization of sciences · Decolonization

14.1 Introduction

The institutionalization of the Social History of Science in Brazil, which began in the 1980s, was not an isolated event. A combination of external and internal factors and the mobilization of scientists boosted the organization of the field. The process began at the end of the 1970s in São Paulo, when some scientists—Shozo

H. M. B. Domingues (✉)
MAST-Museu de Astronomia e Ciências Afins, Rio de Janeiro, Brazil
e-mail: heloisa@mast.br

M. de la P. Ramos-Lara, L. C. Arboleda (eds.), *History of Science in Latin America: The Construction of an Intellectual Field (20th century)*,
https://doi.org/10.1007/978-3-031-75536-1_14

Motoyama,[1] Simão Mathias,[2] Ruy Gama,[3] Maria Amélia M. Dantes[4]—organized themselves to start a postgraduate course in the history of science at the Faculty of Philosophy, Letters and Human Sciences at USP. They soon began working with Latin American science historians. In 1982, at the international congress of the International Union for the History of Science in Bucharest, a group of Latin American historians, including the Brazilian Ubiratan D'Ambrosio,[5] the Mexican Juan José Saldaña and the Colombian Luis Carlos Arboleda met and planned to change the conditions of the discipline on the continent and bring together professionals and students in the field. It was there that the idea of creating the Latin American Society for the History of Science and Technology began to emerge (Dantes, 2007:113). They were all aware that it was necessary to give historical visibility to the sciences of the countries that had emerged from the colonial political process and were still limited by colonialist practices and representations. It was necessary to break down the barriers of ethnocentrism, which "was a tangible reality".[6]

The debates that began opened up questions about local and global science, at the time understood as center/periphery, which created a paradox in the history of the sciences that was being created. The critique of the center/periphery opposition, which discussed the internationalism of the sciences, contributed to boost local histories of the sciences, as part of the international scientific field. The institutionalization of the social history of the sciences in Brazil was concomitant with the formation of the social history of the sciences in many Latin American countries, such as Mexico, Venezuela, Colombia, Argentina, Chile, Bolivia, Peru, as well as Brazil. At the beginning of the 1980s, Juan José Saldaña visited the University of São Paulo as the first president of the newly created Latin American Society for the

[1] Shozo Motoyama (1940–2021), physicist, professor at USP, founder of the postgraduate course at USP, author of a vast body of work on the history of science in Brazil;

[2] Simão Mathias (1908–1991), outstanding chemist, professor at USP, was director of the Institute of Chemistry and, for his work on behalf of the History of Science, is honorary president of the SBHC;

[3] Ruy Gama, architect, professor at the Faculty of Architecture and Urbanism at USP, was director of the Institute of Brazilian Studies at the same university. He has published on the history of technology in Brazil;

[4] Maria Amélia M. Dantes (1944), a physicist by training, was also the founder of the postgraduate course at USP and supervised a hundred masters and doctors in the history of science. Her historical work is internationally recognized.

[5] Ubiratan D'Ambrosio (1933–2021), Mathematician, Professor at UNICAMP, São Paulo, was a historian of Mathematics, established Ethnomathematics and was one of the founders of the Center for Logic, Epistemology and History of Science, at the same university. https://www.unicamp.br/unicamp/noticias/2021/05/13/ Acess: 05/03/2024

[6] Criticism of ethnocentrism came from a long way back, as seen in the notes of Professor Luiz de Castro Faria, when, in 1953, he referred to the scientific expeditions that travelled through Brazil, particularly the anthropological ones. He stated that the travellers' reports had no problem with fidelity. Their descriptions were faithful to what they saw. However, "ethnocentrism was a tangible reality" and the idea guided ethnographic interpretations (Caderno de Notas, 1953; MAST History of Science Archive).

History of Science and Technology (*Sociedade Latino-Americana de História da Ciência e Tecnologia*—SLHCT) helping to instigate the creation of the Brazilian Society for the History of Science (*Sociedade Brasileira de História das Ciências*—SBHC) and, above all, to disseminate the theoretical principles of the social history of science, which gave identity to the history of science. (Dantes, 2007:113) Postgraduate studies at the University of São Paulo began to train hundreds of science historians, who contributed to paving the way for the field in the country and changing the historiography of the sciences.

In 1979, Shozo Motoyama and Mário Guimarães Ferri organized and published the collection *História das Ciências no Brasil* ("Brazilian History of Science"). The book represented an alternative to traditional publications on the history of science in Brazil, such as another collection organized by Fernando de Azevedo in 1956. For Azevedo, as he stated in the Introduction, the sciences in Brazil were "something for foreigners". In addition, the works emphasized the fact that the sciences in Brazil began with universities in the 1930s, considering only the importance of training specialized personnel for the sciences (Azevedo, 1955). Also in 1979, the book *Formação da Comunidade Científica no Brasil* ("Formation of the Scientific Community in Brazil"), by Simon Schwartzman, was published. Its central axis of discussion was the creation of postgraduate courses in the country, which began in 1968. Schwartzman took as his methodological basis the then brand new sociology of the sciences, but his historical axis was the same as Azevedo's, thus reaffirming the outsider nature of the Brazilian sciences (Schwartzman, 1979). From another angle, Motoyama and Ferri's book offered glimpses of a history of the Brazilian sciences. Among the many important chapters, gathered in two volumes, it highlighted the chapter entitled "Institutes of Scientific Research in Brazil", by Maria Amélia Dantes. This work dealt with the first scientific research institutes to emerge in the country, which was a pioneering subject at the time.

Maria Amélia Dantes noted that the institutionalization of the sciences in Brazil took place at the end of the eighteenth century. But this was not the only pioneering aspect of the work, because as well as dealing lucidly with the process of institutionalization of the sciences in the country, Maria Amélia confronted the deep-rooted historiographical view, making it clear that the sciences in Brazil were not the stuff of foreigners. By showing how important scientific production had been in Brazil since the end of the eighteenth century, it opened a deaf struggle against the historiographical monster of Eurocentrism. These works were the beginning of the social history of science in the country. Maria Amélia Dantes' study of the institutionalization of the sciences in Brazil enhanced knowledge and, in 2001, she organized the collection *Espaços de Ciência no Brasil* ("Science Spaces in Brazil") with a group of ex-doctoral students, further increasing the visibility of Brazilian scientific institutions, some of which are now bicentenaries old. This was the case with the Botanical Garden of Rio de Janeiro (1808), the National Museum of Rio de Janeiro (1818) and the National Observatory (1827) (Dantes, 2001).

However, the intensive process of institutionalizing the history of science in Brazil took place in the 1980s. There was a confluence of the group that had started the movement to institutionalize the field in São Paulo with a group of scientists' representative of the country's scientific community. In 1983, a group of scientists,[7] met at the National Observatory; some of them were founders and members of the first board of the Brazilian Society for the History of Science (SBHC). The group comprised the president of the Brazilian Society for the Advancement of Science (SBPC), the president of the Brazilian Academy of Sciences, the Brazil's representative on UNESCO's International Council of Museums and representatives from USP. The aim of the meeting was to discuss the institutionalization of the history of science in the country. They wrote a letter to ask the president of the CNPq—National Research Council (1951)—for a series of political measures that would allow the country to have a great place for scientific knowledge, "because the nation is convinced that science, through technology, can free us from colonialism" In this group's view, science had to become part of Brazilian thinking, which meant its definitive incorporation into the country's culture. They argued that "the true image of a country is given by the culture of its people, which is why they had to fight for science and for institutions allowing to instrumentalize its memory within of national life".[8]

The draft of the letter to the President of the CNPq, written at that meeting, effectively requested that political measures had to be taken to promote the creation of courses and encourage research into the history of science, as well as facilitating the preservation and protection of Brazilian historical-scientific collections, raising awareness among the scientific community at research institutes and universities, as well as making efforts with the Historical Heritage Agency to have architectural and bibliographic collections and scientific instruments listed as heritage sites. They also proposed that the Astronomy Memory Project, already in progress at the National Observatory, should have its work expanded, enabling the creation of a science museum in the headquarters building of the National Observatory (Domingues, 2012). This idea gave rise 2 years later—in 1985, to MAST—the Museum of Astronomy and Related Sciences (Museu de Astronomia e Ciências Afins). This letter was a key instrument in the process of structuring the institutionalization of the history of science in the country. Of the group of signatories to the letter, only Mario Schenberg had lived in exile. However, he, like the others, had lived and acted politically in the post-World War II period, when the so-called scientific neutrality lost its meaning—although they believed otherwise—and science became a government business. At the same time, in the midst of the Cold War, less

[7] The group was constituted by Carlos Chagas Filho (biochemist, then scientific director of CNPq, who went on to become vice-president of SBHC), Fernanda Camargo Moro (archaeologist, Brazilian representative at ICON-UNESCO), Mario Schenberg (physicist, USP, recently returned from exile and a promoter of the popularization of science), Clodowaldo Pavan (biologist, USP, then President of SBPC), Shozo Motoyama (physicist, USP, prof. of the History of Science, then at CNPq), Simão Mathias Chemist, USP, founder and board member of SBHC)) and Maurício Mattos Peixoto (Mathematician, President of the Brazilian Academy of Sciences).

[8] Letter to the President of CNPq, signed by the members of the meeting, cf. note 02. Archive CF, MAST. Letter published in Domingues 2012.

than the communism/capitalism bipolarity, intellectuals were intensely involved in the nationalist movement, turning against colonialism of all kinds (Domingues, 2012).

At the beginning of the 1980s, the military dictatorship, which had lasted since the 1960s, came to an end. The exiles were granted amnesty. At the same time, the economic collapse of "Brazil, [as] the country of the future"—the dictatorship's *slogan*—was imminent. The sciences had been politically rethought since the early 1970s, when Economic Development and Science and Technology Plans were repeated. Those many development plans seemed to have been born on scorched land, which prompted scientists returning to the country to resort to memory and look to the past for links to the present, in order to pave the way for the country's scientific culture.

The development of technology as a driving force for progress was a government policy and a source of criticism from scientists, because it an imported guideline and, in this sense, vulnerable to colonialism. In the same way, the newly instituted postgraduate courses were channeled into training specialists for the market. Carlos Chagas Filho had spoken out at the CNPq against this policy of encouraging imported technology. For him, without investment in the sciences, the most fruitful technological initiatives would be short-lived and would necessarily lead to an era *of "technological colonialism*". The processes of industrialization would remain largely beyond the comprehension of their users. (Italics from the author, Chagas Filho, 1976; Domingues, 2012).

The question of scientific or technological colonialism was not only a historical issue that had rooted ethnocentrism, but also continued to threaten the structures of scientific production in the country. The historical context seemed to push the group to propose a policy for the institutionalization of the history of science in the country. Its members had a vivid memory of the role of science in the aftermath of World War II and had experienced the problems suffered by the country's scientific community in 1964, with the advent of the military dictatorship. In the 1980s, they were experiencing the end of the dictatorship, the return of exiles to the country and, above all, they were faced with an adverse scientific policy based on imported technology and supported by specialized personnel in postgraduate programs with the aim of entering the market. Former exiles, politically persecuted people, or even students, less affected by the movements of the 1960s, who then found themselves in a totally different country in terms of cultural values and the conditions for producing knowledge. In this environment, the history of science found fertile ground for its institutionalization, as it would give visibility to the existing scientific culture in the country, which had little or no communication with the society, and would include social and political values as elements for interpreting the scientific past.

By the end of World War II, at UNESCO, they had raised the issue of political interference in the sciences and had reached the bottom on social issues of scientific activity, by opening the debate on the project of the Scientific and Cultural History of Mankind (SCHM). In Latin American countries, affected by the aftermath of World War II and shaken by the political crises of the 1960s during the Cold War, it awakened a sense of local identity and opened up discussion on the issue of autonomy in the production of knowledge and scientific colonialism, as it will be shown.

In Brazil, the aggravating factor of the [colonizing] policy of encouraging the import of technology, which ignored the local scientific potential, gave more power to the idea of the historical revaluation of the sciences.

14.2 The First Brazilian Institutions of History of Science

When the meeting of scientists at the National Observatory took place, the discussion about formalizing the institutionalization of the history of science was already in progress. As mentioned above, in 1982, a group of representatives from Latin American countries, including Brazil, proposed the creation of the Latin American Society for the History of Science and Technology (SLAHCT). The following year, the group that had met at the National Observatory took part in the creation of the Brazilian Society for the History of Science (SBHC). Professor Simão Mathias was the first president of the SBHC; Carlos Chagas Filho, the first vice-president and perhaps the greatest promoter of the field. Shozo Motoyama was the first general secretary of the SBHC and would go on to write memorable chapters on the history of science in Brazil.[9] The first board of the SBHC also included Milton Vargas, a professor at USP and author of books and articles on the history of technology in Brazil, and historian Francisco Iglésias, a professor at the Federal University of Minas Gerais and then a member of the International Committee for the History of Science. The creation of SBHC was a decisive step in establishing the intense process of institutionalizing the history of sciences in Brazil.

The Brazilian Society for the History of Science has always been representative of the field. In 1985, the first issue of the SBHC's journal, Revista Brasileira de História das Ciências (RBHC), began to circulate, and today it ranks among the top publications in the field. The first SBHC congress was held at MAST—the Museum of Astronomy and Related Sciences—the year after it was created, which established an active and lasting relationship between the two institutions, with MAST, now being their official headquarters. By looking the composition of the society, at that first congress, we may say it was a portrait of the Society that would remain for a long time[6]. With its uninterrupted activities, during all these years the SBHC has maintained the convergence of the field. On its 40th anniversary, the SBHC is holding its 19th National Seminar on the History of Science and Technology, with hundreds of papers expected to be presented. The Society, now affiliated to ICSU (International Unions of the International Council for Science) has 426 members.

MAST, the first institution for the history of science in the country, was created as a research institute of the CNPq, focused on the history of science, education, scientific dissemination and the preservation of documentary collections, and specialized in the safekeeping, conservation and preservation of collections of

[9] Ferri and Motoyama, 1979, 2 vols.; Motoyama, 2004, 2010 (Org.), 2006 (Org.); Motoyama and Nagamini, 2014; etc.

instruments and documentary collections of science and technology, with an emphasis on private archives of scientists and institutional documentary funds. Today, in addition to dozens of scientists' archives, it holds the CNPq archive (1951–1973) and the archives of the Council for the Supervision of Artistic and Scientific Expeditions—which is an archive recognized by UNESCO Memory of the World Programme. The MAST collection includes a set of astronomical observation domes that have been nominated by the International Astronomical Union (IAU) as a candidate for World Heritage status. The institution has maintained the same organization and initial objectives and lately, has created postgraduate courses in collaboration with similar institutions in the areas that concern it: History of Science, Science Communication, Museology.[10] A Master's Degree in the Preservation of Scientific and Technological Collections was established at MAST in 2013. Today, it has published numerous research papers on the history of Brazilian science, which have contributed greatly to institutionalize the social history of science in the country.

Another step towards the institutionalization of the area was taken in 1986, with the creation of the Casa de Oswaldo Cruz (COC—FIOCRUZ), dedicated to the history, dissemination and scientific education, as well as the preservation of the architectural, bibliographic and documentary memory of health. The COC is part of Fiocruz, the largest public health institution in Brazil, and was created with the support and encouragement of Carlos Chagas Filho, thus fitting in with the policy proposal sent to the President of the CNPq a few years earlier. The COC has grown a lot over the years, both in terms of historical research and scientific dissemination. In 1994, the COC launched the journal *História, Ciências, Saúde Manguinhos*, which soon reached the top international levels of publications in the field. In 2000, it set up a postgraduate course in the History of Science, where it has trained numerous specialists in the field. In May 1999, it inaugurated the Museum of Life, with intense activity in the dissemination and teaching of science, and also created a sector for the preservation of institutional memory, together with a documentary archive. It has made a major contribution to the development of scientific culture in Brazil and to the knowledge of the history of science.

As masters and doctors were being formed in the area, from the first postgraduate courses, which received students from all over the country, courses and research centers in the History of Science were established in several states, and research institutes and universities in the country, with emphasis on the groups at the Federal University of Bahia, the Federal University of ABC Paulista, the Federal University of Santa Catarina, the Federal University of Rio de Janeiro, the Pontifical Catholic University of São Paulo—PUC -, etc. The community of historians of science that has been formed since then has come to the present days, contributing to the understanding that cultural diversity is a fundamental component of the process of scientific production. At the same time, it has contributed to redrawing the historiographical

[10] MAST's postgraduate programs were established in History of Science, linked to UNIRIO's Postgraduate Program in History, Science Communication, linked to COC-Fiocruz, and Museology, created with UNIRIO as the first postgraduate program in the area, in Brasil.

map of the sciences to include not only Brazil, but also Latin America, Asia and the East (Gavroglu & Renn, 2007; Domingues, 2012).

14.3 Historiography: A Theory for a Social History of Science

The institutionalization of the social history of science in Latin America in the 1980s represented a real historiographical revolution and a radical change in the cultural vision of science in the region. The transformations in international scientific culture began to be felt in the immediate aftermath of World War II, due to the negative outcome of the war—the atomic bomb—which raised big questions about the sciences.

The creation of UNESCO, within the United Nations Organization (UNO), was the spearhead of projects that would enhance world peace and, instead of ethnocentrism, sought to create a framework of values and knowledge common to all peoples and individuals, in the certainty that this would allow for mutual, international understanding and harmony in all societies (Petitjean & Domingues, 2004:10). The basis they sought was expressed in projects, one of which, the Scientific and Cultural History of Mankind (SCHM), was considered a priority by the first Director General, Julian Huxley, a British evolutionary biologist.

The UNESCO project aimed to draw the mosaic of humanity's cultural diversity, on the understanding that not wars but cooperation between peoples had been the driving force behind humanity's development and progress. The project was directed towards the study of the relationship between science and society, encouraged by Joseph Needham, a biochemist who was appointed head of UNESCO's Natural Sciences Section, and Lucien Febvre, the French well-known historian, who joined the project and was its rapporteur. Unesco planned to replace the old history textbooks that aroused nationalism and wars by a history in which the sciences, as culture, would be the axis of exchanges and interdependence between different human societies (Petitjean & Domingues, 2007:5).

For Febvre, history accumulates and interprets facts, but needs theory to interpret them. The history of mankind, proposed by UNESCO, gave him the ideal opportunity to apply his thoughts on the history of the sciences, the diversity and evolution of civilizations. Febvre had the support of Joseph Needham, and of the first director-general himself, Julian Huxley. In this project, Febvre reasserted his idea of the plurality of civilizations, since any group of human beings, whatever be their means of material and intellectual action on the universe, has its own civilization. He opposed the concept of the plurality of civilizations to what he called "the old conception of a superior civilization, conveyed by the white peoples of western Europe and northern America, which has been incorporated into the facts, as a kind of ideal" (Febvre, 1956:429). He therefore demonstrated his opposition to Eurocentrism, expressing the objectives of UNESCO's SCHM project. For Febvre, "history was the science of the agreement that is negotiated, of that harmony that is established perpetually and spontaneously, in every age, between the diverse and synchronous conditions of human existence; material conditions, technical conditions, spiritual conditions. This is where history meets life" (Febvre, 1992:31–32; Petitjean & Domingues, 2007:9).

However, that moment was not just the end of a war that had decimated lives and violated cultures. It was a moment that held strong nationalist sentiments from the past. In opposition to Febvre's project, there was a project by the Brazilian Miguel Osório de Almeida, also supported by UNESCO. Osório de Almeida proposed a factual history of civilizations, centred on the idea of a history of the progressive development of humanity, in the field of techniques and sciences, culture, spirit and intelligence. He referred to a "civilizing impulse", thanks to which "civilizations die out or are replaced by civilizations that are nascent or in full bloom" (Almeida, 1953:964). He dismissed Febvre's idea that the basis of all civilizations is international cooperation and stated the opposite, that it was necessary to let the facts speak for themselves (Almeida, 1953:964). He ruled out Febvre's idea that the basis of all civilizations is international cooperation and stated the opposite: the need to let the facts speak for themselves (Almeida, 1953: 971). The starting point would be the passage from the primitive state to the civilized state, when history began. Miguel Osório's project, despite having found some support, did not have a huge impact, as well as Febvre's project was also blocked by the discussions that followed at UNESCO, attended by representatives of the member countries.

The consensus for the project of the Scientific and Cultural History of Humanity in the 1950s was a seamless collection of many national histories of local civilizations. In Europe, the years following the war also saw the publication of many general histories of civilization which, like Barker's book, reaffirmed European ethnocentrism.[11] Lucien Febvre, then producing Les Annales, concluded that this historiography "did not consider the rhythm and vision of each civilization. It didn't take into account the awakening of non-European mentalities, in the sense of time, the estimation of values, the appreciation of magnitudes, they would never be the same as those of Westerners." (Febvre, 1954:3; Petitjean & Domingues, 2007:46).

An exception to that criticized historiography came from Samuel Lilley in 1953, in a discussion at the International Union for the History of Science's commission on the Social Relations of Science. The debate on factors internal and external to science sounded fictitious to Lilley because, in his Marxist view, they were complementary factors and the history of science should study their interactions (Lilley, 1953:58–60; Petitjean & Domingues, 2007:45).

[11] Sir Ernest Barker et al., *The European Inheritance*, 3 vols., Oxford, Clarendon Press, 1954; John D. Bernal, *Science in History*, 4 vols., London, Penguin Books, 1954. Charles Singer et al., *A History of Technology*, 8 vols., Oxford, Clarendon Press, 1954–1984. Herbert Butterfield, *The Origins of Modern Sciences*. London, Bell, 1949. Joseph Needham et al., Science and Civilization in China, Cambridge University Press, 1954-. René Taton, *Histoire Générale des Sciences*. Paris, PUF, 1957.

In the 1980s, times were different, but both the theoretical debates and questions raised for historical interpretation maintained a certain continuity. In the post-World War II period, when the so-called scientific neutrality was losing its meaning—even though people believed otherwise—science became a government business. Carlos Chagas Filho repeatedly expressed his anti-colonialism and, in the 1940s, took part in the political movements that resulted in new scientific institutions. This included the National Research Council (currently the National Council for Scientific and Technological Development) in 1951, of which he was one of the creators.

The process of institutionalizing the social history of the sciences in Brazil, more than integrating Brazilian and Latin American historians of science, founded a theoretical-methodological identity, changing the view of the history of science in the country, making it possible to understand the paths of scientific colonialism and, thus, thinking about designing a future free from colonialisms. It revolutionized the sources of research in the history of science; it began to integrate trained historians into its ranks; it brought new and pertinent questions to scientific practice, socializing it.

14.4 The Social History of Science in Brazil

There is no doubt that the history of sciences, as it has been developed in Brazil, has shown the social place of the sciences in the country. At USP, in the early 1990s, the first specialists in the field were graduated. In this course, especially in Maria Amélia Dantes' classes, the Latin American methodological relationships established a few years earlier were felt. Juan José Saldaña's works were required reading in these classes. The journal Quipu, which he created in 1984, was a landmark of the Latin American Society for the History of Science and a wellspring of the theoretical ideas that guided the thinking of the history of science at that time. The journal ran until 2000, but the seeds it planted grew and multiplied. A text by Saldaña published in Quipu in 1986, became a classic of those courses at USP. In it, one sentence summed up the criticism of ethnocentric historiography, previously predominant in Latin America, which was well suited to the new Latin American historians of science to deconstruct and decolonize. Saldaña stated that the history that was made in American countries, "*instead of historicizing the scientific practice of Latin America, sought rather to historicize European science in Latin America*" (Saldaña, 1986:55). He wisely proposed an inversion in the interpretation of the historical role of countries, drawing attention to local practices and their conditions of production in specific political, economic and social contexts.

In 1992, a major congress on the history of science was held in Paris, entitled Sciences et Empires (*Science and Empires)*, under the coordination of Patrick Petitjean, who would soon be in Brazil collaborating on programs at USP. Coordinating the congress with him were Catherine Jami and Anne Marie Moulin, all researchers linked to France's National Council for Scientific Research (CNRS) (Petitjean et al., 1992:4). Dozens of historians of science gathered there to discuss a crucial aspect of international relations in the sciences and the European spread in the world: the question of scientific imperialism. Brazil and other Latin American countries had an active participation. There were also taking part historians of science from Europe, Asia, India, Japan and China.

On the panel entitled Latin American Countries, Maria Amelia Dantes was alongside Juan Jose Saldaña. They were joined by one of Dantes' doctoral student, Ma. Margaret Lopes, Diana Obregón from Colombia, Patricia Aceves from Mexico, Michel Paty from France and Nancy Stepan from the United States, who addressed about eugenics based on Pan-American experiments. None of these works failed to draw attention to the fact that the sciences are international, circulate and are adapted to the peculiarities of each place, taking on local forms (Domingues, 1995:10).

The papers generally drew attention to the relationship between science and local government politics. Juan José Saldaña presented a comparative study between France and Mexico in the nineteenth century, concluding that although the motivations of scientists were similar, scientific activities in Mexico were structured differently to the French model and the results were also different (Saldaña, 1986:153; Domingues, 1995:15). According to Saldaña, the emergence of scientific institutions in Mexico at the end of the nineteenth century was the result of a conscious national policy that generated original scientific work, embedded in a particular culture and forms of organization of funding and social demand, which proved decisive for innovations and the achievement of results.

Likewise, Maria Amélia Dantes referred to the issue of positivism in Brazil, stating that although it cannot be neglected, its influence on scientific or philosophical thought should not be exaggerated, because, just as positivism found many followers in Brazil, it suffered resistance, mainly due to the atheism of certain works, criticized by Catholics (Dantes, 1980:170). Maria Amélia, considering that many positivist works were accepted at the Faculty of Medicine in Rio de Janeiro as legitimate scientific works, concluded that there was no epistemological resistance to positivism. She showed that the dissemination of positivist ideas was original and was changing in the country. History began to be studied through the prism of positivism and the past, as well as the future, were seen as relevant to the western history civilization. In 1881, with the establishment of the positivist church in Brazil, the history of positivism was characterized by a duality of orthodoxy and heterodoxy. According to Maria Amélia, "the followers of the Apostolate were active and participated in the movement for the establishment of the Republic in the country and in public debates on education, the university, slavery and, at the beginning of the 20th century, on vaccination". Although historians of science criticized positivism, Maria Amélia echoed Luís Otávio Ferreira's assertion that positivism was a living part of the history of science in Brazil (Dantes, 1980:171).[12]

[12] Luis Otávio Ferreira defended this idea in his master's thesis entitled *Os Politécnicos: ciência e organização social segundo o pensamento positivista da Escola Politécnica do Rio de Janeiro* (1862–1922), UFRJ, 1988. Cf. He had previously defended the idea that engineers relied on positivism to gain a foothold in the state apparatus.

Although for Latin Americans it may have seemed unquestionable that the sciences were made through exchanges or cultural crossings, the conclusion of the congress organizers was that only two sessions, out of the nine that were held, approached the subject from this perspective: the one on Latin American countries and the one on Asia. The other papers focused mainly on particular disciplines—botany, medicine, technology—and on the work of expeditions and missions, discussing various political and institutional aspects of Europe's colonial and imperial scientific expansion (Petitjean et al., 1992: XI).

During the 1990s, Brazil produced a critical mass of works that showed the strength of the sciences in Brazil since the nineteenth century, in contradiction to what the historiography of science of the mid-twentieth century or Eurocentric historiography still claimed. Silvia Figueirôa, in 1998, in an article discussing the institutionalization of the sciences in Brazil, showed that the first studies in this area, carried out at USP's Graduate School, revealed the history of the development of scientific institutions and practices, particularly in the natural sciences, which were prioritized in Brazil because of their political and economic implications—a sign of the predominance of local incentives over international scientific production. The works showed that sciences were "something for Brazilians" or, perhaps, foreigners, hired by the imperial government and/or provincial governments. It was no longer possible to talk about the backwardness of the sciences in the country, nor to say that they didn't exist or that they were the work of foreigners (Figueirôa, 1998:109). When we look at the subjects of the works studied, the predominance of mining sciences and agricultural sciences, such as botany, zoology—allowed us to get to know the hinterland of the country—or ethnography, which also allowed us to communicate with indigenous peoples, stands out. The works not only showed the intense institutional and political-economic activities of the sciences, but also the engagement of the country's scientific production in the international scientific field, both through publications and participation in international congresses, as highlighted by Figueirôa.[13]

In those early years of the decade, the journal Quipu was a uniting element for Latin American historians of science, and a vehicle for disseminating the history of the work that was being produced. The participation in Quipu of Brazilians who were studying and disseminating the social history of science in the country was intense and representative of the work being done. Brazil was the second country that published the most in Quipu, with 45 articles, after Mexico, which was the journal's headquarters and published 76 articles. The third country that published the most was Colombia, with 20 articles (Silva, 2014: 54). Brazil's extensive participation in Quipu was an evidence of the potential growth of the field. Likewise, Brazilian historians of science constantly attended SLAHCT congresses, and in 1998 it was held in Rio de Janeiro, with participants from several Latin American countries: Argentina, Peru, Bolivia, Colombia, Ecuador, Venezuela, Cuba and

[13] Some of the works listed in Figueirôa, 1992, 1998, Dantes, 1980; Lopes, 1993; Domingues, 1995; Fonseca and Prado, 1996. In addition to these, we should highlight some later works along the same lines: da Motta Barboza, 2002; Marinho, 2008; Ribeiro de Andrade, 1996; Almeida, 2004.

Mexico—the latter took a large delegation, coordinated by Juan José Saldaña (Dantes, 2007).

The historiography of the sciences developed in the 1990s, which covered the Brazilian colonial period and the nineteenth century, showed the dominance of studies on the natural environment. The works highlighted the development of botany, zoology or geology, closely associated with the politics of the countries, but did not ignore the fact that there was a strong imperialist onslaught from European countries, whose aim was to economically exploit the colonies by recognizing their flora and fauna. However, the papers emphasized that the sciences took place on a two-way street, where local scientists played a role that could in no way be neglected, as Juan José Saldaña and Deepak Kumar had rightly pointed out at the same 1992 congress in Paris. Kumar, referring to the relationship between scientists in India and the British in the metropolis, noted that the latter, who had previously been reticent about the work of local scientists, ended up recognizing them and integrating them into their cadres (Kumar, 1992:163; 269).

14.5 Conclusion

In the 1980s, the institutionalization of the social history of the sciences took place in Brazil amid the political context of re-democratization and a serious economic crisis, crossed by the bipolarity of the Cold War and dominated by the ethnocentric and colonialist image of backwardness. Social history was the instrument to combat not only this image, but also political interventionism in the practice of the sciences, which had been going on since at least the World War II. The movement for the institutionalization of the social history of science emerged from the scientific community, based on two issues: the autonomy of research versus scientific ethnocentrism and the connection between science and society. The same issues also mobilized scientists from other Latin American countries.

The meeting between Latin Americans gave impetus to the process of institutionalizing the social history of science in the different countries. It was led by Juan José Saldaña, who not only brought the different groups together, but also formulated the foundations of a theoretical thinking that laid bare ethnocentrism, thus providing the key to understanding the scientific culture shaped in each country.

Nevertheless, the effective institutionalization of the social history of the sciences in Brazil, which was certainly no different from other countries, only came about through the mobilization of local scientists. In Brazil, the institutionalization of the field became a reality thanks to a group of scholars who were aware of the social issues surrounding the development of the sciences in the country. The success of the process of consolidating the institutionalization of the social history of science was due to the persistence of those who created the institutions, those who implemented the postgraduate courses with all the theoretical and methodological strictness that the new area required, and those who maintained and extended the international relations of the field of the social history of science in Brazil. They educated hundreds of specialists and multiplied the capacity for research and staff formation in the area. They provided support and justification for the creation of institutions to preserve collections of science and technology, expanding the communication with the public through both research and the ability to spread science. The cohesion of the field has been maintained by the uninterrupted action of the Brazilian Society for the History of Science and Technology. In other words, the autonomy and persistent action of historians and professionals in the field of the history of the sciences, who trained and worked over the years, ensured the success and continuity of the institutionalization of the social history of the sciences in Brazil. The same can be said of other Latin American countries, confirming Juan José Saldaña's suggestion that breaking the power of scientific colonialism would come from rigorous and serious social histories of science in our countries.

Bibliography

Almeida, M. O. D. (1953). Rapport sur l'histoire scientifique et culturelle de l'humanité. *Cahiers d'Histoire Mondiale*, Neuchâtel, *1*(1), 962–986.

Almeida, M. O. D. (2004). Da Cordilheira dos Andes a Isla de Cuba, passando pelo Brasil: os congressos médicos latino americanos e brasileiros (1888–1929). São Paulo, Fac. Filos., Letras e Ciências Humanas/Universidade de São Paulo (USP), Tese de Doutoramento.

Andrade, A. M. R. D. (1996). *Físicos, mésons e política: a dinâmica da ciência na sociedade.* Universidade Federal Fluminense (UFF), Tese de Doutoramento.

Azevedo, F. (1955). As Ciências no Brasil. Companhia Editora Nacional, 2.

Chagas Filho, C. (1976). Sobre o reforço técnico e administrativo das instituições brasileiras de ciência e tecnologia. Rio de Janeiro. Fundo CNPq; T.4.8.013 (Arquivo de História da Ciência, MAST).

da Motta Barboza, C. H. (2002). *Tempo bom, meteoros no fim do período; uma história da meteorologia em meados do século XIX através das obras de Emmanuel Liais.* FFLCH/USP, Tese de Doutoramento.

da Silva, M. R. Barros. (2014). História e Historiografia Latino-Americana –Quipu 1984–2000. *Revista Brasileira de História da Ciência*, Rio de Janeiro, v. 7, n. 1, p. 47–57, jan | jun.

Dantes, M. A. M. (1980). Institutos de pesquisa científica no Brasil. *En:* Ferri, M. G.; Motoyama, S. (coords.) *(1979–81), História das ciências no Brasil, São Paulo: EPU: Ed. USP, v.2.*

Dantes, M. A. M. (2001). *Espaços de Ciência no Brasil.* Editora Fiocruz.

Dantes, M. A. M. (2007). Integrando o Brasil à América Latina. Um movimento da historiografia dos anos 1980, Org. Ribeiro, Ana Ma. Ribeiro de, *Caminho para as estrelas – Reflexões em um Museu.* Rio de Janeiro, MAST, p. 112–125.

Domingues, H. M. B. (1995). *Ciência: um caso de Política. As relações entre as Ciências Naturais e a Agricultura no Brasil-Império,* São Paulo, Fac. Filos., Letras e Ciências Humanas/ Universidade de São Paulo (USP), tese de doutoramento.

Domingues, H. M. B. (2012). Carlos Chagas Filho: um articulador da História das Ciências no Brasil, Revista *História, Ciência, Saúde, Manguinhos*, https://doi.org/10.1590/S0104-59702012005000001.

Febvre, L. (1954). Sur une nouvelle collection d'histoire, *Annales ESC*, vol. IX, n. 1, janvier, mars, p. 1–6.

Febvre, L. (1956). Civilisation: évolution d'un mot et d'un groupe d'idées, Pour une histoire à part entière. École Pratiques des Hautes Études, SEVPEN, p.482. 1ª edição 1929.

Febvre, L. (1992). Combats pour l'Histoire. Paris, Armand Colin, pp. 20–21. La conférence "Vivre l'histoire. Propos d'initiation" a été prononcée à l'École Normale Supérieure, à l'automne 1941.
Figueirôa, S. S. D. M. (1992). *Ciência na busca do Eldorado: A institucionalização das ciências geológicas no Brasil, 1808–1907*, São Paulo, Fac. Filos., Letras e Ciências Humanas/USP, tese de doutoramento.
Figueirôa, S. F. M. (1998). Mundialização das Ciências e Respostas Locais: Sobre a Institucionalização das Ciências Naturais no Brasil (fins do século XVIII á transição ao século XX), *Asclépio*, Vol. L, 2, p. 107–123.
Fonseca, M. R. D. G., & Prado, M. L. C. (1996). A única ciência é a Pátria: o discurso científico na construção do Brasil e do México (1770–1815). São Paulo, Fac. Filos., Letras e Ciências Humanas/USP, tese de doutoramento.
Gavroglu, K., & Renn, J. (2007). *Positioning the history of science*. Springer.
Kumar, D. (1992). Problems in science administration: A study of the scientific surveys in British India – 1757-1900. In Patrick Petitjean, Catherine Jami e Anne Marie Moulin (Orgs.), *Sciences and empires: Historical studies about scientific development and Europe expansion* (pp. 269–280). Kluwer Academic Publishers..
Lilley, S. (1953). Causes and effects in the history of science. *Centaurus, 3*, 58–72.
Lopes, M. M. (1993). *As ciências naturais e os museus no Brasil no século XIX*. Fac. Filos., Letras e Ciências Humanas/USP, tese de doutorado.
Marinho, P. (2008). *Ampliando o Estado Imperial: Os engenheiros e a organização da cultura no Brasil oitocentista, 1874–1888*. Universidade Federal Fluminense (UFF), Tese de Doutoramento.
Motoyama, S. (2004). *Prelúdio para uma História – Ciência e Tecnologia no Brasil.* São Paulo, EDUSP, p. 520.
Motoyama, S. (2006). *USP 70 anos, imagens de uma história vivida.* (Org.), São Paulo, EDUSP, p. 704.
Motoyama, S. (2010). *Tecnologia e Industrialização no Brasil* (Org.), São Paulo, UNESP, p. 451.
Motoyama, S., & Nagamini, M. (2014). *A engenharia mecânica na Escola Politécnica da USP e suas contribuições para a sociedade brasileira*, São Paulo, EDUSP, p. 208.
Petitjean, P., & Domingues, H. M. B. (2004). International science, Brazil and diplomacy in Unesco (1946-1950). Science, technology & society. *An International Journal Devoted to the Developing World, 9*(1), 29–50.
Petitjean, P., & Domingues, H. M. B. (2007). *1947–1950: quand l'Unesco a cherché à se démarquer des histoires européocentristes : Le projet d'une Histoire scientifique et culturelle de l'humanité*. halshs-00166355f. Acesso-25/03/2024. https://shs.hal.science/halshs-00166355/file/Unesco_et_histoire_de_l_humanite_juillet07.pdf
Petitjean, P., Jami, C., & Moulin, A. M. (1992). *Sciences and empires: Historical studies about scientific development and Europe expansion*. Kluwer Academic Publishers.
Saldaña, J. J. (1986). Marcos conceituais da história das ciências latino americana, El Perfil de la Ciencia en America. Org. Saldaña, J.J., XI Congresso Interamericano de Filosofia. *Cadernos Quipu*, p. 55, Guadalajara, México.
Schwartzman, S. (1979). *Formação da Comunidade Científica no Brasil*. São Paulo.

Heloisa Maria Bertol Domingues Historian of Sciences, Senior Researcher at the Astronomy Museum (MAST), and Director between December 2012 and February 2018. Professor of the Postgraduate Program in History—PPGH-UNIRIO, and the Postgraduate Program in Scientific and Technological Collections—PPACT-MAST. She works on the history of natural sciences and anthropology in Brazil and the history of scientific theories, especially about Darwinism in Brazil. Published several books and articles.

Chapter 15
The Institutionalization of the History of Science and Technology in Mexico

María de la Paz Ramos-Lara

With admiration for an inspiring teacher. Juan José Saldaña. In memoriam.

Abstract The institutionalization of the History of Science in Mexico began in 1964 when Enrique Beltrán established the Mexican Society for the History of Science and Technology (MSHST). It experienced stagnation from 1979 until Juan José Saldaña González became president in 1988. Under his leadership, the society was efficiently reactivated, transforming it from an amateur organization to a professional one. In 1989, he incorporated MSHST as a member of IUHPST/DHST, putting Mexico on the international map very soon. Saldaña was the first Mexican Philosophy and History of Science professional (from France) that promoted this discipline in Mexico, Latin America, and internationally. He always highlighted that the dynamics of Latin American scientific practice deserve recognition in their own right. He was the first president of the Latin American Society for the History of Science and Technology (1982) and the sole editor of both the journals *Quipu* (1984) and *Cuadernos de Quipu* (1986). He professionalized the field at UNAM in 1985. He participated actively in international commissions (UNESCO, IUHPST/DHST, SHOT, ICOHTEC, among others) and founded IASCUD (Mexico, 2000). He made history as the first to host the international congress in a Latin American country (Mexico City, 2001) and also the first Latin American to chair the Secretary General of the IUHPST/DHST from 2001–2005. Actually, the History of Science and Technology is now a well-established and prosperous field in Mexico and Latin America.

M. de la P. Ramos-Lara (✉)
Centro de Investigaciones Interdisciplinarias en Ciencias y Humanidades, Universidad Nacional Autónoma de México, Mexico City, Mexico
e-mail: ramoslm@unam.mx

M. de la P. Ramos-Lara, L. C. Arboleda (eds.), *History of Science in Latin America: The Construction of an Intellectual Field (20th century)*,
https://doi.org/10.1007/978-3-031-75536-1_15

Keywords Mexican Society for the History of science and technology (MSHST) · Juan José Saldaña González · *Quipu* journal · *Cuadernos de Quipu* · LASHST · XXI ICHS · IUHPST/DHST · *Ceromaya* list

15.1 Establishment of the Mexican Society for the History of Science and Technology

The History of Science in Mexico became institutionalized in the twentieth century. The first association in this field was founded by a Mexican biologist and historian, Enrique Beltrán Castillo (1903–1994).[1] Beltrán is considered the first biologist trained in Mexico at what was then called the Facultad Nacional de Altos Estudios (1920–1923), which is today the Facultad de Filosofía y Letras (FFyL) of the Universidad Nacional Autónoma de México (UNAM). Beltrán was an academic who worked in various fields, including protozoology, natural resources, conservation, history of science, sociology, and education. He was a professor at UNAM, the Instituto Politécnico Nacional, and the Escuela Normal Superior. He worked on reviving old Mexican scientific societies and institutions, such as the Sociedad Mexicana de Historia Natural. He designed the Marine Biology Station and the Oceanographic Museum in Veracruz. In 1952, he established and managed the Instituto Mexicano de Recursos Renovables (IMRR), the first civil organization in Mexico dedicated to conservation (Cifuentes, 1992).[2]

Beltrán faced difficulties convening the academic community working in different educational institutions in Mexico City during the 1960s to participate in events on the history of science and technology in Mexico, as there were no professionals yet available. During those years, academics were either scientists or humanists entirely dedicated to their profession and closely linked to their work. Some scholars had written excellent articles or books about the history of science and technology in Mexico. However, their efforts went unnoticed due to the lack of an organization or community to share their work and promote the field. As a result,

[1] I had the pleasure of meeting Dr. Beltrán in the 1990s. At the time, I was finishing my Master's Degree in Sciences (Astrophysics) at the Faculty of Science and beginning my PhD in History at the Faculty of Philosophy and Letters, both of which are part of UNAM. I was awarded the first Dr. Enrique Beltrán Prize on August 31, 1990, for the research work that I completed to obtain my Master's degree. The award was established and granted by the MSHST. This society, together with the University of Puebla, published the laureate work as a book in 1994 titled *Difusión e Institucionalización de la Mecánica Newtoniana en México en el siglo XVIII*, just the year of Beltran's death (Ramos-Lara, 1994). Receiving the award from Dr. Beltran himself was an honor for me. The ceremony was held at the closing of the II Mexican Congress of History of Science and Technology (MCHST) performed in Mexico City in 1990. UNAM Rector José Sarukhán, Eli de Gortari, Patricia Aceves, and Juan José Saldaña (at that time president of the MSHST) were seated at the table of honor.

[2] https://www.biodiversidad.gob.mx/biodiversidad/curiosos/enrique-beltran-castillo (accessed November 2023).

their work remained isolated, with limited institutional and social recognition. Beltrán's first calls had low attendance, but in 1963, personalized invitations sent by himself and his colleagues made the event a success (Beltrán, 1964).

From September 2 to 7, 1963, the First Colloquium on the History of Science in Mexico was held at the Unidad de Congresos of Centro Médico Nacional.[3] This event was dedicated to "The Enlightenment" due to its influence on Hispanic and Latin American culture as part of the discussion on the history of science in Latin America (Beltran, 1964).[4] During the meeting, a proposal to establish the Sociedad Mexicana de Historia de la Ciencia y la Tecnología (SMHCT) was made (Mexican Society for the History of Science and Technology in English—MSHST, hereinafter). This society was established on August 11, 1964, with Enrique Beltrán as its leader. 1969, Beltrán started a specialized publication called *Anales de la Sociedad Mexicana de Historia de la Ciencia y la Tecnología* (Society *Annals*) under MSHST.

The birth of this society marked a new stage in the history of science because of the possibility of implementing group organization strategies to promote this field. Periodic meetings were held in the 1960s to present research progress or results. The proceedings were published in five volumes of the *Anales* in 1969, 1970, 1972, 1974, and 1979. The participation of society members in publishing the *Anales* decreased over time, resulting in immobilization, as evidenced by the spacing between the years of publication (1, 2, and 5 years).

Completing the *Anales* and *Memorias del Primer Coloquio* (First Colloquium Proceedings) was a difficult task due to a lack of commitment from the participants. Most papers had to be transcribed because the authors did not send them as expected. Some of them failed to respond when they were called for review. The irregularity of the *Anales* publication shows the difficulties in convening the meetings. In the last issue, where ten papers were submitted, four were authored by Beltrán. It is worth mentioning that the Consejo Nacional de Ciencia y Tecnología (Conacyt)[5] sponsored *Anales* issue 4 (1974) with the valuable assistance of UNAM. This fact is

[3] Academics from the following institutions participated: UNAM (Centro de Estudios Filosóficos, FFyL, Instituto de Investigaciones Económicas, Instituto de Biología, Escuela Nacional de Medicina, Instituto de Investigaciones Estéticas and Escuela Nacional de Ciencias Políticas y Sociales), El Colegio de México, Instituto Nacional de Antropología e Historia (INAH), Escuela Nacional de Ciencias Biológicas (ENCB) del Instituto Politécnico Nacional (IPN), Instituto Panamericano de Geografía e Historia (IPGH), Comisión Forestal del Estado de Michoacán, IMRR, Instituto de Salubridad y Enfermedades Tropicales, Academia Nacional de Estomatología, Academia Mexicana de Historia, Comité Panamericano de Ciencias Geofísicas, Sociedad Mexicana de Historia de la Medicina, Jardín Botánico "José Celestino Mutis" (Bogotá, Colombia), Secretaría de Educación Pública, Archivo General de la Nación, Sociedad Mexicana de Geografía y Estadística, Biblioteca de la Secretaría de Hacienda y Crédito Público, and Asociación Mexicana de la Historiadores (Beltrán, 1964).

[4] Beltrán received assistance from various individuals in organizing the event, among whom were the renowned Spanish chemistry educator and historian Modesto Bargalló (1894–1981) and the Mexican historian and journalist Arturo Arnáiz y Freg (1915–1980), who was a member of the Mexican Academy of History.

[5] Conacyt is now called Consejo Nacional de Humanidades, Ciencias y Tecnologías (Conahcyt). National Council of Humanities, Sciences, and Technologies in English.

relevant because Conacyt was founded in 1971 to promote the development of science and technology in Mexico and was willing to support the history of science, among many other fields. It is heartening to note that Conacyt was crucial in supporting the journal *Quipu* and extended its sponsorship to numerous events organized by MSHST.

Starting in 1970, at Beltran's age of 67, the *Anales* began publishing the papers of Roberto Moreno de los Arcos (1943–1996). He was a young and outstanding Mexican bibliographer and historian who devoted himself full-time to research on Novo-Hispanic science and technology, as well as the pre-Hispanic history of Mexico. It's worth noting that, unlike most members of the society, Moreno de los Arcos was a full-time researcher in his field. He completed his master's and doctorate studies in History at UNAM and joined the Instituto de Investigaciones Bibliográficas of UNAM as a full-time worker in 1967 (Guzmán Pérez, 1996).[6] He was one of the earliest historians who considered science a pivotal part of Mexico's past (Azuela, 1997). Beltrán served as president of the society until 1979. He was succeeded by Moreno de los Arcos, who also became the director of the Instituto de Investigaciones Históricas at UNAM. Due to his administrative responsibilities and academic research, Moreno de los Arcos was unable to promote activities for MSHST, resulting in its paralysis in the early 1980s.

Beltrán, recognizing the lack of activity in the society, endorsed the nomination of Juan José Saldaña González (1944–2022) for the presidency of the MSHST. Saldaña was the first Mexican to earn a Ph.D. in Philosophy and History of Science from the prestigious University of Paris I (Panthéon-Sorbonne). Beltrán was optimistic about the institutionalization of science in Mexico finally being achieved when Saldaña became president of MSHST in 1988.[7]

Beltran identified specific requirements that must be met to institutionalize a discipline. He described these conditions in 1989:

> In my opinion, a scientific branch must meet certain requirements to be considered institutionalized. First, the field requires a sufficient number of people (critical mass) to hold meetings, form permanent groups, and create affiliations. Second, it must have a permanent journal for publishing the community's work. Third, it must have a specific educational institution that exclusively confers academic degrees, including masters and doctorates, in the field of History of Science. (Beltrán 1989, 84)[8]

After being appointed president of the MSHST, Saldaña presented Beltrán with a long-term plan that included a series of activities and strategies to reactivate the

[6] It is inspiring to learn that Moreno de los Arcos, during his time in the archives, came across a wealth of materials written and published by Novo-Hispanic scientists in the Enlightenment period. His dedication to reproducing and disseminating these materials is a testament to his passion for preserving the history of science in Mexico. We can only imagine the countless hours he spent poring over these documents, driven by a desire to share this valuable knowledge with others. His legacy is a true gift to Mexican and Latin American historians of science.

[7] Saldaña's appointment as president of MSHST is listed as 1987 on his Curriculum Vitae, despite other sources claiming it was in 1988.

[8] Author translated original Spanish text.

society. Saldaña was undoubtedly the best candidate to do it at that time, owing to the following compelling reasons: He presided the Latin American Society for the History of Science and Technology (LASHST) from 1982 to 1988. He founded and managed the journals *Quipu* and *Cuadernos de Quipu*, starting in 1984 and 1986, respectively. He successfully organized multiple events on the history of science across various Latin American countries, the United States, France, and Spain. As a full-time professor and researcher in this field at the FFyL of the UNAM, he had chairs (seminars) to train bachelors, master's, and doctoral students.[9] In 1985, he published the book *Introducción a la Teoría de la Historia de la Ciencia* (UNAM), fundamental for students' theoretical-methodological training (Beltrán, 1989). In addition, he had published several works in the *Cuadernos de Quipu* collection regarding the history of science in Mexico and Latin America,[10] expanding the understanding of scientific developments in these regions (Fig. 15.1).

Saldaña was the president of MSHST from 1988 to 2008. During his tenure, the society shifted from having amateur to professional members. He recruited highly qualified historians of science, including Moreno de los Arcos, and young professionals who were either trained abroad or were studying with him at UNAM. Some of them were added to the MSHST board of directors. In his management, the MSHST played a significant role in legitimizing the field of science and technology in Mexico. He spearheaded annual conference cycles, regional colloquia, and national and international congresses. He established specific chairs in the history

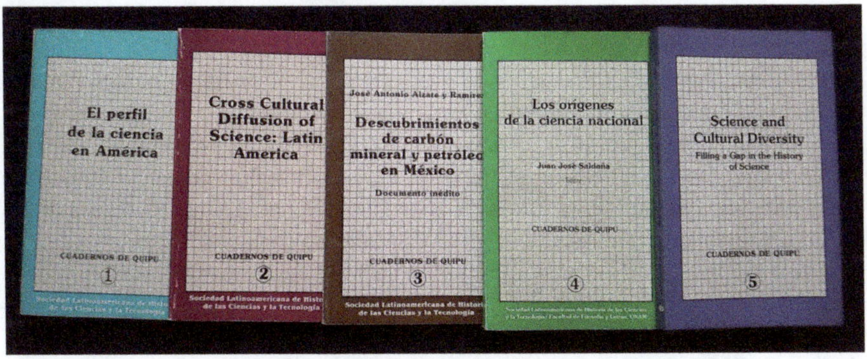

Fig. 15.1 Covers of the five issues of *Cuadernos de Quipu*

[9] Years later, he also included postdoctoral students in his program.

[10] During the 1980s, he published several issues in *Cuadernos de Quipu*: *El perfil de la ciencia en América* (1986), *Cross cultural diffusion of science: Latin America* (1987), *Descubrimientos de carbón mineral y petróleo en México. Documento inédito* del novohispano José Antonio Alzate y Ramírez (1988). In 1992, issue 4, *Los orígenes de la ciencia nacional*, was published, in which I contributed with my first paper entitled "La nueva física y su relación con la actividad minera en la Nueva España" (Ramos-Lara, 1992). Saldaña, Graciela Zamudio, Leonel Rodríguez, and María Lozano Meza published their papers in the same volume. The last three were my classmates in the Saldaña's Seminar.

of science and technology at the postgraduate level at UNAM, and founded prestigious History of Science awards, among other notable achievements. As shown below, Saldaña has been the only president to expand MSHST's functions beyond Mexico to cover Latin America and international domains.

Saldaña integrated the MSHST board of directors by inviting top Mexican experts in history of science and technology, such as Elías Trabulse and Ramón Sánchez Flores (1939–2004). These professionals had specialized publications that were gaining recognition among the Mexican population. Trabulse, a chemist and historian from the Colegio de México, is well-known for his work on the history of science in Mexico, including his remarkable hit collection *Historia de la Ciencia en México* (Trabulse, 1983–1989).[11] Sánchez Flores, historian of technology, who published in 1980 his famous book *Historia de la tecnología y la invención en México* (Sánchez Flores, 1980). Sánchez agreed while Trabulse refused and recommended his student Virginia González Claverán, who authored *La expedición científica de Malaspina en Nueva España: 1789–1794* (González Claverán, 1998).

Two of Saldaña's graduate students, Alberto Saladino and Patricia Aceves, joined the group. Saladino, Ph.D. in Latin American Studies (1988), published his excellent thesis *La ciencia entre los Ilustrados del Nuevo Mundo. Los casos de J. A. Alzate y F. J. Caldas* 2 years later (Saladino García, 1990). Aceves master's thesis, *La difusión de la química moderna en el Real Jardín Botánico de la Ciudad de México*, won the II Beechman Prize awarded by the Beechman Laboratories and the King and Queen of Spain. Table 15.1 contains the names of the Beltran and Saldaña board of directors.

In 1988, Saldaña immediately started organizing the activities of MSHST. These activities included creating the *Boletín de la SMHCT* (*MSHST Bulletin*) and celebrating a monthly cycle of conferences every year. The proceedings of these conferences were published in the *Actas de la Sociedad Mexicana de Historia de la Ciencia y la Tecnología* (*MSHST Proceedings*). His tasks also included organizing national congresses and regional colloquia and searching for resources to publish the proceedings of the events. All these activities cemented MSHST's reputation as a leading organization in Mexico's History of Science and Technology.

The First National Congress occurred in 1988, 9 years after Beltrán's previous event and 25 years after the First Colloquium (1963). The first congress was significantly different from the first colloquium. Beltrán was the honor participant from the first SMHST members generation at the Saldaña I Congress, as various experts of that first generation (Izquierdo, Bargalló, and Somolinos) had passed away.[12] The

[11] Trabulse authored a five-volume encyclopedia on the history of science in Mexico from the sixteenth to the nineteenth century, which includes an index volume (Trabulse, 1983–1989). The work has been highly acclaimed and recognized as a significant contribution to the field. An abridged version of the encyclopedia was published in 1994 (Trabulse, 1994).

[12] Enrique Beltran expertly chaired the organizing committee, while Saldaña served as the Executive Secretary, ensuring a smooth and successful event. The committee was also integrated by Arturo Azuela, Enrique Cárdenas de la Peña, Rául Gío Argáez, Fernando Martínez Cortés, Roberto Moreno, Ramón Sánchez Flores, José Sanfilipo, Juan Somolinos P., Elías Trabulse. The

Table 15.1 MSHST Board of Directors members in 1964, 1979, and 1988

Board of directors	1964	1979	1988
President	Enrique Beltrán	Enrique Beltrán	Juan José Saldaña
Vice President	José Joaquín Izquierdo	Carlos Saenz de la Calzada	Ramón Sánchez Flores
General Secretary	Germán Somolinos D'Ardois	Roberto Moreno de los Arcos	Patricia Aceves
Recording Secretary	Francisco Martínez Cortés	Jorge Álvarez Fuertes	Virginia González Claverán
Treasurer	Samuel Fastlicht	Samuel Fastlicht	Alberto Saladino
			Enrique Beltrán, Honorary Perpetual President

Author's elaboration

number of participants in the First Colloquium and First Congress can be compared using Graph 15.1. The congress in Mexico City was successful, with more than 150 papers and 30 working tables.[13] Two large proceedings volumes were required to publish all the papers (Saldaña, 1989). The congress's open call brought young researchers and prominent intellectuals to the MSHST community.[14] They engaged in direct dialogue, exchanged information on documentary sources, interacted with specialists, and even initiated collaborations.

The renewal of MSHST's membership, innovative reforms, and dynamic activities successfully revitalized this society immediately. Also, several of Saldaña's initiatives, such as creating the Enrique Beltrán Award, were instrumental in strengthening the discipline in Mexico in just a few years. Throughout his 20-year tenure as director of MSHST, Saldaña quantified its contribution to the professionalization process: it oversaw the publication of 15 books, organized around 15 national and international academic conferences, generated approximately 5000 papers, and connected the community through the *Ceromaya* network. Additionally, it supported its members in carrying out their professional projects and empowered them to excel in their professional activities (Vázquez, 2008). After establishing itself as a leading institution, MSHST expanded its influence to the international level. In 2001, the MSHST was designated as a world center for the history of

auxiliary committee included Patricia Aceves, Tamara Azuela, Roberto Jiménez, Adolfo Olea Franco, Lisa Primus, Leonel Rodríguez, and Jaime Vilchis.

[13] Several of his students (including Patricia Aceves, Alberto Saladino, Graciela Zamudio, Emma Yanes, Rafael Rodríguez, Leonel Rodríguez, Porfirio García de León, Elia Patlán Martínez, Patricia Carpy, and María Lozano) participated in the congress, along with other students enrolled in his courses. That year, I was studying for a master in physical sciences; I was not yet interested in the history of science.

[14] Juan Somolinos, Arturo Azuela, Mauricio Beuchot, Anne Staples, Dorothy Tanck, Marco A. Moreno Corral, Omar Moncada, Teresa Rojas Rabiela, Gerardo Sánchez, José Alfredo Uribe, among others, attended it.

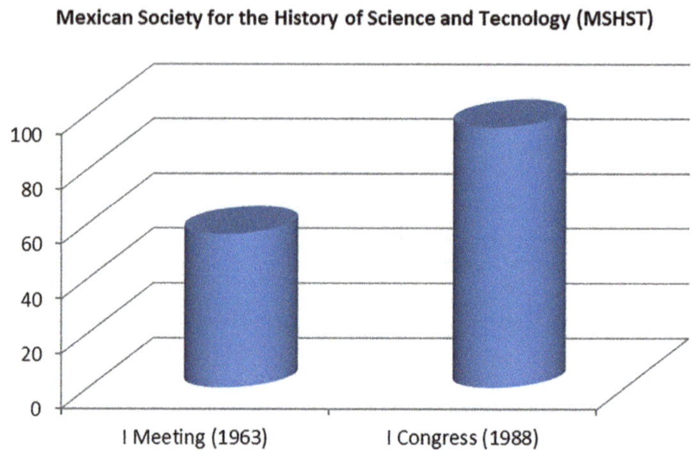

Graph 15.1 Comparison of the number of papers published in the proceedings of the I Colloquium (Beltrán) and I Congress (Saldaña) of the MSHST. (Author's elaboration)

science activities, hosting an international congress, which remains one of its most outstanding achievements.

In 2008, Dr. Ismael Mateos Ledesma (1960–2023) was elected as the new president after Saldaña completed his term as president of MSHST. Ledesma studied Biology and completed his doctorate in Biological Sciences at the Faculty of Sciences of UNAM and his master's degree at Centro de Investigación y de Estudios Avanzados (Cinvestav) of Instituto Politécnico Nacional. He did his postdoctoral studies with Bruno Latour at the Centre de Sociologie de l'Innovation (École Nationale Supérieure des Mines de Paris). Ledesma joined the teaching and research staff at the Facultad de Estudios Superiores "Iztacala" of UNAM. He was a founding director of the School of Biology of the Benemérita Universidad Autónoma de Puebla in 1987. At the beginning of Dr. Ledesma's term, the board of directors of MSHST consisted of Leticia Mayer Celis (vice president), Carlos Ortega Ibarra (general secretary), Jorge Bartolucci Íncico (treasurer), Rolando Neri Vela (recording secretary), Guadalupe Urbán Martínez (first member), and Oliva López Sánchez (second member).[15]

Unfortunately, Ledesma passed away in 2023, just a month and a half after participating in the tribute to Saldaña at the "I Jornada Internacional de Historia de la Ciencia y la Tecnología en América Latina. *Juan José Saldaña González In Memoriam*". The tribute was held on March 23rd and 24th, 2023, at the Aula Magna of the FFyL and the Auditorium of the Centro de Investigaciones Interdisciplinarias en Ciencias y Humanidades (CEIICH) of UNAM. The passing of Ledesma has left MSHST in a problematic situation, but we remain hopeful that it can overcome it, just as it has in the past.

[15] Urbán and Ortega were disciples of Saldaña.

The Association of Historians of Science and Humanities (AHSH) is other organization that brings together many Mexican historians of science. It was founded in March 2007 by Alberto Saladino García, José Omar Moncada Maya, Leonel Rodríguez Benítez, and Luz Fernanda Azuela Bernal, some of whom were participants or graduates of the Saldaña's Seminar. The AHSH has an excellent publishing project called *Saberes Revista de historia de las ciencias y las humanidades*. It is worth noting that the AHSH is a member of the IUHPST/DHST.[16] In 2021, the Seminario Universitario de Historia, Filosofía y Estudios de las Ciencias y la Medicina (SUHFECIM) was established at UNAM, under the coordination of Ana Barahona. This platform serves as a space for worldwide analysis, reflection, research, and dissemination of interactions on topics related to History, Philosophy, Medicine, and Science Studies.[17]

Although there are no specific postgraduate programs in Mexico for the History of Science and Technology, students can pursue research in this field through postgraduate programs at UNAM, Universidad Autónoma Metropolitana (Iztapalapa and Xochimilco), Colegio de México, Instituto Politécnico Nacional, CIDE, Universidad de Puebla, and other Mexican universities. At UNAM, for example, a student can graduate with a research topic in this field in the following postgraduate programs: Philosophy of Science,[18] History, Latin American Studies, "Biological Sciences, Philosophy, Pedagogy, and Philosophy of Science", and others. Currently, Mexico does not have a centralized institute or research center solely dedicated to the study of the history of science and technology. However, despite this challenge, experts continue to seek out and utilize available spaces within various institutions to further their research and knowledge in this field. For example, in 2000, I established the first research program on the History of Science at CEIICH-UNAM, through which I founded two book collections.[19] Other colleagues have also established similar programs in additional UNAM entities and further educational institutions. However, this has led to the scattered distribution of documentary resources across several places, and we need a proper library for the History of Science and Technology.

I am confident that in the near future, Mexico will establish an institution, a library, and a graduate program exclusively dedicated to the History of Science and Technology. However, it is crucial to acknowledge that the field is already

[16] http://www.hch.org.mx/conocenos/ (accessed September 6, 2023).

[17] https://suhfecim.sdi.unam.mx/ (accessed April 2024).

[18] The postgraduate program in Philosophy of Science at UNAM considers it as part of its five fields of knowledge, which include Philosophy of Science, History of Science, Philosophical and Social Studies on Science and Technology, Science Communication, Philosophy of Cognitive Sciences, and Philosophy of Mathematics and Logic of Science.

[19] *Ciencia y Tecnología en la Historia de México* and *Bibliotheca Mexicana Historiae Scientiarum*. The last one is a facsimile edition used to reproduce valuable old scientific books that were published in Mexico. We reproduced the first science book published in America (*Sumario Compendioso*, 1556), the first physics book published in the Americas (*Physica Speculatio*, 1557), and the Mexican flora of the Enlightenment period (*Plantae Novae Hispaniae*).

well-established and continuously developing and enriching itself with the contributions of young professionals trained both within and outside of the country.

15.2 Professionalization and Consolidation of the History of Science in Mexico

Several significant events in Mexico in the 1980s helped advance the History of Science and Technology. One such event was the existence of MSHST, which, although it was paralyzed, was formally established. Additionally, Conacyt showed a willingness to support organizing events and research programs and finance the publication of materials. Several scientists and historians who dedicated themselves to this field were full-time employed by their institutions, resulting in essential publications that significantly contributed to the diffusion of the field in Mexican society. Mexican students trained in Philosophy and History of Science abroad, primarily in France and Spain, returned. Specialized chairs in the History of Science and Technology were established as the Saldaña pioneer Seminary at UNAM. A noticeable trend emerged in Mexico where graduate students in both the exact and biological sciences, as well as humanities, were actively turning towards studying the history of science when Saldaña was beginning his Seminars.

Throughout his career, Saldaña received invaluable support from his students at the Seminar and other colleagues who actively participated in the activities organized by the MSHST. His lectures played a significant role in professionalizing the History of Science in Mexico and helped him achieve his goals. The seminar classrooms were always filled with people eager to learn about theoretical and methodological aspects and develop original research in this field. They were also enthusiastic about participating in the activities organized by the MSHST, whether as audience members, organizers, task assistants, or speakers. As a result, the number of MSHST members increased every year, with new students joining the seminary and people attending events held in regions that had never been visited before. Some seminary students became board members, while others worked as editors of the *MSHST Bulletin,*[20] or helped to organize specific events such as symposia, regional congresses, biannual national congresses, Latin American congresses (held every 3 years), and international events.

Saldaña joined FFyL-UNAM in 1983 as a full-time research professor. His stable position enabled him to implement the idea of a "Seminar" that he had conceived in France, which had become a commitment of the Bucharest Declaration to be developed in several Latin American countries, as we saw in the second section of this book. In 1985, Saldaña started the Research and Thesis Seminar on the History of Science and Technology (from now on Seminar) in the Division of

[20] The *Bulletin of SMHCT* was initially managed by Luz Fernanda Azuela, then by Ana María Carrillo, and later by Carlos Ortega.

Graduate Studies of the Department of History of the FFyL-UNAM. During the 1980s, Saldaña's professorships were the only ones that offered formal theoretical and methodological training in the history of science in Mexico, attracting students from other states of Mexico and even from other countries to listen to his lessons.[21]

The Seminar had two unique aspects for two reasons. Firstly, it was multidisciplinary, and students from various faculties and institutions could pursue research in the history of science in Mexico under the supervision of an expert. Secondly, it was the most renowned and prestigious Seminar at UNAM, facilitating agreements between faculties for students to graduate with a co-director assigned to their faculty. My experience was similar to that of my colleagues. In 1989, I completed the credits required for the Master's Degree in Physics (Astrophysics) at the UNAM Faculty of Sciences. I had to write a thesis for graduation, and I chose the topic of the origins of physics in Mexico, but nobody on the faculty had a doctorate in this field to supervise me. Fortunately, the physicist Dr. Luis de la Peña Auerbach, researcher at the UNAM Physics Institute, suggested that I speak with Saldaña,[22] who agreed to be my supervisor. De la Peña served as my co-director, and thanks to his and Saldaña's guidance, I achieved a master's degree in 1991.[23] Both of them helped me become the first and only Master's graduate in Physics with a specialization in the history of science. The Faculty of Science (UNAM) traditionally considered this field to be reserved for historians and not physicists. For this reason, the Faculty of Science did not accept my doctoral project for being in the history of physics rather than physics, strictly speaking, and its authorities recommended my admission to the FFyL (Ramos-Lara, 2024).

When I enrolled in the History doctorate program at FFyL, I attended Saldaña's Chairs, especially the multidisciplinary Seminar. Professionals from diverse fields of study, such as physics, engineering, mathematics, chemistry, biology, history, medicine, philosophy, veterinary, medicine, sociology, music, education, agricultural sciences, Latin American studies, and geography attended the Seminar. Some attendees were students and research professors from various educational institutions who came to learn about unfamiliar topics, which enriched their research. The Seminar provided an exceptional postgraduate program, which covered an extensive range of topics from the fourteenth century to the twentieth century and included various countries and cultures. The program was different every semester and no repeated. Dra. Clara Carpi, Saldaña's widow, has identified on Saldaña's archive 34 programs taught between 1997 and 2019, and there are ongoing efforts to locate the remaining 20. The FFyL-UNAM is finalizing the publication of a

[21] Among the international students I knew were Luz Marina Duque from Colombia and Guillermo Guajardo from Chile.

[22] De la Peña became acquainted with Saldaña in 1982 at the National Physics Congress organized by the Mexican Physics Society. Saldaña was a plenary speaker and delivered a talk on "The explanation of the polarization of light at the beginning of the 19th century". De la Peña saw Saldaña's ability to explain physics concepts clearly and accurately and assumed that Saldaña studied Physics.

[23] This is where my passion for the subject began.

digital book that will feature the programs and feedback from students (Ramos-Lara & Arboleda, 2024).[24] We attended the seminar for the valuable new material, despite already being accredited. We developed close relationships and collaborations as a result (Fig. 15.2).

Saldaña also taught another Chair called "Historiografía Mexicana de la Ciencia" (Mexican Historiography of Science) that had a consistent content for decades before changing towards the end. The course was particularly interesting as we studied the history of science in Mexico from various fields and time periods, and analyzed texts by historians of science from Mexico, Latin America, and other countries. In the Saldaña classroom, we met some visitor scholars whose works we

Fig. 15.2 Events held in Mexico. Beltrán is above on the right (second from the left) giving me the award, third is Eli de Gortari. On the right in the middle photo is Bernard Cohen participating in Newton in America table. Below I am with Saldaña at my graduation. The three photos on the left show the presentation of the award winning-book. At bottom is the photo of the IV MCHST in Culiacán, Sinaloa, 1994. (author's photos)

[24] The Seminar covered various topics including the History of Provincial Science; History of Science and Scientific Education; History of National Science; Science and Cultural Diversity; Institutions of Scientific Research and Scientific Teaching; Technology and Industry in Mexico in the 19th Century; From Engineering Science to the Beginning of Academic Science; Scientific and Technical Education; Science, Technology and Public Works in Mexico; The State and the Creation of Technical and Technological Culture in Mexico; Images of Science, Technology and the State in Mexico; History of Technology in Mexico, among others.

reviewed in his class. He used to displayed journals (like *Autocrítica*) and information about national, Latin American, and international congresses. Attending an international event presented the opportunity to meet several authors whose work was included in the Seminar's bibliography. For instance, I attended the III Latin American Congress on the History of Science and Technology (LACHST), where I met Bernard Cohen (1914–2003), a famous scientist and historian from Harvard University. We participated in the same discussion panel on Newton in America.[25] Saldaña also invited Roger Hahn from Berkeley University and Ronald Numbers, editor of the prestigious ISIS journal.

The III LACHST occurred in Mexico City from January 12–16, 1992. The congress had a general theme: "America in the formation of a new world: 500 years of scientific exchanges". The Latin American community came together to celebrate the V Centenary of the Columbian Voyage, the bicentenary of the Real Seminario de Minería in Mexico City (considered by some historians as the first House of Science in the New World), and the tenth anniversary of the constitution of LASHST. Additionally, the III Mexican Congress of History of Science and Technology (MCHST) was also included in the event. During the opening ceremony, Saldaña stressed the importance of the congress for Mexicans, mainly due to the presence of researchers from other countries and the opportunity to contemplate the role of science in American and world history. These words were highlighted in his inaugural lecture.

> The topic of the scientific exchanges kept among America and the rest of regions would make possible besides the placing of the American science at its universal dimension. Because, as in the turbulence of a course of water, where several streams interact to form a bed with dams or contours clearly recognizable, that was the way in which American science was formed. The socio historical study of such complexity, with arises both local and international is, therefore, a contribution to the comprehension of American and worldwide science. Since XVI century, America was the laboratory of the westernization of the world. But this was not the only purpose of such process, also and active subject that produces peculiar historic forms of the scientific and cultural practice. The V Centenary gives us effectively the occasion to reflect upon the role of science in the American and worldwide history.[26]

The congress was part of a series of Latin American events that started in Puebla in 1982, followed by Havana in 1985 (I LACHST) and São Paulo (II LACHST) in 1988. The III LACHST had a total of 234 speakers and 196 lectures presented. Graph 15.2 displays that most participants were Mexican, followed by Brazilian and American attendees. The congress was organized by MSHST, LASHST, IUHPST/

[25] Celina Lértora published in Argentina the proceedings entitled "Newton in America". In addition to Cohen's work on Newton, our table referred to the dissemination of the work of this English physicist in several Latin American countries. For instance, Celina Lértora Mendoza presented her findings in Argentina, and Libertad Díaz Molina did the same in Cuba. Similarly, Luis Carlos Arboleda and Diana Soto Arango reported their research in Colombia. María de la Paz Ramos-Lara, Juan Manuel Espinoza, and Patricia Aceves did the same in Mexico (Lértora Mendoza, 1995).

[26] Paragraph taken from the speech pronounced by Juan José Saldaña during the inauguration of the III LACHST (LACHST, 1992).

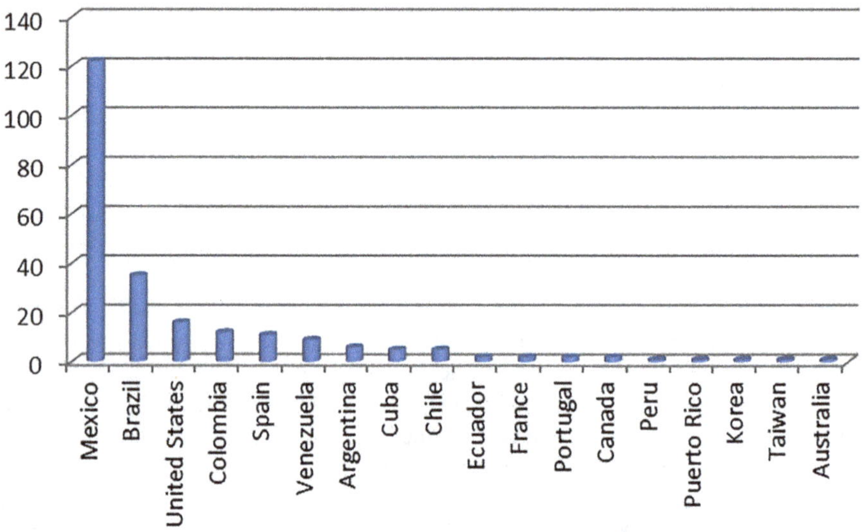

Graph 15.2 Nationality of participants in the III LACHST 1992. (Author's elaboration)

DHST, National Science Foundation, UNAM, Conacyt, and the Consejo Nacional para la Cultura y las Artes. The organizing committee comprised Saldaña (president), Patricia Aceves, Libertad Díaz Molina, María de la Paz Ramos-Lara, Graciela Zamudio, and Alberto Saladino. The five of us were students of Saldaña's Seminar at FFyL-UNAM.[27]

The III LACHST organization demonstrated the significant contribution made by the students of the Saldaña's Seminars to developing the History of Science in Mexico and abroad. Our participation was crucial in organizing congresses, publishing and reviewing papers in *Quipu*, securing the venue, and organizing the 21th International Congress held in Mexico, which will be discussed below. While I do not have the precise number of decent students who regularly attended Saldaña's Seminars each year, I can provide the exact number of the students he supervised until 2018 at UNAM. Between 1985 and 2018, 64 students graduated from the bachelor's, master's, doctoral, and postdoctoral UNAM programs.[28] Given his strengths as a teacher, UNAM authorities recognized and awarded him for his work in training researchers. In 2018, eleven of us worked at UNAM, nine at other national university institutions, and two abroad. Thirteen of us belonged to the National System of Researchers (SNI-Conahcyt) (Fig. 15.3).

[27] All of them are my close friends, and I am proud to say we continue collaborating.

[28] Some graduated already had jobs either inside or outside the academy, while others started their working lives with temporary jobs. During my graduate studies, I worked at the Universidad Autónoma Metropolitana in both Iztapalapa and Azcapotzalco, at the Escuela Superior de Ingeniería Mecánica y Eléctrica of the Instituto Politécnico Nacional, and the Museo *Universum* of the UNAM before being hired by CEIICH-UNAM.

Fig. 15.3 Participants of the III LACHST in Mexico City, 1992. (Photographic archive Juan José Saldaña González)

Although Saldaña Seminar no longer exists, some of Saldaña's disciples are still contributing to training specialists, creating new research areas, establishing collaboration networks, and promoting national and international collective projects.

15.3 *Quipu* Journal and the XXI International Congress of History of Science (ICHS) Held for the First Time in a Latin American Country (Mexico 2001)

The LASHST was founded in 1982 with Saldaña as the president for 6 years. He also edited the *Quipu Revista Latinoamericana de Historia de las Ciencias y la Tecnología*, which was launched in 1984 and had two formats.[29] The First Epoch was in print and ran from 1984 to 2000. The Second Epoch was digital and ran from 2012 to 2014. *Quipu* was a pioneering publication that played a crucial role in enabling communication and fostering the exchange of information among Latin American researchers. In the previous chapter (Chap. 5 of Sect. 5.2), we examined the journal's evolution throughout Latin America. This section will focus on its importance for the community of historians of science and technology in Mexico.

Quipu had a more significant impact on the Mexican academic community than other parts of Latin America, mainly because it was edited and produced in Mexico. This situation made the journal more accessible for libraries and individuals to subscribe to, with reduced shipping costs compared to ordering from abroad. Additionally, Saldaña, the Mexican editor, diligently promoted *Quipu* among his students and other scholars in educational institutions and events he organized in Mexico. Consequently, the largest number of persons and institutions possessing

[29] At the beginning and in other chapters of this book, it is mentioned that Saldaña was also the editor of *Cuadernos de Quipu*. The *Quipu* Journal is available at: https://search.app/ZbYmvvAWekuypJuU6 https://web.archive.org/web/20220804093919 https://www.revista-quipu.com/.

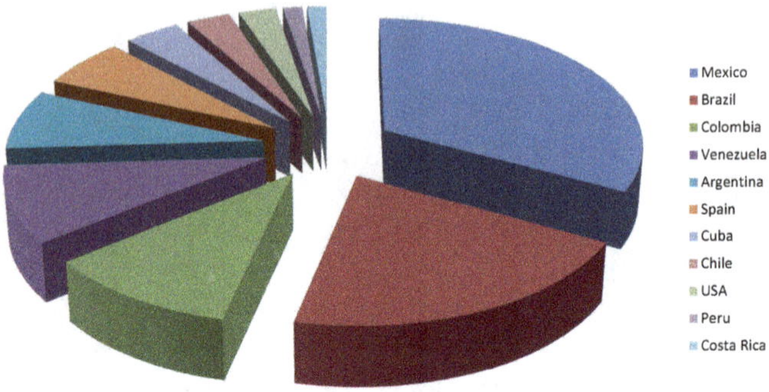

Graph 15.3 Countries of origin of the authors who published in *Quipu* first period (1984–2000). (The graph's lower limit is set at three articles due to its limited size. Consequently, some countries, such as Ecuador, are omitted from the Latin American region). (Author's elaboration)

the entire printed *Quipu* collection are in Mexico. These and other factors significantly increased the readership of *Quipu*, mainly among Mexican scientists and historians of science, which also stimulated their interest in publishing articles in the journal. As shown in Graph 15.3, most authors who published articles in *Quipu* between 1984 and 2000 were Mexican.

Quipu was published every 4 months, featuring Spanish, Portuguese, English, and French articles. It was regularly published for a decade, and its content was almost complete, including articles, reviews, documents, bibliographic essays or bibliographies, testimonials, praise, and the dissemination of international events and recent publications. However, due to the economic crisis in Mexico during the 1990s, Conacyt had to suspend it the subsidy. The journal could only publish some thematic issues subsidized by various Latin American institutions but could not do so between 1995 and 1998. As shown in Graph 15.4, the number of articles and reviews published from 1984 to 2000 indicates 4 years of empty publication due to lack of funding, with its last recovery in 1999 and 2000.[30]

In 2001, Saldaña stopped publishing *Quipu* and stated that it was the responsibility of new generations to continue the task. However, it was not easy as Saldaña had the advantage of having close ties with the Mexican, Latin American, and international communities. In 2011, he decided to revive the project and published the journal digitally under the *Quipu* Second Epoch, but now its copyright was registered under his name. He took advantage of the digital era and the fact that he did not have to pay for printing or material shipments. He implemented a strategy of charging a basic fee for publishing or reading an article, but it did not work for long.

[30] It is worth noting that Saldaña was the only editor of the journal. At the beginning of 2001, he had the crucial task of organizing the ICHS in Mexico. This event, scheduled for July of that year, demanded his undivided attention and efforts.

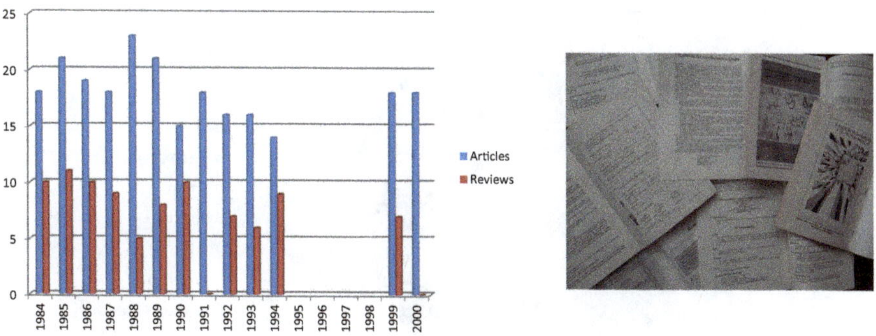

Graph 15.4 Number of articles (left-blue) and reviews (right-red) published in *Quipu* from 1984 to 2000. And information disseminated by Quipu. (Author's elaboration)

One of the reasons for this was that digital journals are more vulnerable to piracy, just like what happened to the Latin American scientific journal *Interciencia.*

Quipu undeniably drew the attention of the Mexican community of historians of science. However, the XXI ICHS held in Mexico in 2001 was a highly significant event for experts in the field and scientists and historians keen to participate in it. We can confirm that *Quipu* ceased publication after providing a comprehensive report on the XXI Congress in the final six pages of its last issue (vol. 13, no. 3, 2000).

The process started in 1989, just 1 year after his MSHST presidency began; he incorporated MSHST into the IUHPST/DHST and became the first Latin American elected as an Advisor to the Executive Council. Its membership was accepted at the IUHPST General Assembly in Munich, Germany, in 1989. On that same occasion, Saldaña proposed that Mexico host the international congress but lost to Spain (Vázquez, 2008, 19). In 1997, Saldaña again requested that Mexico host the international congress at the 20th ICHS in Liège, Belgium. I also attended this congress as a delegate. The delegation from Mexico worked tirelessly to convince the assembly members that Mexico would be the best choice for hosting the next congress. We faced fierce competition from New York (USA) and Beijing (China). We were very excited because winning the venue for the international congress was a triumph, as it was the first time the congress was held outside of the "center countries." We won, and in 2001, Mexico hosted the international congress in Latin America for the first time, demonstrating his exceptional leadership abilities.

Finally, from July 8 to 14, 2001, ICHS took place in downtown Mexico City, where several historic buildings of immense cultural significance served as the venue. Nine hundred seventy-eight lectures were presented, with the participation of 58 countries, 630 in symposia, and 348 in Scientific Sections.[31] The Mexican delegation was the largest, as shown in Graph 15.5. The same happened in the next congress held in China, where the Chinese delegation outnumbered all others.

[31] Book of abstracts 1 Symposia, and book of abstracts 2 Scientific Sections (ICHS, 2001).

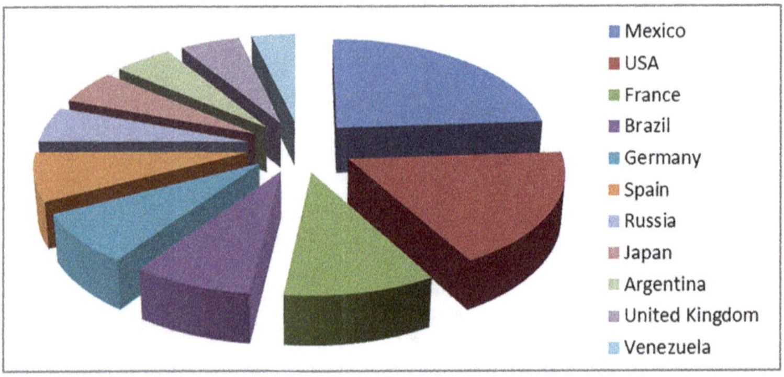

Graph 15.5 Most numerous delegations in the XXI ICHS Mexico City 2001. (This graph considered 19 participants as the lower limit). (Author's elaboration)

The congress had a general theme of *Science and Cultural Diversity*, which was approved by the IUHPST/DHST board during their meeting in Istanbul, Turkey, in 1988, in the same year he became President of MSHST. Saldaña began promoting the congress internationally in the year 2000. The advertisement was written in three languages, and 12,000 copies were distributed across 90 countries (Vázquez, 2008). The lectures were able to be presented in any of the following official languages: English, French, German, Italian, Portuguese, Russian, and Spanish. The organizing committee of the congress was composed as follows:

Executive council: President: Subbarayappa Bidare Venk (India); first Vice-President: Mirko Drazen Grmek (France); second Vice-President: Ekmeleddin Ihsanoglu (Turkey); Secretary General: Robert Halleux (Belgium); Treasurer: Liba Taub (United Kingdom); Assistant Secretary: Fabio Bevilacqua (Italy); Past President: Robert Fox (United Kingdom).

Assessors: Chikara Sasaki (Japan); Mahmoud Hafez (Egypt); Vladimir Kirsanov (Russia); Roman Duda (Poland); Carlos D. Galles (Argentina); Mohammad Baheri (Iran).

International Program Committee: President: Jed Z. Buchwald (USA); Vice-President: Roshdi Rashed (France-Egypt); Kirsti Andersen (Denmark); Luis Carlos Arboleda (Colombia); Mohammad Bagheri (Iran); Michel Blay (France); Xiang Chen (USA-China); Marcos Cueto (Peru); Serguei Demidov (Russia); Ahmed Djebbar (France-Algeria); Liu Dun (China); Gad Freudenthal (France); Carlos Deifino Galles (Argentina); Paulus Gerdes (Mozambique); Roderick W. Home (Australia); Yung Sik Kim (South Korea); Andreas Kleinert (Germany); Deepak Kumar (India); Antonio Lafuente (Spain); Roy MacLeod (Australia); Efthymios Nicolaidis (Greece); Teresa Rojas Rabiela (Mexico); Fumihiko Satofuka (Japan); William Shea (France); Wesley M. Stevens (Canada).

XXI ICHS Local Advising Committee: Mauricio Beuchot, Ana Barahona, Alberto Carabarín Gracia, Martha Díaz de Kuri, Esteban Krotz, Juan Manuel Lozano, Omar Moncada maya, Eduardo Piña Garza, María de la Paz Ramos Lara, Luis Miguel Rionda Ramírez, Leonel Rodríguez Benítez, Ana Cecilia Rodríguez Romo, María

Fig. 15.4 Several photos from the XXI ICHS Mexico City 2001. (Photographic archive Juan José Saldaña González)

Luisa Rodríguez Sala, Teresa Rojas Rabiela, Alberto Saladino García, Gerardo Sánchez Díaz, Elías Trabulse, Rafael Valdés Aguilar, Carlos Viesca Treviño.

Saldaña was appointed as Secretary General of IUHPST/DHST for 2001–2005 due to the success of his work in organizing the congress (Figs. 15.4 and 15.5).

15.4 Final Remarks

It is without a doubt that Juan José Saldaña was a pioneer in the professionalization of the history of science in Mexico. We are deeply saddened by his passing on October 31, 2022, in Mexico City. We would like to use this paper as a tribute to recognize his tireless and invaluable work in consolidating the institutionalization of the History of Science and Technology in Mexico, as well as his significant contributions to Latin America and the international community.

Fig. 15.5 Juan José Saldaña at several international meetings. (Photographic archive Juan José Saldaña González)

Acknowledgments I want to express my gratitude to Dra. Clara Isabel Carpi Navarro, Saldaña's widow, for her unwavering support in locating digital and printed information from the Saldaña archive. I am also grateful to Adriana Cruz Romero and Nicolasa Ramírez from the Department of Information and Documentation of CEIICH-UNAM for their support. My recognition extends to Víctor Méndez, Wendy Gallardo and Miguel Nava from the Audiovisual and Multimedia Production Department of CEIICH for their assistance in digitizing some photos. A special thanks to my colleague and friend Rosi Mendoza for her technical assistance. Lastly, I am deeply grateful to my daughter, Adriana Ivonne Canales, and my husband, Sergio Adrián Canales, for their love and invaluable support in helping me develop and successfully complete this work and the whole book.

Libraries and Archives

National Library of Mexico
Colegio de México Library
UNAM Historical Archive
CEIICH-UNAM Library (printed *Quipu* collection)
Library of the Faculty of Philosophy and Letters, UNAM
Library of the Institute of Geology, UNAM
Library of the Institute of Historical Research, UNAM
Library of the Institute of Philosophical Research, UNAM

References

Actas de la Sociedad Mexicana de Historia de la Ciencia y la Tecnología. (1989). Mexico: Ed. Virginia González Claverán.

Anales de la Sociedad Mexicana de Historia de la Ciencia y la Tecnología (1969, 1970, 1972, 1974, 1979). Mexico; MSHST.

Azuela, L. F. (1997). Roberto Moreno de los Arcos y la Historia de las Ciencias. In memoriam. *Ciencia Ergo Sum, 4*, 119–120.

Bargalló, M. (1955). *La minería y la metalurgia en la América Española durante la Época Colonial: con un apéndice sobre la industria del hierro en México desde la iniciación de la Independencia hasta el presente*. Fondo de Cultura Económica.

Beltrán, E. (1964). *Memorias del primer Coloquio Mexicano de Historia de la Ciencia*. Sociedad Mexicana de Historia Natural (Mexican Society for Natural History).

Beltrán, E. (1989). La historia de la ciencia en México en los últimos cinco lustros (1963–1988). In *Memorias del Primer Congreso Mexicano de Historia de la Ciencia y la Tecnología* (Tomo I) (pp. 21–35). Mexico: MSHST.

Cifuentes L. J. L. (1992). Enrique Beltrán Castillo. In *Nuestros Maestros*. Tomo II. Mexico: Universidad Nacional Autónoma de México, Secretaría General, Dirección General de Asuntos del Personal Académico.

De Gortari, E. (1963). *La ciencia en la historia de México*. México: Fondo de Cultura Económica. https://books.google.com.cu/books?id=Q_t6BwAAQBAJ&printsec=copyright#v=onepage&q&f=false

González Claverán, V. (1998). *La expedición científica de Malaspina en Nueva España : 1789–1794*. Colegio de México.

Guzmán Pérez, M. (1996). Notas Necrológicas. Roberto Moreno de los Arcos. In Memoriam. *Tzintzun, 24*, 158–161.

ICHS. (1985). *XVIIth international congress of history of science*. University of California, Berkeley (31 July–8 August 1985).

ICHS. (1997). *List of participants XXth international congress of history of science* (Liège, Belgium 20–26 July 1997). Centre d'Histoire des Sciences et des Techniques, Université de Liège.

ICHS. (2001). *List of participants XXI international congress of history of science* (Mexico City, Mexico 8–14 July). Also Book of abstracts 1 Symposia, and book of abstracts 2 Scientific Sections.

ICHS. (2005). *List of participants XXII international congress of history of science* (Beijing, China 24–30 July).

Izquierdo, J. J. (1958). *La primera casa de las ciencias en México: El Real Seminario de Minería, 1792–1811*. Ciencia.

LACHST. (1992). *Libro de resúmenes del III Congreso Latinoamericano y III Congreso Mexicano de Historia de la Ciencia y la Tecnología* (12–16 de enero de 1992, Ciudad de México). Mexico: MSHST and LASHST.

LACHST. (1995). *Libro de resúmenes del IV Congreso Latinoamericano de Historia de las Ciencias y la Tecnología* (24–27 de enero de 1995, Cali, Colombia). Cali: LASHST and Universidad del Valle.

Lértora Mendoza, C. A. (compiladora). (1995). *Newton en América*. Memoria del Simposio del III Congreso Latinoamericano de Historia de las Ciencias y la Tecnología (México 1992). Buenos Aires: Ed. Fepai.

Ramos-Lara, M. P. (1992). La nueva física y su relación con la actividad minera en la Nueva España. In *Los orígenes de la ciencia nacional*. México: Sociedad Latinoamericana de Historia de la Ciencia y de la Tecnología, Facultad de Filosofía y Letras (UNAM).

Ramos-Lara, M. P. (1994). *Difusión e Institucionalización de la mecánica newtoniana en México en el siglo XVIII*. Mexico: Sociedad Mexicana de Historia de la Ciencia de la Tecnología (MSHST) and Universidad de Puebla.

Ramos-Lara, M. P. (2024). Viraje de la astrofísica a la historia de la ciencia en el contexto nacional e internacional. In M. P. Ramos-Lara y L. C. Arboleda (Eds.), *La cátedra que profesionalizó la Historia de la Ciencia en México. Contribución de la FFyL-UNAM a América Latina.* Universidad Nacional Autónoma de México (UNAM).

Ramos-Lara, M. P., & Arboleda, L. C. (Eds.). (2024). *La cátedra que profesionalizó la Historia de la Ciencia en México. Contribución de la FFyL-UNAM a América Latina.* Mexico: Universidad Nacional Autónoma de México (UNAM).

Ríos, S. A. (2000). Quipu: una Revista Latinoamericana de la Historia de las Ciencias y la Tecnología. *Biblio 3W. Revista Bibliográfica de Geografía y Ciencias Sociales* 212. https://biblat.unam.mx/es/revista/quipu/

Saladino García, A. (1990). *Dos científicos de la Ilustración Hispanoamericana: J. A. Alzate, y F. J. de Caldas.* México: Universidad Nacional Autónoma de México.

Saldaña, J. J. (1989). *Memorias del Primer Congreso Mexicano de Historia de la Ciencia y la Tecnología (tomos I y II).* MSHST.

Saldaña, J. J. (Ed.). (2003). *Science and cultural diversity. Proceedings of the XXIst international congress of history of science. Volume I Plenary Lectures.* México: Sociedad Mexicana de Historia de la Ciencia y la Tecnología (MSHST) and Universidad Nacional Autónoma de México.

Sánchez Flores, R. (1980). *Historia de la tecnología y la invención en México.* Fomento Cultural Banamex.

Somolinos d'Ardois, G. (1957). *Historia y Medicina: Figuras y hechos de la historiografía medica mexicana.* Imprenta Universitaria.

Trabulse, E. (1983–1989). *Historia de la Ciencia en México.* Fondo de Cultura Económica.

Trabulse, E. (1994). *Historia de la Ciencia en México. Versión abreviada.* México: Conacyt, Fondo de Cultura Económica (última edición 2017). https://www.google.com.mx/books/edition/Historia_de_la_ciencia_en_M%C3%A9xico_versi/LjYqDwAAQBAJ?hl=es&gbpv=1&dq=Historia+de+la+Ciencia+en+M%C3%A9xico.+Versi%C3%B3n+abreviada&printsec=frontcover

Vázquez, Á. (Ed.). (2008). *La construcción en México de una comunidad científica: los historiadores de la ciencia y la tecnología.* MSHST.

María de la Paz Ramos-Lara holds an MSc in Physics and a PhD in History from UNAM. She is a researcher at the Center for Interdisciplinary Research in Sciences and Humanities at UNAM, where she founded the History of Science Department and two book collections: *Bibliotheca Mexicana Historiae Scientiarum* and *Science and Technology in the History of Mexico* (she is currently the editor). She is the head of the Laboratory for Interdisciplinary Research and Complexity Sciences and the Science and Technology Department. She is also a Philosophy of Science Postgraduate Program professor at UNAM and a member of the National System of Researchers (SNI-Conahcyt, Mexico). She was awarded the Dr. Enrique Beltrán Prize, the Sor Juana Inés de la Cruz, and José Joaquín Izquierdo Medals. She has authored three books on Newton's diffusion in New Spain, Engineering training in 19th century Mexico, and the evolution of the Escuela Nacional Preparatoria as a complex adaptive system. She was the facsimile editor of the first science and also physics book published in America. She worked on editing the book about Mexican flora during the Enlightenment. She has coordinated books on the history of science in Mexico, covering topics such as relativity, quantum mechanics, particle accelerators, astronomy, chemistry, medicine, scientific education, and social complexity from an interdisciplinary perspective.

Chapter 16
The Influence of Juan José Saldaña in the Development of the Historical and Social Studies of Sciences and Health in Colombia, 1979–2024

Emilio Quevedo-Vélez and Julieth Nathalia Barrera-Díaz

Abstract The purpose of this text is, in the first place, to analyze the influence of Dr. Juan José Saldaña in the process of building a scientific community in the field of the Social History of Sciences and Technologies in Colombia, and the transition toward Social Studies of Sciences, Technologies and Professions, including Medicine and Public Health. At the second stage, it goes beyond Saldaña's contribution but always takes as a basis the orientations proposed by him. This analysis not only includes the historical understanding of the processes of construction of knowledge and scientific practices, but also of that of construction of new institutional forms of professional practice and their articulation with the undergraduate and graduate curricula of some universities, and their relations with the State and Civil Society.

Keywords Social studies of sciences, health and medicine · Scientific revolutions · Epistemological-history · Anti-evolutionism · Discontinuity · Materialism · Colombia

---

E. Quevedo-Vélez (✉)
Grupo de Estudios Sociales de la Salud, Escuela de Medicina y Ciencias de la Salud, Universidad del Rosario, Bogotá, Colombia

J. N. Barrera-Díaz
Grupo de Estudios Sociales de las Ciencias, las Tecnologías y las Profesiones. Escuela de Medicina y Ciencias de la Salud, Universidad del Rosario, Bogotá, Colombia
e-mail: juliethn.barrera@urosario.edu.co

M. de la P. Ramos-Lara, L. C. Arboleda (eds.), *History of Science in Latin America: The Construction of an Intellectual Field (20th century)*,
https://doi.org/10.1007/978-3-031-75536-1_16

16.1 Immediate Background

16.1.1 *The Beginnings of the Enlightenment and Positivist Histories of Science*

The History of Science, with the pretension of scientific discourse, was born in the enlightened France of the eighteenth century. The first stage of this history, which conceived itself as a science, was framed in the idea of "progress", typical of the eighteenth-century "Enlightenment" (Saldaña, 1989a, 30).

This first model of history of science, of a continuous, progressive, and linear nature, turned hegemonic until the middle nineteenth century and pretended to "apprehend the essence of the scientific revolutions that had taken place in the past: those of Copernicus, Galileo, Bacon, Descartes and Newton" (Saldaña, 1989a, 21).

But at the end of the first third of the nineteenth century, the History of Science took an important leap because of the ideas of the French philosopher and sociologist Auguste Comte. In 1828, he opened the Course on Positive Philosophy, in which proposed the "Law of the three states" of human progress, understanding it conformed by three successive stages or states: the theological or magical-religious state; the metaphysical or speculative state; and, the positive or scientific state (Larroyo, 1979, XXXVI–XXXVII; Comte, 1979, 33–34).

According to Comte, the Positive state was the one of positive science. That is to say, the positive tool for knowing and managing the world, which would progressively eliminate theological and metaphysical explanations and fill the empty spaces on the board of objective and positive knowledge. According to Michel Fichant, for Comte the past of science was no longer science but the history of its process of constitution (Fichant, 1975, 58–61).

This vision of the history of science was clearly "internalist". It analyzed and explained the historical process of science based on the ideas of the scientists and researchers themselves, as a purely intellectual, autonomous, and independent reality. This reality had no connection to the social, cultural, and economic-political elements and movements that have intervened in the processes of scientific and technological production in a specific social formation.

In this context of positivism, O. Tannery proposed the need to separate the history of science from philosophy and to understand it as part of general history. This proposal had an important influence on a generation of historians of science headed by George Sarton and Aldo Mieli (Motoyama, 1987, 45). Sarton, of Belgian origin, was one of the most significant representatives of this group of "historians of science" and worked constantly for the constitution of it as a discipline, independent, specialized, and foreign to any metaphysical concern (Fichant, 1975, 63).

Thus, the program of a positivist history of science considered science to be one and unique, becoming continuous and linked to the experimental method. Therefore, the unity of science resided in its methods (all of them variations of the experimental method) and its common goals. The history of science was thus the history of how the ordering of results and methods into theories was organized (Sarton, 1950).

This position had, as a result, the reinforcement of *internalism* (Quevedo-Vélez, 1993a, 25). This conception of the History of Science predominated until well entered the twentieth century.

Since the mid-nineteenth and early twentieth centuries, the first works of History of Science began to appear in Latin America, largely oriented from the perspective of the enlightenment idea of "progress" and with a descriptive, cumulative, and linear approach. This History of Science, and its variants, studied by other authors throughout this book, will be hegemonic in Europe, the United States, and Latin America until the 1930s (Lértora Mendoza, 1986b, Motoyama & Guimaraes, 1980).

16.1.2 The Second International Congress on the History of Science and Its Consequences: Internalism Versus Externalism

In 1930, European historians of science convened the Second International Congress on the History of Science (Huerga Melcón, 2004, 680–681). The Russians saw in that Congress the opportunity they were waiting for to present to the Western world their progress in the study of science, scientific planning, and science and technology policy. Therefore, the Russian Stalinist government organized a delegation of scientists who, in turn, were historians of science, to participate in the Congress.

This II Congress met in London from June 29 to July 4, 1931, and the presence of the Russian delegation gave the event an unexpected turn, since the Russian physicist Boris Hessen presented a paper, using the approaches and methods of Dialectical and Historical Materialism. From this perspective, Hessen affirmed that Newtonian physics was a scientific response to the problems that the Industrial Revolution had posed to British scientific thought. In addition, he concluded that it was then necessary to begin a new history of the sciences: a social history of the sciences, based on an understanding of the economic-political relations between the material processes of social reproduction of societies and the natural scientific explanations of the material and social world (Hessen, 1985, 14).

The followers of the internalist-positivist conception closed ranks around a clear opposition to these Marxist thesis and devoted themselves to highlighting the intellectual aspect of science, emphasizing the autonomy of scientific thought without concerning social conditions (Motoyama, 1987, 46).

Nevertheless, some younger British historians took up the Russian delegation's approaches. Notable were the works resulting from this process, written by John D. Bernal (Bernal, 1976a, b), Joseph Needham (Needham, 1964), and other authors. This influence will appear in Latin America, but later, through the work of the Cuban José López Sánchez (López Sánchez, 1950, 1987, 1997, 1999) and the Mexican Elí de Gortari (De Gortari, 1980a), among others.

Another alternative path resulted because of the presence in the Congress' auditorium of the young American sociologist Robert K. Merton, a candidate for a doctorate at Harvard University. His interest in the debate would give rise to his doctoral thesis in Sociology (Merton, 1938), in which he studied the influences of society on the development of science and its institutionalization in seventeenth century England, but from the perspective of Sarton's positivist History of Science and the Talcott Parsons' American functionalist sociology, among others (Merton, 1984). This thesis configured as the inaugural piece of an *externalist*, positivist-functionalist and anti-Marxist trend, of North American imprint, which auto-called itself Sociology of Science (Merton, 1949, 1977), and which would play an important role in the Social Studies of Science in the following decades.

New advances appeared in the discipline in Latin America, between 1940 and 1950, during the decades after the II London Congress. Most of them, were still under the influence of the Enlightenment traditions, but re-read from a positivism strengthened by the post-Congress debates (Ortiz & Pyenson, 1985, Lértora Mendoza, 1986b, 139–140, Papp, 1985, 123–128, Azevedo, 1955, 1971, Beltrán, 1984, López Sánchez & Torriente Brau, 1972, Vessuri, 1984, Valdizán, 1928, 1944).[1]

16.2 The Scientific Revolutions of the First Half of the Twentieth Century and the New European Trends in the History of Sciences

In France, during the 1930s and 1940s, new proposals for the practice of the history of science emerged.[2] The preceding histories of science had emerged as a result of the reflection on the birth of modern science, in the 16th and 17th centuries. But these new proposals were the consequence of the reflection on the revolutions that took place in the actual history of science, at the beginning of the twentieth century: the development of non-Euclidean geometries, the theory of relativity, the beginnings of microphysics, and so on (Lecourt, 1973, 24–25).

In this atmosphere, some new researchers began to work in these perspectives. One of the first ones was the French philosopher Gaston Bachelard, that interpreted Einstein's physics as an *epistemological rupture* with previous scientific thought and proposed the notion of "discontinuity" to explain this rupture (Bachelard, 1934), thus, constructing a new epistemology that confronted the *continuist* principle of the positivist History of Science (Lecourt, 1973, 26). This new epistemology posted that "scientific knowledge" was produced in rupture against "common knowledge", displacing it (Bachelard, 1981). This led to a historical epistemology, which would account for that displacement (Bachelard 1973b, 148–164); or, what his disciple, Georges Canguilhem, would later call, an *epistemological history of the sciences* (Canguilhem, 1975a, 9–23, 2009).

---

[1] I will not sound in this subject because it appears in other chapters of this book.

[2] As another chapter of this book studies the influence of new French epistemology in Latin American History of Sciences, I will just touch here some general spots, in order to have further space to speak broadly about its influence in Colombia, where it was very important.

This epistemological history did not believe in the existence of a Science, with a capital letter, recognizing that there is not one, but many sciences, each one of them with its intellectual object of study, whose formulation marked the beginning of its history and defined its methods. However, its statement that science had begun with Galileo and that all historical knowledge about science, before him could be considered as non-science or ideology, was still a positivist lag in Bachelard's Historical Epistemology (Bachelard, 1973c).

In parallel, Alexandre Koyré, a Russian philosopher of Jewish origin, who had moved to Paris in 1934, argued that Galileo's work configured it self as a "scientific revolution" because it had changed the way of looking at the physical world, using mathematics as a tool of analysis (Koyré, 1940). This new orientation allowed him to overcome the positivist facet of the Bachelardian approach. According to Koyré, the history of science should focus on the understanding of the revolutionary processes, or rupture processes, of the different ways of doing science through different historical processes (Koyré, 1973a).

A few years later, between 1945 and 1950, and as a projection of Robert K. Merton's doctoral thesis, the so-called *Sociology of Science* (Merton, 1938, 1945, 1949) began to take shape and consolidate, especially in English-speaking countries (Motoyama, 1987, 45). Merton elaborated an approach in which he considered science as a social institution, in which scientific communities worked, with a characteristic ethos, and subjected it to an analysis based on structural functionalism (Merton, 1945: 366–405).

In the 1960s, a new model for the history of science and technology emerged, contributed by the American physicist, philosopher and historian of science, Thomas S. Kuhn (Kuhn, 1962, 1975). He proposed the alternation of *extraordinary science* and *normal science through* time. Taking from grammar the concept of *Paradigm*, which means "a typical way of doing something", Kuhn proposed, dynamizing Koyré's ideas, that science was a succession of paradigms, that is, typical ways of doing science, in given historical times, which are born, develop, decay, and disappear by *scientific revolutions*. *Normal science* would be that practiced and developed by the accumulation of knowledge within the theoretical-methodological model that characterizes a given paradigm. *Extraordinary* or *abnormal science* is that which begins to answer the extraordinary questions that arise in the context of the dominant paradigm and that can not be solved by the paradigmatic scientists. As these types of questions and problems multiply and the paradigm becomes increasingly incompetent the moment of paradigmatic rupture arrives. The *scientific revolution* and the old paradigm begin to bump into each other, until its replacement by the new paradigm. Thus, the extraordinary science ends up becoming the normal science of the new paradigm (Motoyama, 1987, 47).

In these same decades from 1960 to 1970, and especially in the context of the May 1968 Revolution in Paris, the French perspective of the epistemological history of sciences, promoted by Bachelard and his followers, was reread from the standpoint of historical materialism by Louis Althusser, Michel Fichant, Michel Pécheux and Dominique Lecourt. These thinkers took up Bachelard's *discontinuist* epistemology but added a materialist tinge, and constructed an epistemological history of the sciences, based on political economy. Upon this new epistemology, they initiated critiques on French epistemological internalism, functionalist Sociology of Science and Leninist empiricist theory of knowledge (Althusser, 1970b, 1974a, 1975a; Fichant, 1975; Fichant & Pécheux, 1975; Lecourt, 1973, 1975, 1982, 1974; Althusser, 1975b).

16.2.1 The Reception of Positivist Sociology and History of Science in Latin America and Colombia 1960–1979

While all this was happening in Europe, since the 1960s, and within the framework of the developmental ideas and policies of the Alliance for Progress, in Latin America and Colombia began to rise a change of perspective on the importance of the social role of science and technology as instruments for development.

In this context, the Brazilian government, interested in bringing its country into the circle of developed countries, introduced policies to implement science and technology.[3] An important synthesis of the historical process of science in Brazil was described in Simon Schwartzman's book: *Formação da comunidade científica no Brasil* (Schwartzman, 1979).

In the same context, the First Mexican Colloquium on the History of Science and Technology was held in Mexico City, on September 2–7, 1963 (Beltrán, 1964). Because of this Colloquium, the Mexican Society for the History of Science and Technology was created and formally established 1 year later on August 11, 1964 (Beltrán, 1984, 21).

In Colombia, in 1960, a curricular reform was implemented in the Sociology academic program of the National University of Colombia and a new period of Sociology began in the country. The rationale for this change was set out in a document written by the Sociology Faculty professors of the National University (Professors of the Department of Sociology, 1969). "Sociologist Darío Mesa [was] the most influential personality during that period [...]" (Cataño, 1986, 41–42). With his guidance, national sociologists began to seek new fields of work and research.

Thus, within the framework of development tendencies and policies, the Sociology of Science began its penetration in the country, during the early 1970s, based on the ideas of Merton and Talcott Parsons. On the other hand, the German Sociology of Knowledge began to influence students through the work of Max Weber, frequently cited by the sociologist and Professor Dario Mesa in his classes at the National University. These, and the works of his disciple, Gabriel Restrepo, were the vehicles of these influences on groups of Colombian sociologists, who began to take an interest in externalist historical-sociological studies of science and technology in the country.

[3] These processes are studied in other chapters of this book.

Eight years later, on November 20, 1968, President Carlos Lleras Restrepo created the Colombian Fund for Scientific Research and Projects "Francisco José de Caldas" (COLCIENCIAS). This initiative was part of his State reform program, consistent with the developmental policy of the Alliance for Progress to promote the development of science and technology as tools to ensure industrial growth and consequent economic productivity. President Carlos Lleras Restrepo attached the Fund to the Ministry of National Education (Plata, 2013, 66–68).

Within the framework of this policy, COLCIENCIAS organized the First Colloquium on the History of Science in Colombia, at the Quirama Campus, in the city of Rionegro, Antioquia, from September 25 to 28, 1969, to discuss the state of science in Colombia. The aim of the Colloquium was the construction of "the general history of sciences that had been most cultivated in the country […] and the way that should be pointed out for further and more complete research in this important aspect of the cultural history of Colombia" (Colciencias, 1970). Because of this Colloquium, a volume entitled *Apuntes para la Historia de la Ciencia en Colombia (Notes for the History of Science in Colombia)* was published, in which 8 papers were collected, written by different scientists and science professionals. However, all these articles had a descriptive and internalist character and, although they contributed important data, they did little to explain the processes.

Eight years later, in 1978, a new book entitled *Science and Technology in Colombia* appeared, also published by COLCIENCIAS, with the collaboration of the Colombian Institute of Culture. The introduction to this text was explicit to define the role of science and technology as tools for development. There is a "[...] growing conviction that science and technology have become major factors in the economic and social process, to such an extent that the differences between developed and underdeveloped countries are largely attributed to the disparities in their stock of knowledge" (Chaparro & Sagasti, 1978, 7).

According to the compilers of the book, the works gathered in this volume had "the objective of giving a global and synthetic vision of the evolution and situation of science and technology at that time in Colombia" (Chaparro & Sagasti, 1978, 9). However, in general, the historical works were, either revisions of the previous publications, or were new works, but, with very few exceptions, almost all of them had been written within historiographical traditions already out of fashion.

Nevertheless, it is important to highlight the work of Francisco Sagasti because, although it was not a historiographic study based on documentary sources, he did attempt to elaborate a first global outline of the History of Science in Latin America. At least, he showed knowledge of the most advanced bibliography on the subject (Sagasti, 1973), produced up to that moment in the continent and in the United States, which already showed symptoms of a new methodological and critical approach to the discipline in the American continent (Sagasti, 1977).

In contrast with these intellectual postures, during this decade, the first manifestations of the Bachelardian influence appeared in Colombia, from two main sources. Luis Carlos Arboleda, professor at the Universidad del Valle, Cali, and mathematician, with a doctorate in History and Teaching of Sciences at the Alexander Koyré Center in Paris, under the guidance of René Taton, was the first one to publish an article, in 1977, with a clear internalist but discontinuist-materialist orientation. In this paper, he gathered the approaches of several European authors of the time, such as Gaston Bachelard, Georges Canguilhem, Alexandre Koyré, Ludovico Geymonat, Pierre Raymond, Nicolas Bourbaki, Louis Althusser and Suzanne Bachelard, among others, who formed part of the discussion that at that time was taking place between the different trends and research groups in Europe (Arboleda, 1977, 101–122).

Secondly, in 1979, a text on the history of Renaissance algebra was published. It was written by Carlos Eduardo Vasco, professor of the Mathematics Department of the Universidad Nacional, in which he manifested a discontinuist and genetical constructivist conception of the history of science, based on the ideas of Jean Piaget, among others (Vasco Uribe, 1985). Vasco also took, as a conceptual framework for his history of Renaissance algebra, several of Althusser's structuralist Marxist approaches to Bachelard, particularly the concept of "spontaneous philosophy of scientists" (Althusser, 1975a), together with Pierre Raymond's methodological approach to the history of science (Raymond, 1973, 1975).

16.2.2 An Innovative Pedagogical Experience: The Seminar on Philosophy and History of Science (SFHC, by its name in Spanish: Seminario de Filosofía e Historia de las Ciencias) at the Colombian School of Medicine (ECM: Escuela Colombiana de Medicina), Bogota, February 1979

With the creation of the new Colombian School of Medicine, in 1979, an innovative pedagogical experience was born in the field of teaching the History of Sciences and Medicine to medical students. The architect of this initiative was Professor Emilio Quevedo V. who, since his undergraduate studies of Medicine and specialty in Pediatrics, had been interested in history as a tool for understanding medical knowledge and practice. While he was attending his regular medical studies, he studied, on his own account and personal interest, the historical epistemology of Gaston Bachelard (1978) and the epistemological history of Georges Canguilhem (1971, 1975b). He also read the archeological history of knowledge of Michel Foucault (Foucault, 1975, 1976, 1977), and the scientific revolutions of Alexandre Koyré (1940, 1973a, b). Aside, he also studied the discontinuist-materialist epistemology of Louis Althusser (1969, 1970a, b, c, 1974a, b, 1975a, b, 1977a, b; Althusser & Lacan, 1970; and Dominique Lecourt 1971, 1973, 1974, 1975; Lecourt et al., 1974). More over, he began to study, again on his own, Henry E, Sigerist's History and Sociology of Medicine and of Public Health (Sigerist, 1974).

Within this background, during an informal meeting, Professor Quevedo proposed to the authorities of this new School of Medicine to introduce in their medical curriculum a seminar on epistemology, history of science, and medicine, the which must cross the entire curriculum and should serve as a beacon for the undergraduate student to understand the historical-epistemological and social processes that lay at the origin, basis, and development of the sciences and medical practices, which he would be studying and learning in the other components of his curriculum (Quevedo-Vélez, 1990b).

The directors of this School crazily accepted the proposal and, from the first to the tenth semester of the curriculum, Profesor Quevedo, with help of the Sociologist Armando Borrero. Began to implement, within the scope of this new Seminar, teaching work based on research in this field. The theoretical-methodological perspective of this seminar was built from the discontinuist and non-positivist historical epistemological perspective, based on the works of the authors mentioned above, Bachelard, Canguilhem, Foucault and Koyré, but from the re-reading that Canguilhem, Althusser and Lecourt had done, turning it into a discontinuist and materialist epistemological history, confronting the empirio-criticist Leninist approach (Quevedo-Vélez, 1990a, 232–235, 238–239).

It is important to emphasize that the curricular structure of that Seminar was very particular. For the first time in the history of medical education, a space like this, for research and teaching, that crossed the entire medical career, was introduced inmedical education. The main objective of this Seminar was "[...] to allow and encourage the development of a critical, active and conscious attitude towards the different scientific and technical contents that students received during their training. This entailed, as a consequence, the possibility of integrating, structurally and organically, not mechanically, the contents generically called social and humanistic with the bio-clinical and medical contents" (Quevedo-Vélez, 1984a, 13).

To achieve these objectives, the Seminar promoted in both, students and teachers, the "reflection and research on the origin, development and problems of the knowledge (concepts) that serve as a basis for medical practice, as well as reflection and research on the different social aspects that make up medical practice itself" (Quevedo-Vélez, 1984a, 13).

In this sense, this reflection and research were carried out in different areas:

- That of the historical process of production and chaining of concepts and theories within the various medical and social sciences.
- That of the different aspects that constitute the daily work of the physician, as an art and as a technique, and of the historical process of its constitution.
- That of the interrelation of these concepts, theories, and practices with scientific and technical fields, not strictly speaking biological, and with other forms of knowledge, thought, and activities different from the sciences, such as philosophy, art, literature, cinema, opera, and religion, etc.
- That of the historical determinants (economic, social, political, etc.) that allow a global understanding of scientific production as part of the productive processes, in general, and of social practices, as well as their implementation.

The seminar covered 10 of the 12 semesters of the curriculum and had a weekly intensity of 4 hours. It was located in the Psychosocial Area of the curriculum but in close relation with the other subjects, both in the same area and in the Bioclinical and Community Medicine areas. Likewise, it was programmed in the same three levels in which the curriculum was articulated (primary, secondary, and tertiary prevention), which were integrated to the three levels of care of the National Health System (NHS). "Because of its intensity, its sustained coverage, and its connection within the curriculum, this seminar constitutes an original and novel integrative experience in the Colombian university" and, why not say it, in the world (Quevedo-Vélez, 1984a, 14).

16.2.3 The First Latin American Meeting of Historians of Science and the Creation of the Latin American Society for the History of Science and Technology (LSHCT), August 1982

In 1981, the XVI International Congress on the History of Science happened in Bucharest (Romania), 3 years after the opening of this Seminar of the Colombian School of Medicine. Within the framework of this event, an informal meeting of the seven Latin Americans participating in the Congress occurred. This meeting was the starting point for the development of a Latin American program for research and teaching in the field of Social History of Science and its institutionalization[4] (Galles, 2002).

Because of this meeting, and thanks to the efforts of Juan José Saldaña and his group, the First Latin American Meeting of Historians of Science was held in the city of Puebla, Mexico, August 23-26, 1982, and it was attended by 200 people from 16 countries, and 120 scientific papers were presented (Galles, 2002).

Within the framework of this meeting, was launched the proposal to create a Latin American Society for the History of Science and Technology (LSHCT), because of the initiative that had arised in the Bucharest Congress of 1981. As a consequence of this meeting and of the creation of the Latin American Society, the publication of Quipu Latin American Journal of the History of Science was proposed and its first issue appeared in January 1984.

The creation of the LSHCT and the journal Quipu were an important step towards the development and institutionalization of the new trends and conceptions that had been developing in Europe and the United States and towards the organization of a community of historians of science in Latin America. This institution quickly became the catalyst for these new trends in the region.

The first president of this Society was Dr. Juan José Saldaña, who trained in France as a historian of science and had brought to Latin America a very rich intellectual background, composed of all those trends that were swarming in the European debates on the history of science after the Second Congress. There for, the LSHCT began to promote the creation of national associations of history of science

[4] This event has been described in other chapter of this book.

and technology in Latin America. In this way, the Society became the engine for the development and growth of this scientific specialty.

Incidentally, the Quipu Journal, in its volume II, published a rather extensive article, 20 pages long, showing the structure, objectives and working methods of the recently created Seminar of Philosophy and History of Science of the Colombian School of Medicine in Bogota. This publication was the first national and international presentation of the Seminar in society and especially to historians of science (Quevedo-Vélez, 1984c, 223–243).

16.2.4 Workshop on the Teaching of Epistemology and History of Science and Medicine in the Training of Physicians and Health Professionals, Colombian School of Medicine, Bogotá (ECM), August 1983

The second outstanding event was the workshop on the Teaching of Epistemology, History of Science and Medicine in the Education of Physicians and Health Professionals. This event was organized by the Seminar on Philosophy and History of Science (SFHC) of the Colombian School of Medicine, on August 10, 11 and 12, 1983, four and a half years after its creation. It was held at the headquarters of the Colombian School of Medicine, in Bogotá, as a previous and complementary event to the First National Conference on History of Medicine, organized by the recently created Colombian Society for the History of Medicine and the Colombian School of Medicine, which was held on August 13 and 14 of the same year.

This workshop had as an important antecedent the celebration of the *Lanceros Workshop of* the Seminar of Philosophy and History of Sciences-SFHC, a closed workshop, which was held at the Lanceros Hotel in the city of Paipa, Boyacá, Colombia (hence its name), on May 19 and 20, 1983. During this activity, a balance was made of the achievements reached up to that moment in the Seminar, both from the teaching and research point of view. It intended to better define the different paths towards the future and to specify more precise and adequate methodologies, both for the pedagogical work, and for the research that was actually being carried out (Quevedo-Vélez, 1984a, 11–18).

This Lanceros Workshop allowed the preparation of the topics that would be presented and discussed publicly, 3 months later, at the Workshop on the Teaching of Epistemology, the History of Science and Medicine in the Training of Physicians and Health Professionals, which, as mentioned above, was held on August 10–12, August. Professor Agustín Albarracín Teulón, researcher at the Spanish National Research Council and disciple of Professor Pedro Laín Entralgo, and Professor Felipe Cid of the Autonomous University of Barcelona, director of the Museum of the History of Medicine in that city and disciple of Georges Canguilhem, participated as special guests. This invitation responded to the fact that the professors and students of the Colombian School of Medicine (ECM) had already begun to realize that Historical Epistemology and Social History of Science were fundamental to understanding the knowledge of the medical sciences (Quevedo-Vélez, 1990a), and that it was also necessary a historical and sociological study of the techniques with which both medical and sanitary practices had been built, that is, first as *Tekné* and then as a profession, as it was insinuated in one of the papers presented by Emilio Quevedo in that Workshop (Quevedo-Vélez, 1990b, 289).

In this workshop, the teaching and research experiences of the Philosophy and History of Sciences Seminar were shown, discussed, and contrasted with similar experiences presented by the national and international speakers who attended the event. A group of students from various semesters, the professors and coordinators of the seminar, the dean of the Colombian School of Medicine, and several guest professors, both from that school and from other Colombian universities, participated in these discussions. This was the case of Luis Carlos Arboleda, from the Universidad del Valle; Luis Enrique Orozco, from the Universidad de los Andes; Roberto Serpa Flórez, from the Universidad Industrial de Santander; and, the foreign guest professors Agustín Albarracín and Felipe Cid. In addition, Professor Francisco Guerra, from the Universidad de Salamanca in Spain, Dr. Eduardo Estrella and Dr. Rodrigo Fierro, from the Universidad Central de Quito, attended the seminar. Eduardo Estrella and Rodrigo Fierro were going to participate in the forthcomming First Conference on the History of Medicine, during the following days. The papers read and discussed during this event, were published in 1990 by the Colombian School of Medicine, in the book Proyecciones de un Programa, which was part of the Medical Education Collection, Volume III (Autores Varios, 1990).

As a result of this Workshop, the group of researchers linked to the SFHC of the ECM began to build a series of academic and research relationships. On the one hand, with the Spanish groups represented at the Workshop by Albarracín, Cid, and Guerra. The SFHC group already had prior knowledge of the works of Pedro Laín Entralgo, director of the Arnau de Vilanova Institute for the History of Medicine in Madrid (Laín Entralgo, 1961, 1964, 1976, 1982), the institute where Albarracín was working. These students had also studied Cid publications (1974a, b, 1978). With them, during the days of the Workshop, extensive conversations were held on the research project that the Seminar group was designing on José Celestino Mutis and the Bourbon reforms in the New Kingdom of Granada, within the framework of Enlightenment Medicine.

On the other hand, the SFHC group stablished relations with the Ecuadorian group, with whom the group had not had contact before. They also stablished relations with the Colombian researchers Luis Carlos Arboleda, from the Universidad del Valle, and Diana Obregón, from COLCIENCIAS. They, since the beginning of the year had been working on a proposal of a project on the Social History of Science in Colombia, which was being promoted by COLCIENCIAS, with the support of the Regional Program for Scientific and Technological Development in the Organization of American States (OAS). Arboleda and Obregón's attendance at this meeting boosted the participation of Néstor Miranda and Emilio Quevedo as researchers in this COLCIENCIAS project, as will be seen below.

16.3 The Definitive Institutionalization of the Social History of Science in Colombia

16.3.1 The Magna Instauration*: The COLCIENCIAS Project, February 1983*

In February 1983, COLCIENCIAS, with the financial support of the Regional Program for Scientific and Technological Development of the Organization of American States, began to work on the design and implementation of a large project called Social History of Science in Colombia. The first stage of the research began in March 1983 and was coordinated by Carlos Eduardo Vasco, on behalf of the Colombian Society of Epistemology, and Diana Obregón, on behalf of COLCIENCIAS. For the development of this project, "an interinstitutional and interdisciplinary team was organized. It was composed of specialists in the different sciences under study and social scientists, linked to the main universities in the country" (Quevedo-Vélez, 1993b, 13).

Initially, the researchers participating in this project, and their research topics were: Luis Carlos Arboleda, History of Mathematics; Jorge Arias de Greiff, History of Astronomy; Armando Espinosa, History of Geology; Olga Restrepo, Natural History and Botany; Jesús A. Bejarano, History of Agricultural Sciences; Gabriel Poveda Ramos, History of Engineering and Techniques; Regino Martínez-Chavanz, History of Physics; José Luis Villaveces, History of Chemistry; Emilio Quevedo Vélez, History of Medicine (17th and 18th centuries); Néstor Miranda, History of Medicine (nineteenth century); Gabriel Restrepo and Olga Restrepo, History of Social Sciences (later joined by Salomón Kalmanovitz, Rodrigo Parra Sandoval and Telmo Eduardo Peña Correal). The teams began their work and met periodically to comunicate their partial results and discuss them with the other participants.

16.3.2 The Seminar on Methodology for the Social History of Science in Latin America, November 1983

During the first week of November 1983, the coordinators of the COLCIENCIAS project, at the request of the participating researchers, organized an International Seminar for the study of the Methodology of the Social History of Science in Latin America, as part of the activities of the project funded by COLCIENCIAS and the OAS. The need to hold this seminar arose from the realization, on the part of the researchers, that all the participants in the Project had different approaches to the history of science and that only a few had approaches to the new trends of the Social History of Science. For this reason, in some of the first meetings, the researchers concluded that, to develop the Project, it was necessary to promote a discussion on the new trends in Sociology and Social History of Science that were being debated in Europe and the United States. Luis Carlos Arboleda, Diana Obregón and Emilio

Quevedo, who had joined the recently created LSHCT, were already participating in the project and they proposed to organize an international seminar with researchers from different Latin American countries, who were also linked to this Society, and who formed a group that was becoming a spearhead of these new trends in the continent. The idea was to discuss, with the participation of historians of science from several Latin American countries, what the methodology for a social history of science should be.

Once the dates of the seminar were set, researchers from Mexico, Venezuela, Peru, and Spain were invited to participate. The LSHCT had a very active representation and participation in the development of the Seminar, and it was possible to "discuss the advances that up to that moment had been achieved by the different Ibero-American academic groups, especially in the definition of the theoretical and methodological instruments of the history of science in the Region" (Quevedo-Vélez, 1993b, 13).

The seminar included presentations to stimulate discussion by Carlos Eduardo Vasco and Diana Obregón, project coordinators; Joseph Hodara, Israeli sociologist and journalist, who was working on the development of science in Latin America; Juan José Saldaña, president of the LSHCT; Ubiratan D'Ambrosio, from the University of Campinas in Brazil; Ruy Gama, from the University of Sao Paulo; Hebe Vessuri, researcher at the Venezuelan Institute of Scientific Research (IVIC); Néstor Miranda and Emilio Quevedo, from the Colombian School of Medicine in Bogotá; Pierre Raymond, professor at the Pontificia Universidad Javeriana; Antonio Lafuente and José Luis Peset, from the Spanish National Research Council (CSIC); Ernesto Yepes del Castillo, historian of Peruvian medicine; and Luis Carlos Arboleda, professor at the Universidad del Valle.

In his inaugural lecture, *Hacia una crítica histórica, teórica y metodológica de la historiografía latinoamericana de las ciencias*, Juan José Saldaña, after criticizing the history of science practiced by historians who had participated in the First Ordinary Meeting of the Mexican Society for the History of Science and Technology (SMHCT), and thinking about a Latin American History of Science, made the following considerations:

This Latin American history of the sciences:

1. "It will not be an internalist history.
2. "It will not be a continuist history.
3. "It will not be an externalist history.
4. "It will not be a history of contributions.
5. "It will not be a history of the 'universal' sciences.
6. "It will not be a positivist history of the sciences.
7. "It will not be a history detached from the epistemology and sociopolitics of the sciences.
8. It will not be a recursive history (Saldaña, 1983, 343–344).

Saldaña's intervention in this seminar marked a turning point in the history of science in Colombia, because, from that moment on, the History of Science in Colombia began to stop being the collection of "an ordered and cumulative set of true knowledge [...] and the gallery of martyr heroes who, anointed with genius and through a life of sacrifice, made important discoveries" (Forero Pineda, 1993). Indeed, this Seminar created the conditions for the group of researchers linked to the LSHCT and, especially, the intellectual guidance of Juan José Saldaña, its president, to become the illuminating beacon of the path towards the Social History of Sciences in Colombia. This influence was able to bear fruit because it found a previously fertilized soil: the group that COLCIENCIAS had assembled for this project already had, on the one hand, sociologists linked to the sociology of science proposals of Merton and Weber, as was the case of Olga and Gabriel Restrepo; on the other hand, historians of science and professions close to the French epistemological history of Bachelard, Canguihem, Althusser and Lecourt, such as Luis Carlos Arboleda, Néstor Miranda and Emilio Quevedo; and, the latter, close to the school of the Spanish History of Medicine of Pedro Laín Entralgo and his disciples, and the German-American school of the Social History of Medicine of Henry Sigerist. This, in addition to other scientists who, like Jorge Arias de Greiff, Armando Espinosa, and Regino Martínez, already had experience in the history of their respective sciences in Colombia. For example, in the case of Arias de Greiff, who had already done several works on Francisco José de Caldas, since 1960.

In this Seminar, Emilio Quevedo presented the paper, *The Relationship between the History of Sciences and Techniques in Medicine*, which expanded what he had been proposing since the previous meetings, and in which he showed that for the historical study of the professions, the History of Sciences was not enough, since the professions had a practical component, of a technical nature, which did not consist simply in a knowledge materialized in the capacity for technical and instrumental action. It was insisted that all those preventive, diagnostic, therapeutic, and rehabilitative capacities of which medical practice was composed, developed historically in the interaction of social and human relations articulated to the processes of social production, which implied the daily economic-political reproduction of a given social formation (Quevedo-Vélez, 1983). However, since most of the members of the COLCIENCIAS research team were scientists and did not know the internal details of medical practice, they were unable to see the social dimensions of the professions, that Quevedo was posting.

From then on, and throughout the 4 years of the project, the group of historians who worked on it maintained constant interaction with the international researchers who had participated in the seminar and with their research groups. Some of them were linked to the LSHCT and many of them frequently attended the Latin American Congresses on the History of Science and Technology, which the LSHCT organized every 3 years. In the cases of the History of Sciences group of the Universidad del Valle and the Seminar of Philosophy of the History of Sciences of the Colombian School of Medicine, this relationship was constant and fruitful. They were perhaps the only two groups that maintained the orientations proposed by Saldaña, although with variations because of the disciplines of study and the evolution of each group.

The results of the COLCIENCIAS History of Science Project were published individually in the journal *Ciencia Tecnología y Desarrollo*, between the end of

1985 and the beginning of 1986. Despite the heterogeneity of its results, due to the diversity of the academic orientations and backgrounds of the participants, the project achieved a first methodological and conceptual approach to the problem of the Social History of Science in Colombia, with the possibility of influencing national scientific and technological policy.

After the end of the project, the influence of this group of researchers on the others was decisive: the project became "the institutional space where the new international trends in the Social History of Science began to be received, acclimatized and domesticated, and we can say that most of the groups that are still working have been doing so under the parameters and paths that emerged from the discussions within this team of researchers and from the concrete results of the project" (Quevedo-Vélez, 1993b, 39).

Finally, the three general orientations that determined the project were the sociological study of science, within the framework of the Mertonian paradigm; the epistemological history of sciences, with a Bachelardian approach, in its Althusserian and Lecourtian re-reading; and that of the new social history of sciences and technologies, contributed by the LSHCT, which attempted to integrate all these elements in a theoretical-methodological framework of its own. The results of the project were finally published in their entirety in the book series entitled *Historia Social de la Ciencia en Colombia*, in 10 volumes, between 1993 and 1996, under the academic coordination of Carlos Eduardo Vasco, Diana Obregón and Luis Enrique Orozco, and the scientific and editorial direction of Emilio Quevedo (Quevedo-Vélez, 1993–1996).

In this way, the Social History of Science in Colombia entered to form part of this discipline in Latin America, participating and contributing to the integration of these three conceptual and methodological traditions that emphasized the importance of understanding the existing relationships between the national and social context, and the meaning of science at each given moment (MacLeod, 1982).

From another integrating perspective, which makes possible a history of science and technology proper to countries outside the "main stream" trends of scientific and technological development, it was "a history of science that, because of its discontinuist historical-epistemological tradition, recognizes the importance of the analysis of the internal development of scientific concepts and their appearance through epistemological ruptures or scientific revolutions, but which, because of its discontinuist materialist vision, accepts that this internal development from one group of theories and concepts to others is a productive process that is part of the general production and social reproduction processes. Likewise, because it assimilates sociology, it understands science as a social activity and its process of organization as a process of institutionalization" (Quevedo-Vélez, 1993b, 40). At the end of this Seminar, Saldaña proposed to Emilio Quevedo to organize and direct the Section of History of Medicine of the LSHCT,

16.4 The Consequences of the COLCIENCIAS Project: The Developments of the New Social History of the Sciences in Colombia

16.4.1 The Jornadas de Historia de la Medicina Hispano-Andina, *Quito, April 23–27, 1984*

Several processes resulted from this COLCIENCIAS project. In the first place, a certain continuity was established between the presentations and discussions that had taken place the previous month, August, at the Workshop on the Teaching of Epistemology and History of Science and Medicine, held at the ECM. This workshop motivated the first invitation of Emilio Quevedo, his group of professors-researchers and his outstanding students, as foreign speakers, to an international event: the First Spanish-Andean Conference on the History of Medicine and the First International Course on the History of Spanish-Andean Medicine, organized by the Quito group, with the collaboration of the Spanish group. These conferences were held in Quito, Ecuador, from April 23 to 27, 1984, and were attended by several Spanish historians of medicine, including Agustín Albarracín, José Luis Peset, Delfín García Guerra, Elvira Arquiola, Luis Montiel and Diego Gracia Guillen, among others. During these conferences, friendships and cooperation processes were established among the participating groups.

16.4.2 Modern Science and the New World: First Meeting on the History of Science and Technology of the Iberian and Ibero-American Countries, Madrid, September 25–28, 1984

The combination of the results of the workshop held at the Colombian School of Medicine, the methodological seminar of the COLCIENCIAS project, the Quito Conference, and the membership of the Latin American Society for the History of Science and Technology, resulted in the invitation to Luis Carlos Arboleda and Emilio Quevedo, as international speakers at the First Meeting on the History of Science and Technology of Iberian and Ibero-American Countries, organized by the Spanish group and held in Madrid from September 25 to 28, 1984. This meeting was attended by other invited Latin American speakers: Ruy Gama, from Brazil; Jorge Arias de Greiff, from Colombia; Celina A. Lértora Mendoza, from Argentina; and Juan José Saldaña, Carlos Viesca Treviño and Elsa Malvido, from Mexico. In this meeting Arboleda read a paper on *Dificultades estructurales de la profesionalización de las matemáticas en Colombia* (Arboleda, 1985); and Quevedo, a paper on *La Ilustración y la enseñanza de la Medicina en la Nueva Granada* (Quevedo-Vélez, 1985). In these papers, both gave an account of the research that each one had been conducting in their groups, based on the new trends in the history of sciences and, in the case of Quevedo, on the conjunction of the new orientations of the social history of sciences, linked to the Sociology of professions and to the methods of analysis of the new Spanish history of medicine, materialized in the works of Pedro Laín Entralgo and his disciples, which Quevedo had discussed with Albarracín and Cid, during the said Workshop held at the Colombian School of Medicine the previous month.

This Madrid meeting served as a space for Latin American researchers to interact with each other and with different working groups in Spain. It also cleared the way for the Colombian School of Medicine Group to build a closer relationship with Spanish medical historians of a new type, strengthening future collaborations, especially with José Luis Peset, Antonio Lafuente, Delfín García Guerra, and Elvira Arquiola, who, in their research, were already going beyond the internalism of Laín Entralgo. This experience was aimed at overcoming the isolation that had characterized Latin American historians of medicine and science. The publication of the works presented there came out the following year under the title *La Ciencia Moderna y el Nuevo Mundo* (Peset et al., 1985).

16.4.3 First Latin American Seminar on Alternatives for the Teaching of the History of Science and Technology, Cali, November 1984

Also, because of the COLCIENCIAS project, and framed within the LSHCT strategies, the Latin American Seminar on Alternatives for the Teaching of the History of Science and Technology was held in Santiago de Cali, Colombia, from November 4 to 10, 1984. This event was jointly organized by the LSHCT (based at the National Autonomous University of Mexico), the Seminar on History of Sciences (of the Faculty of Sciences of the University of Valle, in Cali) and with the support of the Colombian Institute for the Promotion of Higher Education (ICFES). The seminar was held within the framework of an LSHCT strategy that consisted of discussing these problems of teaching the History of Science at the universities where they worked. Also, to train teachers in this subject and to produce teaching texts.

"The Seminar was convened to identify, stimulate and reinforce the experiences of teaching the History of Science in these countries and to foster the exchange of ideas, research programs and comparative studies in this discipline" (Arboleda, 1986b, 2).

In this event participated all historians of science who had been carrying out academic and professional activities on this topic in Colombia. A total number of 40 researchers participated, including international visitors. Some members of the newly created LSHCT were also present in this meeting, such as: Juan José Saldaña, current President of the LSHCT; Ubiratan D'Ambrosio, who at that time was the secretary of the Brazilian Society for the History of Sciences; and Jorge Núñez, who later would became the president of, what was constituted years later as the Society for the History of Sciences of Cuba (Arboleda, 1986b).

Before beginning the discussions that were to take place in this seminar, in the Inaugural Conference, entitled *General Balance of the History of Science in Latin America*, Juan José Saldaña, in his capacity as president of the LSHCT, resumed his role as intellectual guide of the group, stating that two reasons explained why, up to that moment, we still did not have a History of Science in Latin America. Quoting the Mexican researcher Enrique Beltrán, he said that the first was the isolation and scarce contact among historians of science in the continent (Beltrán, 1984). Saldaña, going further, stated that the second reason was the limited space that the History of Science had occupied up to that time in the culture of each of the Latin American countries. In this sense, he remarked that this fact had already been pointed out by different Latin American researchers such as Shozo Motoyama, from Brazil (Motoyama & Guimaraes, 1980); Eli de Gortari, from Mexico (De Gortari, 1980b, 149); and others. Referring to the debates and discussions held the previous year in the framework of the past Seminar on Methodology of the Social History of Science, organized in Bogotá by COLCIENCIAS, he stated that this second cause had also been recognized by several of the historians attending that event. He concluded that "both reasons, the isolation of Latin American countries and the isolation of historians of science and technology within their own countries, make the activity that has been being carried out in the field of the History of Science and Technology an adventure and a discovery" (Saldaña, 1986, 3–4).

Based on these reflections, Saldaña concluded that "Latin American integration is something yet to be built, although at present there are several signs that an integrating process has begun [...] in the face of an essentially unfair and aggressive structure of international relations, Latin American countries are beginning to find that they have much to gain by associating and making historically true the similarities between them and the tendencies to bring together our present and our future" (Saldaña, 1986, 4). And, he wondered if what was beginning to happen in the political and economic spheres could also occur in the sciences and culture; and he affirmed that "the comparative, regional history of our scientific-technical evolution could help to understand the tendencies that constituted the present moment and, even more, to unravel the role that the sciences can play in the future and the conditions that need to be met for this" (Saldaña, 1986, 4–5). Likewise, he insisted that what would be achieved in this seminar would imply a modification effect, "from didactics to the very role of teaching in our countries. By the way, this is an effect on present affairs and not a small one" (Saldaña, 1986, 5).

The discussions held at this Seminar were based on the experiences of other international centers led by Professors J.C. Guédon, Canada; F. Bevilacqua and G. Bellodi, Italy; and H.W. Schütt, from the Federal Republic of Germany. Based on these experiences, the Latin American participants discussed their own experiences in one of the three workshops held, depending on the topic:

- The History of Science in the curricula, especially in the Faculties of Science and Engineering.

- Different alternatives for the dissemination of the history of science in the region, and
- Training of Latin American researchers in the History of Science.

These debates were moderated and coordinated by the professors of the Seminar on History of Sciences of the Universidad del Valle: Ángel Zapata, Ramiro Tobón, and Luis Carlos Arboleda. Among the most important conclusions of this Seminar was the recommendation to the LSHCT to promote the implementation of a package of regional and subregional programs on the three fronts mentioned above" (Arboleda, 1986b, 1). All this should lead, in the medium term, to the creation of a Latin American Center for *Research and Postgraduate Studies*, whose dissemination organ should be the QUIPU Journal. This should give rise to a Scientific School in this academic area, which would become autonomous and internationally competitive (Arboleda, 1986b, 1–2).

In addition to the workshops, three symposia and a series of conferences were held to show the variety of interests and new approaches to these studies in Latin America:

- "Epistemology and History of Science, with emphasis on the discussion of the object and method of historical research in the Region,
- History of Science and Ethnoscience of pre-Hispanic Cultures, and
- Sciences in Colombia at the time of the botanical expedition of José Celestino Mutis (End of the XVIII century)".

The Cycle of Conferences brought together studies that were already underway at that time, such as the history of pre-Columbian mathematics in Colombia (Víctor Albis); the numerical representation of some pre-Columbian cultures (Diego Pareja); the Enlightenment and teaching of mathematics in New Granada (Luis Carlos Arboleda); and the History of Physics in Argentina (Celina A. Lértora). Lértora also gave a lecture with methodological proposals for the study of primary sources in the history of science (Lértora Mendoza, 1986a). For his part, Arboleda presented a program for the training of Latin American researchers in the history of science at the third workshop mentioned above (Arboleda, 1986a). It is important to highlight Ubiratan D'Ambrosio's paper on an alternative proposal from Ethnoscience for the history and teaching of science, with an outline of the fundamental principles of the new field of ethnomathematics that was then being constituted in the international arena under his guidance (D'Ambrosio, 1986). Emilio Quevedo could not travel to this Seminar, but he sent a paper, which was read by one of his colleagues (Quevedo-Vélez, 1984b).

Within the framework of this meeting, the Colombian Society for the History of Science and Technology was constituted, under the presidency of Jorge Arias de Greiff; the vice-presidency of Regino Martínez; the executive secretary of Luis Carlos Arboleda; and with control and auditing positions, headed by Diana Obregón and José Luis Villaveces. Although this Society had an ephemeral life, it was part of the strategies of the LSHCT to promote and shape the development of Societies in different countries.

From that moment on, the Cali group began to produce the *Boletín de Historia de las Ciencias*, which had a dozen issues during the decade. Finally, this Seminar favored the consolidation of the Seminar on the History of Science and Technology of the Universidad del Valle (Arboleda, 1986a, 2).

16.4.4 Consolidation of the Research Line on the History of Science and the Seminar on the History of Science and Technology at Universidad del Valle, 1984

In this 1984 Cali Seminar, the Research Line on the History of Sciences of the Universidad del Valle was consolidated as part of the strategies of the LSHCT. All this generated a ferment of activities in the field of History of Sciences, particularly in Cali, which led to the constitution of the Seminar on History of Sciences, in which Luis Carlos Arboleda participated in the area of mathematics, Armando Espinosa in the area of geological sciences, the physicist Álvaro Perea of the Faculty of Sciences, Ramiro Tobón in physics, Pedro Rovetto in medical sciences and Alfonso Claret Zambrano in the history and teaching of biology. But the most visible and important head of the leadership of this Seminar was Professor Angel Zapata, who was an engineer, a humanist and an intellectual of great value, trained at the University of Antioquia in Medellin, and who had been the creator of different academic programs at the Universidad del Valle and, in particular, the Chemistry Program in the Faculty of Sciences and Chemical Engineering (Arboleda, 2024).

Taking as a reference the Seminar of Philosophy and History of Sciences, which had been initiated by Emilio Quevedo at the Colombian School of Medicine in Bogota since 1979, a very interesting initiative was produced, within the framework of this Line of Research, which was new in the country. The idea was to create a humanistic training platform, based on the history of sciences, for all the careers offered by the Universidad del Valle. In this way, students could fill the Humanities and Philosophy credits of their undergraduate courses by participating in the newly consolidated History of Science Seminar. This teaching strategy, which continued for several years, was organized in the faculties of Science, Engineering, Education, and Humanities. The teaching texts of this Seminar were published by the Colombian Institute for the Promotion of Higher Education-ICFES in 1990. This Institution had been promoting, since 1984, this type of approach to introduce the study of the history of science in the training of professionals in science and engineering, humanities (Arboleda, 1990).

16.5 Other Significant Activities of the LSHCT for the Development of the History of Science in Latin America and Colombia

16.5.1 The First Latin American Congress on the History of Science and Technology, July 21–25, 1985

The Latin American Congresses on the History of Science and Technology were one of the fundamental activities of the LSHCT. They met once every 3 years and almost always had a theme that brought researchers together. The First Latin American Congress on the History of Science and Technology was held in Havana, Cuba, July 21–25, 1985. This was a second opportunity for a massive meeting of Latin American historians of science and technology to exchange new ideas and theoretical and methodological approaches to research and teaching.

To this Congress, Emilio Quevedo traveled accompanied by a group of undergraduate students from the Colombian School of Medicine, who had actively participated as co-investigators of the projects of the Seminar of Philosophy and History of Science of the Colombian School of Medicine. He also served as Coordinator of the History of Medicine Section of the Congress and read two presentations and a free paper, which were the result of two different research projects within the SFHC: the research project on the history of the Bourbon educational reforms in Colonial New Granada (Quevedo-Vélez & Hernández de Alba, 1985; Quevedo-Vélez & Zaldúa, 1986) and the research project of the history of diseases and epidemics (Quevedo-Vélez, et al., 1986a, b).

16.5.2 Latin American Database Project for the History of Infectious Diseases and Health Education, January 1986 to December 1988

In 1986, the Ministry of Culture of Spain, the Latin American Society for the History of Science and Technology, the Colombian Society for the History of Science and Technology and the Colombian School of Medicine signed an agreement to carry out a project for the construction of a Latin American database to collect, classify and systematize the existing documents in each Spanish-speaking country to support a history of infectious diseases and health education. Emilio Quevedo, as director of the History of Medicine Section of the LSHCT and as professor-researcher and director of the SFHC of the Colombian School of Medicine, played the role as director of the Project for Colombia. The project ran from January 1986 to December 1988, and collected, classified and systematized documents on the subject found in the National Library of Colombia, the General Archive of the Nation, the Luis Angel Arango Library and the Central Library of the National University of

Colombia in Bogota; in the Library of the University of Cartagena; in the Historical Archive of Antioquia; in the Central Library of the University of Antioquia in Medellín; in the Historical Archive of Cauca; and, in the Library of the University of Cauca, in Popayán. The results were published (Quevedo-Vélez, 1989b) and physical electronic copies of the database were donated to the General Archive of the Nation and the National Library of Colombia, with the commitment that these institutions would give copies of the database to the other libraries and historical archives in the country.

16.5.3 Second Latin American Seminar on Alternatives for the Teaching of the History of Science and Technology, São Paulo, February 24–26, 1987

As part of the fundamental activities of the LSHCT, under its sponsorship and with the organizational support of the Brazilian Society for the History of Science (SBHC), the II Latin American Seminar on Alternatives for the Teaching of the History of Science and Technology was held in the city of São Paulo, Brazil, from February 24 to 26, 1987, to make a general assessment of the state of science and technology teaching in Latin America.

As the Brazilian authors will refer to this Seminar in their respective chapters of this book, I only want to highlight, particularly, a few facets that touch on the Colombian participation in it.

In the Presentation of this Seminar, and taking up the thread of what was discussed at the First Seminar on the subject held in Santiago de Cali, Colombia, in March 1984, Juan José Saldaña reminded and emphasized that "the teaching of the history of science and technology is more than an academic or pedagogical matter; it is also the path that today's societies are hopefully following for the formation of a new scientific culture and their modernization [...] The undoubted scientific and technological progress, of which our era is heir to, is only comprehensible and, consequently, only manageable, through the understanding of its earlier forms. [...] To teach science is also to teach its past and its mechanisms of evolution" (Saldaña, 1989b, 1).

In this Seminar, Luis Carlos Arboleda participated with a central paper on *Social History and the Formation of a Scientific Culture* (Arboleda, 1989), which gave rise to a series of debates due to the complexity of the topics discussed and the conceptual and methodological approach used. One of the most critical topics of discussion was the type of demands, both theoretical-methodological, and the empirical documentation that the social history of science approach poses to research on the development of Latin American science, and the appropriation and use of the approach in the professional training of historians, scientists, and the public. Notwithstanding its complexity, this rich discussion gave rise to positive analyses that proved important for the future development of the discipline, particularly in Colombia.

For his part, Emilio Quevedo presented a paper on *Alternatives for Teaching the History of the Professions* (Quevedo-Vélez, 1989a), which broadened and deepened a topic already touched upon in previous Seminars and Congresses. Thus, he considered fundamental to discuss the differences in the treatment of the social history of professions as compared to the social history of sciences. His presentation gave rise to positive comments, but also some critical comments, partly because the history of the professions itself is not yet well configured.

These two papers demonstrated the need to continue these debates to better shape these social histories of sciences and professions.

16.5.4 Second Latin American Congress on the History of Science and Technology, São Paulo, June 1988

From June 30 to July 4, 1988, the Second Latin American Congress on the History of Science and Technology was held in the city of São Paulo, Brazil. At the close of his inaugural lecture, *Science and State in Latin America*, Juan José Saldaña stated that "we must do much more research on our [Latin American] science to construct the conceptual framework that will allow us to make accurate forecasts of the future [...] the structural character of the Science-State relationship in Latin America leads me to propose a very clear alliance between scientists and democratic States [...] so that science and the social effort that sponsors it, can be oriented towards a 21st century Latin America that must be more prosperous and, above all, more just" (Saldaña, 1988, 25). But already in the Presentation of the First Latin American Seminar on Alternatives for the Teaching of the History of Science and Technology, held in Cali in 1986, Luis Carlos Arboleda had emphasized the need to promote research programs and comparative studies in the Social History of Science (Arboleda, 1986a, 2) and its importance in overcoming the isolation of Latin American researchers (Saldaña, 1986, 3–4). These two strategies already pointed to the dream now proposed by Saldaña of anchoring the sciences and the State to achieve prosperity and justice.

These previous slogans had been taken up by the members of the Seminar of Philosophy and History of Sciences (SFHC) of the Colombian School of Medicine and, consequently, they presented, to this second Latin American Congress, a paper entitled *Colonial Medical Education in Latin America: a double process of subject construction* (Quevedo-Vélez & Mora, 1989), which did a first attempt for a comparative study of the medical education reforms implemented by the Bourbon State in Spain, the New Kingdom of Granada, the Royal Audience of Quito, the Viceroyalty of New Spain, and by the Braganza family in colonial Brazil, showing precisely the inter-actions of science, education and political and economic affairs in the Metropolis and in the colonial territories, seeking similarities and differences between the programs, the socio-political actors, the State policies in the colonial period, and the mechanisms of subjection implemented by the State to guarantee social reproduction. New concepts were introduced into the analysis, such as Néstor A. Braunstein's "double process of social subjection" (Braunstein, 1980) and Althusser's "Ideological Apparatuses of the State" (Althusser, 1974a).

16.6 Change of Direction and Incorporation of New Approaches to Research in the Seminar of Philosophy and History of Science of the Colombian School of Medicine, 1988–1994

16.6.1 Participation in the Health Sector Study, 1988–1989

While Emilio Quevedo was a professor and researcher at the Colombian School of Medicine, and because in front of the main door of that institution was the headquarters of the Colombian Institute for the Promotion of Science and Technology (COLCIENCIAS, today Minciencias: Colombian Ministry of Science, Technology and Innovation), the Director of that Institute asked the leaders of the Colombian School of Medicine to allow Professor Quevedo, in his free time and without remuneration, to act as Advisor to the National Health Research Program of COLCIENCIAS. In effect, this Advisory began to operate in 1986 and continued until 1988. This resulted in a proposal to reform and implement the Colciencias National Program for Scientific and Technological Development in Health (Quevedo-Vélez et al., 1986a). While working on this subject, at the end of 1988, Professor Quevedo coincided, in some of the meetings of the Health Program of COLCIENCIAS, with the sanitary specialist Francisco José Yepes Lujan, who was beginning to work on the design of the Sectoral Health Study, a government project that was to be financed by the World Bank and executed by the Ministry of Health and the National Planning Department. After listening to some of Quevedo's postings on the relationship between the National Health Research Program and the problems of the national health policy, Yepes, confidentially explained Quevedo, the whole Program of the Sectoral Study and proposed that he could, in his capacity as an officer at the Ministry of Health, ask the Colombian School of Medicine (ECM) to give him an unpaid leave of 1 year so that, within the framework of this Study, he and part of his research group could be paid to do a research on the History of Decision Making in Health in Colombia from 1886 to 1986 (a hundred years of Public Health in Colombia). After making the necessary arrangements, the ECM granted Quevedo and five other of his researchers, a leave of absence to form a research team, since this was a major work that had not been done so far.

Thus, from November 1, 1988, to July 31, 1989, Quevedo acted as principal researcher and coordinator of the research group of the study *Socio-historical and anthropological analysis of decision-making in health in the last one hundred years in Colombia,* within the framework of the Health Sectorial Study and under the orders of the Ministry of Health. Professors Néstor Miranda Canal, Consuelo

Mariño Samper, Mario Hernández Álvarez, Hugo Cárdenas and Carolina Wiesner joined the group as co-investigators. As a result, a large document was delivered at the end of the project, of which a schematic version was published as the first chapter of the book that gathered all the results of the Sectoral Study (Quevedo-Vélez et al., 1990a), and which was later published as a separate book (Quevedo-Vélez et al., 1990b).

This change of direction implied adjusting the parameters of the theoretical-methodological approach of the research group, to combine the approaches of the Social History of the Sciences and the History and Sociology of the Professions, with the new approaches to the study of Public Health and public policies in health, and especially, the approaches of Henry Sigerist (de Arellano, 1981; Restrepo Chavarriaga & Villa, 1980; Rosen, 1985; Vasco Uribe, 1988; Sigerist, 1935, 1937, 1938, 1941, 1946, 1960, 1974, 1981a, b).

16.6.2 Retirement of Emilio Quevedo from the Colombian School of Medicine and His Transfer to COLCIENCIAS, 1990–1994

In August 1989, Emilio Quevedo and his team rejoined the Colombian School of Medicine and reassumed their work at the SFHC of that institution. During the research process in the Health Sector Study, Quevedo became interested in the history of public health, as a body of knowledge and practices, thus, as a profession. But also in health policies, both national and international, as part of the professional field of public health. But especially, he focused on the role played by the Rockefeller Foundation in that field in Latin America. Upon returning to work at the SFHC, he began to design a research project on this topic that would consist of two phases: first, the study of the role of the Rockefeller Foundatión in the development of Colombian Public Health and Health Policies; and second, a comparative study on the role played by this Foundation in the other Latin American countries.

But, in August 1990, the neoliberal president, César Gaviria Trujillo went into power in Colombia and named the economist Clemente Forero Pineda as the new director of COLCIENCIAS. During his government, Gaviria Trujillo entrusted Forero Pineda with the reform of that institution and the creation and organization of the National System of Science and Technology. At the end of the first semester of 1991, Forero Pineda and Quevedo knew one each other, and Forero Pineda proposed Quevedo to join the new team that, under his direction, was given the task of reforming COLCIENCIAS. Quevedo was a medical Doctor that hold a PhD in Social Studies of Science and Technology, had experience in medical curriculum reforms, was working as a university teacher and researcher in History of Sciences and Medicine, History of Health Systems, and was interested in national and international Health policies. Thus, Forero Pineda decided to appoint Quevedo as the head of the Social Development Division of the Colombian Institute for Scientific and Technological Development-COLCIENCIAS, since July 1, 1991. In this Division, were located three big national research programs: Health Sciences, Education Sciences and Social Sciences, and Forero Pineda thought that Quevedo had the perfect profile to reorganize these three programs in the frame of the COLCIENCIAS proposed Reform. This appointment was done by means of a Special Cooperation Agreement, signed by COLCIENCIAS and the Colombian School of Medicine, which would be in force until December 31 of that same year.

Thus, Quevedo's historical project focused on the Rockefeller Foundation's health programs overseas had to be temporarily postponed and, in his spare time, Quevedo continued to expand his comparative studies on the Bourbon's health reforms in Latin America.

Thus, Quevedo entered to participate in the process of the COLCIENCIAS reform and the creation of the National System of Science and Technology. This reform emerged in the context of the new Colombian Constitution of 1991. Quevedo then joined the Forero Pineda management team made up of university professors with a research background and experience in the design and management of public policies. This team shared the ideal of a strategic alliance between the State and the scientific community for nation-building within the principles of the new Constitution.

At the end of the initial agreement, Forero proposed to Quevedo that he resign from the Colombian School of Medicine and become a permanent employee of the institution he was leading. The proposal was tempting and Quevedo accepted the challenge. Later, within the framework of the reform, the research programs of COLCIENCIAS were redesigned and Quevedo was appointed head of the National Program of Sciences and Technology of Health, from January 1 to December 31, 1992.

16.6.3 *Publication of the Collection* Historia Social de la Ciencia en Colombia *and the Construction of the COLCIENCIAS Social Studies of Science and Technology Seminar as a Tool to Guide Science and Technology Policy in Colombia, 1992*

In 1992, in addition to performing the tasks related to the organization of the new Sciences and Technology of Health Program, with the permission of the director Forero Pineda, he dedicated some time to the publication of the ten volumes of the collection *Historia Social de la Ciencia en Colombia* (Quevedo-Vélez, 1993–1996), which collected all the results of the project.

In addition, during that time it was a happy coincidence that Quevedo and Arboleda were both appointed officers of COLCIENCIAS. Once there, they decided to organize an informal seminar on Social Studies of Science and Technology for all middle and high-ranked officers (the director, deputy directors, heads of Research

Programs, and assistant professionals of these programs) of the institution. In this seminar, which met for three hours once a week, they began to study the new works of Thomas Kuhn (Kuhn, 1972, 1975, 1978, 1989), Karen D. Knorr-Cetina, Michael Mulkay, Barry Barnes, Daryl Chubin, H. M. Collins, Bruno Latour, and Steve Steve Latour. Collins, Bruno Latour, and Steve Woolgar, compiled a book that Arboleda had brought from New York in 1990, entitled *Science Observed* (Knorr-Cetina & Mulkay, 1983) and other related bibliographies that were obtained and related to the Edinburgh and Paris "Strong Program in Sociology of Knowledge" (Knorr-Cetina, 1983; Woolgar, 1988, 1991; Latour, 1983, 1987; Latour & Woolgar, 1979). These authors were the ones who were standing out in Europe and the United States for their various attempts to erase the disciplinary boundaries between the social sciences to consolidate a new transdisciplinary approach that soon was known as the Social Studies of Science, Technology and Medicine (SSSTM). These researchers began to align themselves into two groups or tendencies within this approach: the so-called Strong Program and Weak Program of the SSSTM, whose difference was in how strongly or how weakly they were based on a relativistic position with respect to truth and objectivity of science.

16.6.4 The Colombian Science and Technology Observatory Program, COLCIENCIAS, January 1993 to March 1994

This bibliography was closely related, in general, to science policy and was very much in line with what Luis Carlos Arboleda and Emilio Quevedo wanted to set in motion in COLCIENCIAS: the proposal to create a Permanent Program of Social Studies of Science, Technology, and Medicine that, from the institution, would support research that would serve as a space to consolidate a new view of science in Colombia. This proposal, arround which Quevedo and Arboleda had talked to Clemente Forero Pineda since they both joined the institution, was consolidated, but it required some time to mature. Finally, the proposal was accepted by the director and the Management Committee of the Institution, at the end of 1992. Therefore, Quevedo, who was the one who had promoted it, was hired again as head of the Colombian Science and Technology Observatory Program of COLCIENCIAS, from January 1, 1993, to March 1, 1994. The name of the program was taken from the title of Karen Knorr-Cetina and Michael Mulkay's book, *Science Observed*, which had inspired them to create the Seminar for COLCIENCIAS' officials (Knorr-Cetina & Mulkay, 1983).

Throughout 1993, Quevedo worked on the design and implementation of the Program and on the call for proposals to choose the research projects with which the work of observing various aspects of science in Colombia would begin. Unfortunately, in February 1994, when the Program was already set up and the first set of projects had been approved, the director Clemente Forero Pineda, withdrew the assigned budget, to pass it to another program that, according to the National Government, was of a higher priority. As a sign of protest, dissent, and disenchantment, Quevedo resigned from his position at the Institution, effective February 28.

By that time, Quevedo had already configured the first part of his project on the Rockefeller Foundation and the transition from Hygiene to Public Health in Colombia (Quevedo-Vélez, 1994a), and with it, he won a call for a teaching-researcher position at the Faculty of Medicine of the National University of Colombia, Bogotá, and, on March 1, he went to work there.

16.7 IV Latin American Congress of History of Science and Technology, Cali, January 24–27, 1995

Between January 24 and 27, 1995, the Sixth Latin American Congress on the History of Science was held at the Universidad del Valle, which proceedings were published in May 1997, under the title *Nationalism and Internationalism in the History of Science and Technology in Latin America* (Arboleda & Osorio, 1977), since this was the central theme of the Congress. Precisely, this theme led to the formation of a preparatory organizing committee for this congress, and a series of efforts were made in the southwestern region of Colombia for its preparation and organization. The process of preparing the event had repercussions at the national level, and work was done with researchers who had participated in the Latin American Seminar on Alternatives for the Teaching of the History of Science and Technology, held in November 1984, in Cali (Arboleda, 2024).

This topic was in vogue and was part of the debates among the researchers of the different Colombian groups of History of Science and Technology. The members of the LSHCT also joined the process. It was a matter of establishing new theoretical-methodological proposals, epistemological, sociopolitical and cultural approaches, which came from the Social Studies of Science, but, at the same time, from National Histories of scientific disciplines, particularly mathematics and medicine, which in Colombia were the most advanced in that sense. The Social Studies of Science and Health had already produced very mature works, as the result of research, with an epistemic, conceptual and methodological robustness that were very well received in this Latin American Congress, which later, with the publication of the proceedings, had a wider circulation, becoming significant contributions to the development of the Social History of Latin American Sciences (Arboleda, 2024).

At this Congress, Quevedo, who had been working with the Congress organizing comitee, presented two papers: one, entitled *¿Políticas de salud o políticas insalubres? De la higiene a la salud pública en Colombia en la primera mitad del siglo XX* (Quevedo-Vélez, 1996), in which he announced the use of new theoretical-methodological approaches in his new research project on the Rockefelller Foundation and Public Health in Colombia, and in which he had been working for a year and a half at the National University; and, another, entitled *José Celestino*

Mutis ante la higiene pública: un oráculo periférico preso en un paradigma metropolitano contradictorio (Quevedo-Vélez, 1994b), which was not published in the first volume of the proceedings of the Congress, but was later published by the Universidad del Rosario, Bogotá, in the proceedings of another event to which he presented it again. This study represented the closing of the project on Illustrated Medicine in the Colony and Medicine in the Bourbon Reforms, to which he had dedicated many years. But it was a very special closure because it was his farewell to the colonial period, but with a new approach. It was an experiment to see Mutis from the Kuhnian viewpoint of the Scientific Revolutions and the Scientific Paradigms, in contrast with the epistemological and social history that he had used in previous works on the subject. It was also an effort to integrate Kuhn's new type of internalism with the discontinuist materialist models of Canguilhem, Althusser, Lecourt, and Koyré, and with the proposals of the Social History of Public Health, in which he was beginning to immerse.

16.8 Epilogue

I would have many more things to tell, but this book does not gives me more room. I would only like to emphasize that, despite all the time that Arboleda and Quevedo worked as a team, at the end, life took them through different, but similar, academic settings. On the one hand, Arboleda, ended setting at a strong university, with a firm trajectory (Universidad del Valle, Cali), where he managed to find a suitable niche for his work and, after all the vicissitudes described here before, he has been able to devote his last years to the professional teaching of the history of mathematics, his native field, without having to change his parent university, despite several strategic temporary departures (Arboleda, 2024). As Arboleda himself states,

"The history of mathematics has been introduced more professionally in our country, moving beyond amateur levels through institutional efforts that have been incorporated into curricula and in the training of researchers at the Universidad del Valle. This journey began with the Faculty of Sciences and the Department of Mathematics, where the first graduate theses in Colombia on the history and epistemology of mathematics were completed. Subsequently, the Institute of Education and Pedagogy (now the Faculty) further expanded the focus on universal themes of mathematics by incorporating the study of the development of mathematical and scientific cultures in Colombia. The creation of the Doctorate of Education in the Faculty of Education and Pedagogy at the turn of the millennium marked the beginning of a third line of historical and mathematical education, with a focus on exploring the historical, epistemological, and philosophical aspects of mathematical practice to better serve teacher training" (Arboleda, 2024).

On the other hand, Quevedo, despite having found a first space that allowed him to start and establish himself in the field of the Epistemological and Social History of Science and Medicine, combining teaching with research, the structural precariousness and isolation of the Colombian School of Medicine, which was a school-university that did not have, at least at that time, any other faculty or associated school, with a minimum of contact with the national reality, had to look for other spaces to develop his intelectual and Scholar program. During his time in COLCIENCIAS, he linked the History of Sciences and Medicine with science, professions, and science and health policies, broadening his "intellectual and methodological landscape". Then in his transcourse at the National University (1994–2009), he opened important and enriching paths and perspectives. There he re-created the History of Medicine Research Center, that turned into his academic headquarter, from where, with his coleagues and students, he developed his project on the Rockefeller Foundation in Colombia and in the world (Quevedo-Vélez, et al. 2004; Quevedo-Vélez, 2000, 2001, 2005). Also, during that time, there he started his project on the History of Medicine in Colombia, although without the support of the National University, but since the project was funded by a private entity, Tecnoquímicas, he was able to start developing it in his free time, with the support of a team of high-quality researchers and the search and use of new bibliographical resources that contributed with new frameworks and spotlights that expanded their intellectual views and research and interpretation methods. In this way, this research team understood professions as labor activities of human collectives (social actors) that daily participate in social, cultural, economic, and political processes, trying to solve people's problems. To do it, they need to use specialized bodies of knowledge and technical or technological abilities and, they do it working in specialized institutions. In these, professionals practice their professions, guarantee the reproduction of the professions, the knowledge, the practices, the professional autonomy, and create social relationships with the State and the Civil Society. The History of Medicine, as a profession, is not only the history of medical sciences or techniques. It has to have in account all these elements said above (Freidson, 1986; Freidson, 2001; Abbott, 1992; González Leandri, 1999a, b; González Leandri, 2008; Johnson, 1972; Johnson et al., 2005; Burrage & Torstendahl, 1990; Parsons, 1939; Turner, 1987, 1999). These History of Medicine Research Center also allowed new ways of teaching and understanding Social Studies of Health and Medicine. In december 2008, Quevedo was forced to legally retire from National University of Colombia, after 30 years of doing teaching and conducting research.

But at the end of December, he was called by the Dean of the Faculty of Medicine of the Rosario University of Bogotá, a private university, who invited him to join these university to continue with his teaching and research programas in Social Studies of Health and Medicine. Thus, in February 2009, Quevedo was again begining to organice a new academic space, at the Universidad del Rosario: the Group of Social Studies of Sciences, Technologies and Professions, a research and teaching group, with the participation and cooperation of old coleagues and new recently graduated studentes. There he was able to finish his major work, the *History of Medicine in Colombia* (as a profession), in four volumes (Quevedo-Vélez, et al. 2007, 2008, 2010, 2013a) and strengthen its program of comparative studies of Medicine and Public Health in Latin America (Quevedo-Vélez, 2011; Quevedo-Vélez, et al. 2013b, 2023; Núñez-Gómez et al., 2021), But, in addition, at the Universidad del Rosario he was invited to actively participate in the design and implementation of a new curricular reform of the Medicine Academic Program, articulating and supporting it in the Social Studies of Science, Medicine and Health, a field that there was called Medical Socio-Humanism (Quevedo-Vélez, 2012). Today, Quevedo continues to work in this institution as a full time Professor Emeritus and researcher. He also is the coordinator of the curricular axis of Medical Sociohumanism, that supports the entire structure of the medical curriculum in that Universty. In April 26, 2016, Quevedo was awarded as Reaearcher Emmeritus by Colciencias, today Minciencias (Colombian Ministry of Science, Technology and Innovation), and continues his research and publishing activities.

Bibliography

Abbott, A. (1992). *The system of professions: An essay on the division of expert labor*. Chicago-London: University of Chicago Press. Original edition, 1967.

Althusser, L. (1969). *Para leer El Capital*. Segunda edición en Español ed. México: Siglo XXI Editores, S. A. Original edition, 1967.

Althusser, L. (1970a). *Ideología y Aparatos Ideológicos de Estado*: Ediciones Los Comuneros.

Althusser, L. (1970b). *La Revolución Teórica de Marx*. Qunta edición en Español ed. México: Siglo XXI Editores, S. A. Original edition, 1965.

Althusser, L. (1970c). *Sobre el trabajo teórico: dificultades y recursos*. Barcelona: Cuadernos Anagrama.

Althusser, L. (1974a). Ideología y aparatos ideológicos de Estado (*Notas para una investigación*). In L. Althusser (Ed.), *Escritos* (pp. 105–170). Barcelona: Editorial Laia.

Althusser, L. (1974b). *Para una crítica de la práctica teórica. Respuesta a John Lewis*. Buenos Aires: Siglo XXI Argentina editores, S. A.

Althusser, L. (1975a). *Curso de filosofía para científicos, El curso fue dictado entre 1967 y 1968*. Barcelona: Editorial Laia.

Althusser, L. (1975b). *Lenin y la filosofía*. 2a edición en español ed, *Serie Popular*. México: Ediciones Era, S. A. Original edition, 1969.

Althusser, L. (1977a). Defensa de Tesis en Amiens. In L. Althusser (Ed.), *Posiciones*. México: Grijalbo.

Althusser, L. (1977b). *Posiciones*. Barcelona: Grijalbo.

Althusser, L., & Lacan, J. (1970). *Freud y Lacan/El objeto del psicoanálisis*. Barcelona: Anagrama.

Arboleda, L. C. (1977). Notas sobre el objeto y el método de la historia de las ciencias. *Revista Universidad del Valle, 3–4*, 101–122.

Arboleda, L. C. (1985). Dificultades estructurales de la profesionalización de las matemáticas en Colombia. In J. L. Peset, J. J. Saldaña, J. S. Catalá, R. Á. Peláez, A. Lafuente, & M. Sellés (Eds.), *La Ciencia Moderna y el Nuevo Mundo* (pp. 27–38). Madrid: Consejo Superior de Investigaciones Científicas/Sociedad Latinoamericana de Historia de las Ciencias y la Tecnología.

Arboleda, L. C. (1986a). "Presentación." In *Seminario Latinamericano sobre Alternativas para la Enseñanza de la Historia de las Ciencias y la Tecnología*, edited by Serie memorias de eventos cientificos colombianos N° 31. Bogotá: Editora Guadalupe/ICFES/Universidad del Valle.

Arboleda, L. C. (Ed.). (1986b). *Seminario Latinoamericano sobre Alternativas para la Enseñanza de la Historia de las Ciencias y la Tecnología. Universidad del Valle, Cali, 4–10 de noviembre de 1984 Serie memorias de eventos científicos colombianos N° 31*. Bogotá: Editora Guadalupe, Instituto Colombiano para el Fomento de la Educación Superior-ICFES, Universidad de Valle.

Arboleda, L. C. (1989). "Historia Social y Formación de una Cultura Científica." In Milton Vargas, Shozo Motoyama, Julio Katinsky, Ana María Goldfarb, Marta Belmonte Barbosa and Mauro

Rolla Duarte (Eds.), *Número especal: Seminário Latino-Americano sobre Alternativas de Ensino da História da Ciencia e da Tecnologia. Revista da Sociedade Brasileira de História da Ciência* (pp. 11–18). Sâo Paulo, Brasil.

Arboleda, L. C. (1990). *Historia General de las Ciencias, Seminario de Historia de las Ciencias.* Cali: Universidad del Valle-ICFES.

Arboleda, L. C. (2024). *Conversaciones con Luis Carlos Arboleda sobre la Historia Social de la Ciencia en Colombia.* edited by Entrevistador: Emilio Quevedo V.

Arboleda, L. C., & Osorio, C. (1977). *Nacionalismo e Internacionalismo en la Historia de las Ciencias y la Tecnología en América Latina.* Santiago de Cali, Colombia: Universidad del Valle. Memorias del IV Congreso Latinoamericano de Historia de las Ciencias y la Tecnología, 24 al 27 de enero de 1995.

Autores Varios. (1990). *Proyecciones de un Programa. Taller sobre la Enseñanza de la Epistemologia, la Historia de las Ciencias y de la Medicina en la formación del Médico y el Profesional de la Salud.* Edited by Jaime Escobar Triana and Carolina Wiesner Ceballos. Vol. 3, *Colección Educación Médica.* Bogotá: Escuela Colombiana de Medicina.

Bachelard, G. (1934). *Le nouvel sprit scientifique.* Paris: Presse Universitaires de France.

Bachelard, G. (1973a). *El compromiso racionalista.* Buenos Aires: Siglo XXI Editores, S.A.

Bachelard, G. (1973b). La actualidad de la historia de las ciencias. In G. Bachelard (Ed.), *El compromiso racionalista* (pp. 148–164). Siglo XXI.

Bachelard, G. (1973c). *La filosofía del no* (p. 1940). Buenos Aires: Amorrortu. Original edition.

Bachelard, G. (1978). *El racionalismo aplicado.* Buenos Aires: Editorial Paidos.

Bachelard, G. (1981). *La formación del espíritu científico: Contribución a un psicoanálisis del conocimiento objetivo.* México: Siglo XXI Editores. Original edition, 1938.

Barnes, B. (1972a). *Sociology of Science: Seleted readings.* London-Baltimore: Penguin Books.

Barnes, B. (1972b). *Sociology of science: Selected readings.* London-Baltimore: Penguin Books.

Bateman, A. (1972). *Páginas para la historia de la ingeniería colombiana.* Biblioteca de Historia Nacional. Bogotá: Editorial Kelly.

Beltrán, E. (1964). *Primer Coloquio Mexicano de Historia de la Ciencia.* México, D. F.: Sociedad Mexicana de Historia Natural/Sociedad Mexicana de Historia de la Ciencia y la Tecnología.

Beltrán, E. (1984). La Historia de la ciencia en América Latina. *Quipu. Revista Latinoamericana de Historia de las Ciencias y la Tecnología, 1*(1), 7–23.

Bernal, J. D. (1976a). *Historia Social de la Ciencia.* Cuarta edición ed. Vol. Vol 1/La ciencia en la historia, *Serie Universitaria Historia/Ciencia/Sociedad 9.* Barcelona: Ediciones península. Original edition, 1954.

Bernal, J. D. (1976b). *Social History of Science* (4th ed.). Vol 1/La ciencia en la historia, *Serie Universitaria Historia/Ciencia/Sociedad 9.* Barcelona: Ediciones península. Original edition, 1954.

Bernal, J. D. (1976c). *Historia Social de la Ciencia.* Cuarta edición ed. Vol. Vol 2/La ciencia en nuestro tiempo, *Serie Universitaria Historia/Ciencia/Sociedad 10.* Barcelona: Ediciones península. Original edition, 1954.

Bernal, J. D. (1976d). *Social history of science* (4th ed.). Vol 2/La ciencia en nuestro tiempo, *Serie Universitaria Historia/Ciencia/Sociedad 10.* Barcelona: Ediciones península. Original edition, 1954.

Braunstein, N. A. (1980). *Psiquiatría, teoría del sujeto, psicoanálisis (hacia Lacan).* México: Siglo XXI Editores.

Burrage, Michael, and Rolf Torstendahl. 1990. Professions in theory and history: rethinking the study of the professions. London/Newbury Park/New Delhi: Sage Publications. London/ Newbury Park/New Delhi.

Canguilhem, G. (1971). *Lo normal y lo patológico.* 1a edición en español ed. Buenos Aires: Siglo XXI Argentina, S. A. Original edition, 1966.

Canguilhem, G. (1975a). L'objet de l'histoire des sciences. In G. Canguilhem (Ed.), *Études d'histoire et de philosophie des sciences* (pp. 9–23). Paris: Librairie Philosophique J. Vrin. Original edition, 1968.

Canguilhem, G. (1975b). *La formación del concepto de reflejo en los siglos XVII y XVIII* (Juan Rovira, Trans.). *Avance Teoría*. Barcelona: Juan Lliteras Editor. Original edition, 1955.

Canguilhem, G. (2009). El objeto de la historia de las ciencias. In G. Canguilhem (Ed.), *Estudios de historia y de filosofía de las ciencias* (pp. 11–26). Buenos Aires: Amorrortu editores. Original edition, 1968.

Cataño, G. (1986). *La Sociología en Colombia*. Bogotá: Plaza & Janes.

Chaparro, F., & Sagasti, F. (1978). *Ciencia y Tecnología en Colombia*. Bogotá: Editorial Escala.

Cid, F. (1974a). Evolución de la Historia de la Medicina. In *Reflexiones sobre la Historia de la Medicina. Cuadernos Anagrama* (pp. 46–66). Barcelona: Editorial Anagrama.

Cid, F. (1974b). *Reflexiones sobre la Historia de la Medicina, Cuadernos Anagrama*. Barcelona: Editorial Anagrama.

Cid, F. (1978). *Breve historia de las ciencias médicas*. Barcelona: Espaxs.

Colciencias, República de Colombia. (1970). *Apuntes para la Historia de la Ciencia en Colombia, Documentación e Historia de la Ciencia en Colombia*. Bogotá: Servicios Especiales de Prensa.

Comte, A. (1979). *Curso de Filosofía Positiva*. México: Editorial Porrúa, S. A. Original edition, 1830.

D'Ambrosio, U. (1986). Etnociencia: Alternativa para la Historia y la Enseñanza de las Ciencias. In *Seminario Latinoamericano sobre Alternativas para la Enseñanza de la Historia de las Ciencias y la Tecnología. Universidad del Valle, Cali, 4–10 de noviembre de 1984* edited by Luis Carlos Arboleda. Bogotá: Instituto Colombiano para el Fomento de la Educación Superior-ICFES.

de Arellano, A. B. R. (1981). The politics of public health in Puerto Rico: 1926-1940. *Revista de Salud Pública de Puerto Rico, 3*, 35–58.

de Azevedo, F. (1955). *As Ciencias no Brasil*. São Paulo: Melhoramentos.

de Azevedo, F. (1971). *A Cultura Brasleira*. 2 Vols. São Paulo: Melhoramentos/Editora da USP. Original edition, 1943.

De Gortari, E. 1980a. *La Ciencia en la Historia de México, Tratados y manuales Grijalbo*. México, D. F.: Editorial Grijalbo S. A. Original edition, 1979.

De Gortari, E. (1980b). Problemas de la historia de la ciencia y la tecnología. In *Reflexiones históricas y filosóficas de México*. México: Editorial Grijalbo.

Fichant, M. (1975). Idea de una historia de las ciencias. In M. Fichant & M. Pécheux (Eds.), *Sobre la historia de las ciencias* (pp. 46–126). Buenos Aires: Siglo XXI Argentina Editores.

Fichant, M., & Pécheux, M. (1975). *Sobre la historia de las ciencias*. 2a edición ed. 1 vols. Buenos Aires: Siglo XXI Argentina Editores.

Figueiroa, Silvia Fernanda de Mendonça. (n.d.). s/f. A puzzle of refracted images: Preliminary reflections on the relation between semiotic and history of sciences. Manuscript, kindly provided by the author.

Forero Pineda, C. (1993). *Folleto de presentación de la obra Historia Social de la Ciencia en Colombia*. edited by República de Colombia. Instituto Colombiano para el Desarrollo de la Ciencia y la Tecnología (Colciencias). Bogotá: Instituto Colombiano para el Desarrollo de la Ciencia y la Tecnología (Colciencias)/Departamento nacional de Planeación.

Foucault, M. (1975). *El nacimiento de la clínica. Una arqueología de la mirada médica*. 2a edición en español ed. México: Siglo XXI Editores. Original edition, 1963.

Foucault, M. (1976). La crisis de la medicina o la crisis de la antimedicina. *Educación Médica y Salud, 10*(2), 152.

Foucault, M. (1977). El nacimiento de la medicina social. *Revista Centroamericana de Ciencias de la Salud, 6*, 89–108.

Freidson, E. (1986). *Professional powers: A study of institutionalization of formal knowledge*. Chicago/London: University of Chicago Press.

Freidson, E. (2001). Teoría de las profesiones. Estado del arte. *Perfiles Educativos, 23*(93), 28–43.

Galles, C. (2002). VI Congreso Latinoamericano de Historia de las Ciencias y la Tecnología.

González Leandri, R. (1999a). *Curar, persuadir, gobernar. La construcción histórica de la profesión médica en Buenos Aires, 1852–1886*. Madrid: CSIC.

González Leandri, R. (1999b). *Las Profesiones. Entre la vocación y el interés corporativo. Fundamentos para su estudio histórico*. Madrid: Editorial Catriel.

González Leandri, R. (2008). José María Ramos Mejía. Médico, intelectual y funcionario del Estado (1870-1900). In A. Álvarez & A. Carbonetti (Eds.), *Saberes y prácticas médicas en la Argentina. Un recorrido por historias de vida* (pp. 95–133). Mar del Plata: Universidad de Mar del Plata-Eudem.

Hernández de Alba, G. (1966). Contribución a la Historia de la Medicina Colombiana. In R. M. Briceño & G. Hernández de Alba (Eds.), *De Hipócrates a Pasteur/Contribución a la Historia de la Medicina Colombiana* (pp. 95–259). Bogotá: Biblioteca Schering Corporation USA.

Hessen, B. (1985). *Las raíces socioeconómicas de la Mecánica de Newton*. La Habana: Editorial Academia. Original edition, 1931.

Huerga Melcón, P. (2004). El Congreso de Londres de 1931. *Llull: Revista de la Sociedad Española de Historia de las Ciencias y de las Técnicas, 27*, 679–703.

ICFES. (1982). *Historia y Epistemología de lasl Ciencias. Ciclo de Conferencias*. Medellín: ICFES; Universidad Nacional, seccional Medellín; Universidad de Antioquia; Universidad Pontifica Bolivariana; Biblioteca Pública Piloto.

Jaramillo Uribe, J. (1977). *La personalidad histórica de Colombia y otros ensayos*. Vol. Este trabajo recoge varios ensayos publicados entre 1952 y 1977. Bogotá: Instituto Colombiano de Cultura.

Jaramillo Uribe, J. (1980). *Manual de Historia de Colombia* (Vol. 3). Bogotá: Instituto Colombiano de Cultura.

Jaramillo Uribe, J. (1982). *El pensamiento colombiano en el siglo XIX*. Bogotá: Editorial Temis.

Johnson, T. (1972). *Professions and power*. London: Macmillan.

Johnson, T., Larkin, G., & Sacks, M. (Eds.). (2005). *Health professions and the state in Europe* (p. 1995). London: Routledge. Original edition.

Knorr-Cetina, K. (1983). The ethnographic study of scienctific work: Towards a constructivist interpretation of science. In K. Knorr-Cetina & M. Mulkay (Eds.), *Science observed. Perspectives on the social study of science* (pp. 115–140). London: SAGE Publications.

Knorr-Cetina, K., & Mulkay, M. (1983). *Science observed. perspectives on the social study of science*. London: SAGE Publications.

Koyré, A. (1940). *Études galiléennes*. Paris: Hermann.

Koyré, A. (1973a). *Estudios de historia del pensamiento científico*. México: Siglo XXI Editores. Original edition, 1973.

Koyré, A. (1973b). Perspectivas de la historia de las ciencias. In A. Koyre (Ed.), *Estudios de historia del pensamiento científico* (pp. 377–386). México: Siglo XXI Editores.

Kuhn, T. S. (1962). *The structure of Scientific revolutions*. Chicago: University of Chicago Press.

Kuhn, T. S. (1972). Los paradigmas científicos. In B. Barnes, T. S. Kuhn, & R. K. Merton (Eds.), *Estudios sobre sociología de la ciencia* (pp. 79–102). Madrid: Alianza Editorial. Original edition, 1963.

Kuhn, T. S. (1975). Prefacio. In T. S. Kuhn (Ed.), *La estructura de las revoluciones científicas* (pp. 9–19). México: Fondo de Cultura Económica. Original edition, 1962.

Kuhn, T. S. (1978). *La revolución copernicana*. Barcelona: Ariel. Original edition, 1957.

Kuhn, T. S. (1989). *¿Qué son las revoluciones científicas? y otros ensayos*. Barcelona: Paidos/Universidad Autónoma de Barcelona/Instituto de Ciencias de la Educación.

Laín Entralgo, P. (1961). *La historia clínica. Historia y teoría del relato patográfico*. 2a edición ed. Barcelona: Salvat Editores, S. A. Original edition, 1950.

Laín Entralgo, P. (1964). *La relación médico-enfermo. Historia y teoría*. Madrid: Revista de Occidente.

Laín Entralgo, P. (1976). *Historia Universal de la Medicina* (Vol. 7). Barcelona: Salvat Editores. Original edition, 1972.

Laín Entralgo, P. (1982). *El diagnóstico médico. Historia y teoría*. Barcelona: Salvat Editores, S. A.

Larroyo, F. (1979). Estudio introductivo: la filosofía positiva de Augusto Comte. In A. Comte (Ed.), *La Filosofía Positiva*. México: Editorial Porrúa, S. A.

Lastres, J. B. (1951a). *Historia de la Medicina Peruana. Vol. 2: La Medicina en el virreinato*. Lima: Universidad Mayor de San Marcos-Imprenta Santa María.

Lastres, J. B. (1951b). *Historia de la Medicina Peruana. Vol. 3: La Medicina en la República*. Lima: Universidad Mayor de San Marcos-Imprenta Santa María.

Lastres, J. B. (1951c). *Historia de la Medicina Peruana. Vol. 1: La Medicina Incaica*. Lima: Universidad Mayor de San Marcos-Imprenta Santa María.

Latour, B. (1983). Give me a laboratory and I will raise the world. In K. Knorr-Cetina & M. Mulkay (Eds.), *Science observed. perspectives on the social studies of science*. London: SAGE Publications.

Latour, B. (1987). *Science in action*. Cambridge: Harvard University Press.

Latour, B., & Woolgar, S. (1979). *Laboratory life*. London: SAGE Publications.

Lecourt, D. (1971). Prólogo. In G. Canguilhem (Ed.), *Lo normal y lo patológico* (pp. VII–XXX). México: Siglo XXI, Editores.

Lecourt, D. (1973). *Para una historia crítica de la Epistemología*. Buenos Aires: Siglo XXI Argentina Editores.

Lecourt, D. (1974). *Ensayo sobre la posición de Lenin en filosofía*. Primera edición en Español ed. Buenos Aires, Argentina: Siglo XXI Argentina Editores. Original edition, 1973.

Lecourt, D. (1975). *Bachelard o el día y la noche. Un ensayo a la luz del materialismo dialéctico*. Barcelona: Editorial Anagrama.

Lecourt, D. (1982). *Para una crítica de la Epistemología*. 4a. ed. ed, *Colección mínima N° 72*. México: Siglo XXI Editores.

Lecourt, D., Lysenko, T. D., & Aragón, L. (1974). *El "Caso Lysenko"*. Barcelona: Cuadernos Anagrama.

Lértora Mendoza, C. (1986a). Fuentes primarias para el estudio de la Historia de las Ciencias en especial en América Latina. In *Seminario Latinoamericano sobre Alternativas para la Enseñanza de la Historia de las Ciencias y la Tecnología. Universidad del Valle, Cali, 4–10 de noviembre de 1984* edited by Luis Carlos Arboleda. Bogotá: Instituto Colombiano para el Fomento de la Educación Superior-ICFES.

Lértora Mendoza, C. (1986b). Los estudios de historia de la ciencia en Argentina. *Quipu. Revista Latinoamericana de Historia de las Ciencias y la Tecnología, 3*(1), 135–147.

López Sánchez, J. (1950). *Vida y obra del sabio médico habanero Tomás Romay y Chacón*. La Habana: Instituto del Libro.

López Sánchez, J. (1987). *Finlay. El hombre y la verdad científica*. La Habana: Editorial Científico-Técnica.

López Sánchez, J. (1997). *Cuba, medicina y civilización. Siglos XVII y XVIII*. Instituto Cubaño del Libro/Editorial Científico-Técnica.

López Sánchez, J. (1999) Carlos J. Finlay. His life and his works. La Habana: Editorial José Martí.

López Sánchez, J., & de la Torriente Brau, Z. (1972). *Bibliografía científica cubana (1790–1748)*. La Habana.

MacLeod, R. (1982). Changing perspectives in the social history of science. In R. Spiegel, D. J. De Solla, & Price. (Eds.), *Science, technology and society, a cross-disciplinary perspective*. London/Beverly Hills: Sage.

Merton, R. K. (1938). *Science, technology and society in seventeenth century England*. Edited by Osiris. Vol. 4 pt 2, *Studies in the History and Philosophy of Science, and on the History of Learning and Culture*. Bruges: St. Catherine Press,. Original edition, 1938.

Merton, R. K. (1945). Sociology of knowledge. In G. Gurvitch & W. E. Moore (Eds.), *Twentieth century sociology* (pp. 366–405). New York: Philosophical Library.

Merton, R. K. (1949). *Social theory and social structure*. New York: The Free Press.

Merton, R. K. (1977). *La sociología de la ciencia*. 1a edición ed. 2 Vols. Madrid: Alianza Editorial.

Merton, R. K. (1984). Prefacio (1970). In R. K. Merton (Ed.), *Ciencia, tecnología y sociedad en la Inglaterra del siglo XVII* (pp. 31–37). Madrid: Alianza Editorial. Original edition, 1938.

Motoyama, S. (1987). Un análisis de la historia de la ciencia en el contexto latinoamericano. In A. Lafuente & J. J. Saldaña (Eds.), *Nuevas tendencias en historia de las ciencias*. Madrid: Consejo Superior de Investigaciones Científicas.

Motoyama, S., & Guimaraes, M. (1980). *História das Ciências no Brasil* (Vol. 2). São Paulo: Editora Pedagógica e Universitaria.

Needham, J. (1964). La ciencia y la sociedad en oriente y en occidente. In J. D. Bernal (Ed.), *La ciencia de la ciencia* (pp. 201–232). México: Editorial Grijalbo.

Núñez-Gómez, María Camila, Maria Catalina Sánchez-Martínez, & Emilio Quevedo-Vélez (2021). Viruela en Santiago, Concepción y Santafé: Comparación de las estrategias ilustradas (1782–1807). *Revista Ciencias de la Salud* 19 (Número Especial):1–22.

Ortiz, E., & Pyenson, L. (1985). José Babini-Eloge. *Isis, 76*, 568.

Papp, D. (1985). Mi camino hacia la historia de la ciencia y de la medicina. *Quipu. Revista Latinoamericana de Historia de las Ciencias y la Tecnología, 2*(1), 123–128.

Parsons, T. (1939). The professions and social structure. *Social Forces, 17*, 457–467.

Pérez Arbeláez, E. (1983). *José Celestino Mutis y la Real Expedición Botánica del Nuevo Reino de Granada*. Bogotá: Editorial Bolívar.

Peset, J. L., Saldaña, J. J., Catalá, J. S., Peláez, R. Á., Lafuente, A., & Sellés, M. (Eds.). (1985). *La Ciencia Moderna y el Nuevo Mundo*. Madrid: Consejo Superior de Investigaciones Científicas/ Sociedad Latinoamericana de Historia de las Ciencias y la Tecnología.

Plata, J. J. (2013). Colciencias cuarenta años. Aprendizajes organizacionales y retos en las sociedades del conocimiento. In M. Salazar (Ed.), *Colciencias cuarenta años. Entre la legitimidad, la normatividad y la práctica*. Bogotá: Observatorio Colombiano de la Ciencia y la Tecnología/ Universidad Nacional de Colombia/Universidad Colegio Mayor de Nuestra Señora del Rosario.

Profesores del Departamento de Sociología, Universidad Nacional de Colombia. (1969). "Neocolonialismo y Sociología en Colombia: un intento de respuesta." Ponencia presentada al IX Congreso Latinoamericano de Sociología, México (Manuscrito).

Quevedo-Vélez, E. (1983). Las relaciones entre la Historia de las Ciencias y de la Técnicas en la Medicina. *Ciencia, Tecnología y Desarrollo, 7*(3), 379–387.

Quevedo-Vélez, E. (1984a). Seminario de Filosofía e Historia de las Ciencias. Taller de Lanceros-Paipa mayo 19 y 20 de 1983. In *Reflexiones sobre un Programa* (pp. 11–47). Bogotá: Escuela Colombiana de Medicina.

Quevedo-Vélez, E (1984b). Importancia de la investigación de la Historia de las Ciencias y de la Medicina en la formación del médico general y del especialista. *Primer Seminario Latinoamericano sobre Alternativas para la Enseñanza de la Historia de las Ciencias y la Tecnología*, Santiago de Cali, Colombia, 4–10 de Noviembre (Manuscrito).

Quevedo-Vélez, E. (1984c). Papel de la Historia de las Ciencias en la Integración Curricular de la Escuela Colombiana de Medicina. *QUIPU-Revista Latinoamericana de Historia de las Ciencias y la Tecnología, 1*(2), 223–243.

Quevedo-Vélez, E. (1985). La Ilustración y la enseñanza de la Medicina en la Nueva Granada. In J. L. Peset, J. J. Saldaña, J. S. Catalá, R. Á. Peláez, A. Lafuente, & M. Sellés (Eds.), *La Ciencia Moderna y el Nuevo Mundo* (pp. 85–110). Madrid: Consejo Superior de Investigaciones Científicas/Sociedad Latinoamericana de Historia de las Ciencias y la Tecnología.

Quevedo-Vélez, E. (1989a). Alternativas para la Enseñanza de la Historia de las Profesiones. *Revista da Sociedade Brasileira de História da Ciência, 3*, 91–103.

Quevedo-Vélez, E. (1989b). Proyecto para construir una base de datos sobre fuentes documentales para la historia de las Enfermedades Infecciosas y la enseñanza sanitaria en Colombia. In J. P. Preciado, J. O. López, I. P. Escobar, & P. G. H. Ramírez (Eds.), *Hacia el rescate de nuestra memoria colectiva. Programas de Archivos en Colombia* (pp. 69–78). Tunja: Magister en Historia-UPTC.

Quevedo-Vélez, E. (1990a). La Historia de las Ciencias y sus relaciones con el conocimiento médico. In *Proyecciones de un Programa, Colección Educación Médica*. Bogotá: Escuela Colombiana de Medicina.

Quevedo-Vélez, E. (1990b). Papel de la Historia de las Ciencias en la integración curricular de la Escuela Colombiana de Medicina. In J. E. Triana & C. W. Ceballos (Eds.), *Proyecciones de un Progarama* (pp. 289–320). Bogotá: Escuela Colombiana de Medicina. Original edition, 1983.

Quevedo-Vélez, E. (1993a). Los Estudios Histórico-Sociales sobre las Ciencias y la Tecnología en América Latina y en Colombia: Balance y Actualidad. In E. Quevedo (Ed.), *Historia social de la ciencia en Colombia* (pp. 19–86). Bogotá: COLCIENCIAS/Tercer Mundo Editores.

Quevedo-Vélez, E. (1993b). Presentación. In E. Quevedo-Vélez (Ed.), *Historia Social de la Ciencia en Colombia* (pp. 13–16). Bogotá: Colciencias/Tercer Mundo Editores.

Quevedo-Vélez, E. (1994a). El transito desde la higiene hacia la salud pública en Colombia en el contexto de las interacciones internacionales. In *Proyecto de investigación*. Bogotá: Universidad Nacional de Colombia. Facultad de Medicina.

Quevedo-Vélez, E. (1994b). José Celestino Mutis ante la higiene pública: un oráculo periférico preso en un paradigma metropolitano contradictorio. In E. Varios (Ed.), *El humanismo de Mutis: Proyección y vigencia* (pp. 91–112). Bogotá: Universidad del Rosario-Colciencias.

Quevedo-Vélez, E. (1996). ¿Políticas de salud o políticas insalubres? De la higiene a la salud pública en Colombia en la primera mitad del siglo XX. In L. C. Arboleda & C. Osorio (Eds.), *Nacionalismo e Internacionalismo en la Historia de las Ciencias y la Tecnología en América Latina. IV Congreso Latinoamericano de Historia de las Ciencias y la Tecnología* (pp. 195–215). Cali: Universidad del Valle.

Quevedo-Vélez, E. (2000). El tránsito desde la higiene hacia la salud pública en América Latina. *TIERRA FIRME, Revista de Historia y Ciencias Sociales, 72*(18), 611–661.

Quevedo-Vélez, E. (2001). *A coffee with sugar, please*: La Fundación Rockefeller y la salud pública en Centroamérica, hacia una mirada comparativa. *Colombia, Ciencia & Tecnología, 19*(3), 24–36.

Quevedo-Vélez, E. (2005). No one knows for whom he is actually working! The indirect role played by the Rockefeller Foundation in the shift from poor law medical relief to the National Health Service in England, through the London School of Hygiene and Tropical Medicine (1913–1948). In I. Borowy & W. D. Gruner (Eds.), *Facing illness in troubled times. Health in Europe in the interwar years* (pp. 365–408). Frankfurt am Main: Peter Lang. Europäischer Verlag der Wissenschaften.

Quevedo-Vélez, E. (2011). In U. del Rosario (Ed.), *Programa de Estudios Histórico-Comparativos y Transdisciplinares sobre la historia de la medicina y de la salud pública en América Latina.* Bogotá: Universidad del Rosario, Grupo de Estudios Sociales de las Ciencias, las Tecnologías y las Profesiones. Archivo del GESCTP.

Quevedo-Vélez, E. (2012). Comprensión histórico-crítica del proceso salud-enfermedad: base para una renovación curricular en medicina. In G. A. Quintero (Ed.), *Educación Médica. Diseño e implementación de un curriculo basado en resultados del aprendizaje* (pp. 269–314). Bogotá: Editorial Universidad del Rosario.

Quevedo-Vélez, E, and Gonzalo Hernández de Alba. (1985). Las reformas universtarias borbónicas en el Nuevo Reyno de Granada: El plan de Moreno y Escandón. *Ponencia presentada en el I Congreso Latinoamericano de Historia de las Ciencias y la Tecnología Primer Congreso Latinoamericano de Historia de las Ciencias y la Tecnología*, La Habana, 21–25 de julio.

Quevedo-Vélez, E., & Mora, O. L. (1989). La Educación Médica Colonial en Latinoamérica: un doble proceso de sujetación. In *Actas del Segundo Congreso Latinoamericano de Historia de las Ciencias y la Tecnología* (pp. 190–208). Sao Paulo: Editorial Stella.

Quevedo-Vélez, E., (Ed.). (1993–1996). *Historia Social de la Ciencia en Colombia* (Vols. 10). Bogotá: Colciencias/Tercer Mundo Editores.

Quevedo-Vélez, E., & Zaldúa, A. (1986). Antecedentes de las reformas médicas del siglo XVIII y XIX en el Nuevo Reino de Granada: una polémica entre médicos y cirujanos. *Quipu. Revista Latinoamericana de Historia de las Ciencias y la Tecnología, 3*(3), 311–334.

Quevedo-Vélez, E., Durán de Pardo, M. C., & Mesa, G. (1986a). *Programa Nacional de Investigación en Salud.* Bogotá: Colciencias.

Quevedo-Vélez, E., Echeverry de Pardo, G., & Gómez, M. C. (1986b). Brujería, Enfermedad Mental, Mujer y Sociedad. *Revista Colombiana de Psiquiatría, 15*(4), 367–396.

Quevedo-Vélez, E., Mario Hernández, A., Néstor Miranda, C., Mariño, C., Cárdenas, H., & Wiesner, C. (1990a). La Salud en Colombia. Análisis socio-histórico. In F. J. Yepes (Ed.), *Estudio Sectorial de Salud. Documento General* (pp. 111–217). Bogotá: Ministerio de Salud/Departamento Nacional de Planeación.

Quevedo-Vélez, E., Mario Hernández, A., Néstor Miranda, C., Mariño, C., Cárdenas, H., & Wiesner, C. (1990b). La Salud en Colombia. Análisis Sociohistórico. In F. J. Yepes (Ed.), *Estudio Sectorial de Salud*. Bogotá: Ministerio de Salud-Departamento Nacional de Planeación.

Quevedo-Vélez, E., Borda, C., Eslava, J., García, C., Guzmán, M., Mejía, P., & Noguera, C. (2004). *Café y gusanos, Mosquitos y Petróleo. El tránsito de la higiene hacia la salud pública en Colombia, 1873–1953*. Bogotá: Universidad Nacional/Instituto de Salud Pública/Centro de Historia de la Medicina.

Quevedo-Vélez, E., Germán Enrique Pérez R., Néstor Miranda C., Juan Carlos Eslava C., Mario Hernández A., Lisa Priscila Bustos J., María Fernanda Durán, Marcela García, Carolina Manosalva R., Elquin Morales L, Diana Farley Rodríguez, & Carlos Vladimir Villamizar. (2007). *Historia de la Medicina en Colombia*. 5 vols. Vol. 1. Prácticas médicas en conflicto, 1492–1782. Bogotá: Tecnoquímicas; Grupo Editorial Norma.

Quevedo-Vélez, E., Germán Enrique Pérez R., Néstor Miranda C., Juan Carlos Eslava C., Mario Hernández A., Lisa Priscila Bustos J., María Fernanda Durán, Marcela García, Carolina Manosalva R., Elquin Morales L, & Diana Farley Rodríguez. (2008). *Historia de la Medicina en Colombia*. 5 vols. Vol. 2. De la Medicina Ilustrada a la Medicina Anatomoclínica, 1782–1865. Bogotá: Tecnoquímicas; Grupo Editorial Norma.

Quevedo-Vélez, E., Germán Enrique Pérez R., Néstor Miranda C., Juan Carlos Eslava C., Mario Hernández A., María Cristina Acosta B., Laura del Pilar Cadena Afanador, Juan Carlos Ferrero, Abel Martínez Martín, Orlando Mejía Rivera, Hugo Portela, Pedro Rovetto, Rita Sierra M., Jairo Solano Alonso, Lisa Priscila Bustos J., Claudia Margarita Cortés G., María Fernanda Durán, Marcela García, Carolina Manosalva R., Elquin Morales L, Diana Farley Rodríguez, Marlín Téllez, and Manuel Vega. (2010). *Historia de la Medicina en Colombia*. 5 vols. Vol. 3. Hacia una profesión liberal, 1865–1918. Bogotá: Tecnoquímicas-Grupo Editorial Norma.

Quevedo-Vélez, E., Hernández, M., Cortés, C., & Eslava, J. C. (2013a). Un modelo para armar: Una propuesta metodológica para abordar el estudio comparativo de la historia de la salud pública, de las profesiones de la salud y de sus relaciones de doble vía con la sociedad. *Revista Ciencias de la Salud, 11*(3), 295–321.

Quevedo-Vélez, E, Germán Enrique Pérez R., Néstor Miranda C., Juan Carlos Eslava C., Mario Hernández A., Carolina Manosalva Roa, María Cristina Acosta B., Laura del Pilar Cadena Afanador, Orlando Mejía Rivera, Hugo Portela, Pedro Rovetto, Rita Sierra M., Jairo Solano Alonso, Lisa Priscila Bustos J., Claudia Margarita Cortés G., María Fernanda Durán, Marcela García, Giovanna Matiz, Elquin Morales L, Diana Farley Rodríguez, Marlín Téllez, and Manuel Vega. (2013b). *Historia de la Medicina en Colombia*. 5 vols. Vol. IV. De la práctica liberal a la socialización limitada, 1918–1977. Bogotá: Tecnoquímicas-Grupo Editorial Norma.

Quevedo-Vélez, E., Nuñez-Gómez, M. C., Ardila-González, J. L., Hernández, Q., & Sánchez-Martínez, M. C. (2023). Las reformas ilustradas de la educación médica en tres colonias americanas, siglo XVIII y XIX. *Bulletin of Spanish Studies, 100*(8), 1147–1176. https://doi.org/10.1080/14753820.2024.2326379

Raymond, P. (1973). *Le Passage au Matérialism*. Paris: Maspero.

Raymond, P. (1975). *L'Histoire et les Sciences*. Paris: Maspero.

Restrepo, G. (1981). Elementos teóricos para una historia social de la ciencia en Colombia. *Ciencia, Tecnología y Desarrollo, 5*(3).

Restrepo Chavarriaga, G., & Villa, A. (1980). *Desarrollo de la salud pública colombiana*. Medellín: Universidad de Antioquia/Escuela Nacional de Salud Pública.

Rosen, G. (1985). *De la policía médica a la Medicina Social*. México: Siglo XXI.

Rosselli Quijano, H. (1968). *Historia de la Psiquiatría en Colombia*. 2 Vols. Vol. 1. Bogotá: Editorial Horizontes.

Sagasti, F. (1973). Underdevelopment, science and technology: The point of view of the underdeveloped countries. *Science Studies, 3*, 47–59.

Sagasti, F. (1977). Esbozo histórico de la ciencia en América Latina. In F. Chaparro & F. Sagasti (Eds.), *Ciencia y Tecnología en Colombia* (pp. 13–36). Bogotá: Instituto Colombiano de Cultura/Colciencias.

Salas Olano, E. (1894). *Historia de la Medicina en Chile. Con importantes documentos sobre la medicina de nuestros predecesores*. Santiago de Chile: Imprenta Vicuña Mackenna.

Saldaña, J. J. (1983). Hacía una crítica histórica, teórica y metodológica de la historiagrafía latinoamericana de las ciencias. *Ciencia, Tecnología y Desarrollo, 7*(3), 333–344.

Saldaña, J. J. (1986). Balance General de la Historia de las Ciencias en América Latina. In U. del Valle (Ed.), *Seminario Latinoamericano sobre alternativas para la Enseñanza de la Historia de las Ciencias y la Tecnología*. Bogotá: Editora Guadalupe/Universidad del Valle/Universidad Autonoma de México/ICFES. Original edition, 1984.

Saldaña, J. J. (1988). Ciencia y Estado en América Latina. In U. D'Ambrosio (Ed.), *Anais do 2° Congresso de História da Ciência e da Tecnologia* (pp. 17–25). Sao Paulo: Nov Stella.

Saldaña, J. J. (1989a). Estudio sobre las fases principales de la evolución de la historia de las ciencias. In J. J. Saldaña (Ed.), *Introducción a la teoría de la historia de las ciencias* (pp. 21–78). México: Universidad Nacional Autónoma de México. Original edition, 1985.

Saldaña, J. J. (1989b). Presentación. In M. Vargas, S. Motoyama, J. Katinsky, A. M. Goldfarb, M. B. Barbosa, & M. R. Duarte (Eds.), *Número especal: Seminário Latino-Americano sobre Alternativas de Ensino da História da Ciencia e da Tecnologia. Revista da Sociedade Brasileira de História da Ciência*.

Santos Filho, L. (1947). *História da Medicina no Brasil (do século XVI ao século XVI)*. Sao Paulo: Editora Brasiliense Ltda.

Sarton, G. (1950). Est-il possible d'enseigner l'histoire des sciences? *Archives Internationales d'Histoire des Sciences, 10*.

Schwartzman, S. (1979). *Formação da comunidadecientífica no Brasil*. Río de Janeiro: FINEP.

Sigerist, H. E. (1935. L'histoire de la médecine et la sociologie medicale. *X Congreso Internacional de Historia de la Medicina*, Libro de Actas, Madrid.

Sigerist, H. E. (1937). *Socialized medicine in the Soviet Union*. New York: W. W. Norton.

Sigerist, H. E. (1938). Socialized medicine. *The Yale Review*, 463–481.

Sigerist, H. E. (1941). *Medicine and human welfare*. New Haven: Yale University Press.

Sigerist, H. E. (1946). *Civilización y Enfermedad*. México: Fondo de Cultura Económica.

Sigerist, H. E. (1960). From Bismarck to Beveridge: Developments and trends in social security legislation. In M. J. Roemer (Ed.), *Henry E. Sigerist on the history of medicine*. New York: MD Publications.

Sigerist, H. E. (1974). *Historia y Sociología de la Medicina*. Bogotá: Editado por el Dr. Gustavo Molina.

Sigerist, H. E. (1981a). *Hitos en la historia de la salud pública*. México: Siglo XXI Editores. Original edition, 1956.

Sigerist, H. E. (1981b). Los modelos cambiantes de la atención médica. In H. E. Sigerist (Ed.), *Hitos en la historia de la salud pública* (pp. 85–98). México: Siglo XXI Editores. Original edition, 1956.

Soriano Lleras, A. (1966). *La Medicina en el Nuevo Reino de Granada durante la Conquista y la Colonia*. Bogotá: Universidad Nacional de Colombia/Imprenta Nacional.

Turner, B. (1987). *Medical power and social knowledge*. London: SAGE.

Turner, B. (1999). Profesiones, conocimiento y poder. In C. De la Cuesta Benjumea (Ed.), *Lecturas básicas en sociología de la medicina*. Medellín: Universidad de Antioquia.

Valdizán, H. (1928). *Apuntes para la Biblioteca Médica Peruana*. Lima: Imprenta Americana.

Valdizán, H. (1944). *Historia de la Medicina Peruana*. Lima: Ediciones Hora del Hombre.

Vasco Uribe, C. E. (1985). *El álgebra renacentista*. 2a Edición ed. Bogotá: Departamento de Matemática y Estadística. Universidad Nacional de Colombia. Original edition, 1979.

Vasco Uribe, A. (1988). *Estado y Enfermedad en Colombia*. Medellín: Editorial Universidad de Antioquia.

Vessuri, H. (1984). *Ciencia Académica en la Venezuela Moderna*. Caracas: Fondo Editorial Acta Científica Venezolana.

Woolgar, S. (Ed.). (1988). *Knowledge and reflexivity: New Frontiers in the sociology of knowledge*. London: Sage Publications Ltd.

Woolgar, S. (1991). *Ciencia: abriendo la caja negra*. Barcelona: Anthropos-Editorial del Hombre. Original edition, 1988.

Zuluaga, O. L. (1987). *Pedagogía e Historia: la historicidad de la pedagogía, la enseñanza un objeto de saber*. Bogotá: Ediciones Foro Nacional por Colombia.

Emilio Quevedo-Vélez MD, PhD Physician, pediatrician, PhD in Social Studies of Science. Professor Emeritus and Director of the Social Studies of Sciences, Technologies and Professions Reseach Group, School of Medicine and Health Sciences, Universidad del Rosario, Bogotá, Colombia. Emeritus Researcher of the Ministry of Science, Technology, and Innovation of Colombia.

Magister Julieth Nathalia Barrera-Díaz Nurse, Magister in Public Health Nurse, MSc. in Public Health. Temporal Professor at the School of Medicine and Health Sciences, Universidad del Rosario, Bogotá, Colombia.

Chapter 17
Intellectual Collaboration at the Early Stage of the Social and Historical Studies of Science and Technology in Venezuela

Hebe Vessuri and Isabelle Sánchez-Rose

Abstract We are concerned with a specific type of knowledge, namely academic knowledge, produced in one of the world's regions in and about specific institutions dedicated to research and teaching. In the 1980s and 1990s, a group of scholars from several Latin American countries agreed to participate in the creation of the field of history of science and technology in Latin America. In 1982, the Latin American Society for the History of Science and Technology (LASHST) came into being and in 1984 the *Quipu* journal was founded, serving as the official journal of the Society. We review the Venezuelan participation in *Quipu* from issue 1 in 1984 to its demise in 2001.

Keywords LASHST · Intellectual collaboration · Social bonds · Networking · Knowledge circulation · Venezuela

17.1 Introduction

Throughout the history of modernity, multiple experiences have accelerated the circulation of knowledge. In complex forms, knowledge, places, and people continue to be transformed during the process of circulation. What remains stable or changes cannot be decided theoretically but must be empirically studied (Keim et al., 2023). Here we are concerned with a specific type of knowledge, namely academic knowledge, produced in one of the world's regions in and about specific institutions dedicated to research and teaching.

H. Vessuri (✉)
Venezuelan Institute of Scientific Research (IVIC), Macarao, Venezuela

I. Sánchez-Rose
Center of Development Studies (CENDES), Universidad Central de Venezuela, Caracas, Venezuela

M. de la P. Ramos-Lara, L. C. Arboleda (eds.), *History of Science in Latin America: The Construction of an Intellectual Field (20th century)*,
https://doi.org/10.1007/978-3-031-75536-1_17

In the 1980s and 1990s, a group of scholars from several Latin American countries agreed to participate in the creation of a field of reflection and research, that of the history of science and technology in Latin America. In 1982, the Latin American Society for the History of Science and Technology (LASHST) came into being,[1] chaired by Juan José Saldaña as a result, among other things, of his enthusiasm and convening capacity. Soon after, in 1984 the journal *Quipu* was founded, immediately becoming the Society's scientific dissemination organ offering the novelty of its Latin American scope. With the journal as a unifying element, this network frequently met at different venues aiming to develop the activity of research on history of science, including its contemporary phase, in the region. At the time not much discrimination was made whether practitioners were historians or sociologists, economists, or political scientists. The multidisciplinary "umbrella" seemed broad enough to cover them all. They identified themselves as researchers in the broad field of the humanities and social sciences interested in how the processes of implantation and development of scientific-technical activity and its role in their society took place in countries that by then were known as "developing" or "underdeveloped", distant from the world centers of knowledge production and validation.

At the time policy considerations related to the construction of national scientific communities and national research institutions were common. Science was an essential component of development policies, so much so that a book like *Science and Politics in Latin America* (Herrera, 1971) became a bestseller, and science policies were an integral part of the novel 'dependency theory' that became common in the region. To some, science looked as being too important to be left to natural and "hard" scientists alone, being guided by the inner conviction that attention should be paid not just to the "internalist" history of science but also to its external conditions, the history of politics and social dynamics with which research grew inextricably linked in the world at large. Research was required to identify instantiations of academic knowledge and provide adequate conceptualization for them, isolate, analyze, and situate knowledge spaces and actors, identify channels of knowledge exchanges between regions, determine historically limited moments or sequences, and remain sensitive to the power dynamics at play (Keim et al., 2023).

When the LASHST came into being, Hebe Vessuri, co-author of this chapter, already knew Juan José Saldaña. She lived and worked in Venezuela, exiled from Argentina. She had recently launched Venezuela's first graduate program on Science and Technology Policy, which also turned out to be the first in Latin America. That happened at the Center for Development Studies (CENDES) of the Central University of Venezuela, in Caracas, more than 40 years ago. In those circumstances, she was looking to relate to the initiatives she was beginning to learn about, from neighboring fields in countries in the region and beyond. Vessuri, like most participants in the regional knowledge adventure, was very much aware that place matters in the world of knowledge, both for understanding its production and

[1] Sociedad Latinoamericana de Historia de la Ciencia y la Tecnología (SLHCT), original name in Spanish.

reception, a belief that at the beginning of the 2000s Livingstone would synthesize in a widely read book (2003). International collaborations grew everywhere, offering opportunities for additional research beyond the limits of national borders, while publications were already controlled by powerful "gatekeepers" in all academic disciplines.[2]

17.2 Growth Through Networking of an Intellectual Shared International Space

During this growth period, Latin American meetings multiplied in Brazil, Colombia, Argentina and Mexico being crucial for inter personal communication and intellectual exchange. One contact led to another and intellectual concerns, ideas, and exchanges, rapidly expanded in different directions, while both the networks and the journal together reflected participants expectations' of assuming a position as a Latin American school or collective for the study of the history of science.[3] The personal dimension of scientific relations created conditions for perceived and shared views and approaches. Research questions fostered by personal bonds through the region's knowledge space contributed to some critical deconstruction of their imbrications with power relations in works that would later crop up in the literature.

We understand personal bonds as "a social relationship in which two or more people establish a shared and distinctive body of interpersonal norms, worldviews, rules, and expectations that are not predetermined by their group or institutional background" (Rodriguez Medina & Vessuri, 2021). The social bonds knitted around *Quipu* journal and the LASHST were the basic units of social relations that created and stabilized a shared intellectual space in the history and social study of science in the region and beyond. Perceptions of trust, credibility, supportiveness, and influence were important qualities that grew from the successful interactions in the shared journal/ association setting. Personal bonds among actors of a varied nature shaped connections, local and international, influencing many social scientists' careers and production. In a world where knowledge production was (and still is) unevenly distributed, there were occasional episodes of tensions and disagreements among participants in events, but the challenge and achievements of connectivity and levelled dialogue prevailed. Conversations between researchers of different origin, from both North and South, helped produce a more propitious flatter environment for collaboration.

[2] Vessuri, H., J.-C. Guédon & A. M. Cetto, in a paper of 2013, visited the evolution of the gap between national and international publishing and the undue inequalities this created.

[3] Something similar is suggested by Barros da Silva (2014) when she refers to the shared expectation of authors to make of the journal "o campo de uma demarcação".

There was shared awareness of the simultaneous importance of ensuring an international dialogue that would lower communication barriers in a upward (for some) and downward (for others) double process of levelling scholarly communication between world centers and peripheries, diminishing asymmetries. It is known that scientific conferences provide transient spaces for face-to-face (and more recently also virtual) encounters between scholars. At conferences researchers present and discuss research results, work in progress or ideas in the making; they socialize with their colleagues and establish new contacts. The links in the regional network between scholars like Lewis Pyenson, Jean-Claude Guédon, and Yakov Rabkin from Canada; Ubiratam d'Ambrosio, Shozo Motoyama, Maria Amelia Dantes, Regis Cabral from Brazil; Luis Carlos Arboleda, Olga Restrepo Forero and Emilio Quevedo from Colombia; Alicia Lértora and Alfredo Kohn Loncarica from Argentina; Antonio Lafuente, José Sala Catalá and Horacio Capel, from Spain; Angel Ruiz from Costa Rica; Eduardo Estrella from Ecuador, and numerous other colleagues from the journal's host country Mexico as well as from other countries, often resulted in intellectual collaborations and friendships unrelated to the local or global structure of academic labor.[4] Retrospectively, the interplay between the personal and the international emerges time and again in the historiographical record of the period.

Actual encounters that favored interaction, though, were not restricted to the region's boundaries. In this respect the International Congress on the History of Science in Berkeley in 1985, in which for the first time there was an organized Latin American presence, was particularly interesting. There was a Symposium called *Cross Cultural Transmission of Natural Knowledge and its Social Implications: Latin America*, published by Saldaña in 1988. In the preceding World Congress in Bucharest in 1981, some ten Latin American historians participating on an individual and disaggregated way, were surprised to find that topics dealt almost exclusively with the history of European science with hardly anything about research in other latitudes. They decided to articulate their activities on a future regional meeting, writing a Bucharest Declaration (1981) (Arboleda, 2021). But this was not the only contact point with the international world. A growing number of meetings that took place in different places attracted Latin American participants, often through the mediating role of a scholar from the North.

One such mediator was Everett Mendelsohn, from Harvard, who was a relevant figure in the Berkeley Congress. An outstanding researcher and teacher some of his features were common to some Latin Americans identified with institution and bonds-building. While Mendelsohn was engaged in the history of biology, he was also interested in social aspects of science, science policy, politics and the relationship between science and warfare. A former defender of peace in Vietnam, he was

[4] We mention only a few because our memory fails us but we dare to provide this initial incomplete list of the universe of relevant individuals hoping that other contributions improve it towards the completion of the intellectual map of people involved in that seminal stage. For relevant publications of some of these scholars during that period, see the References at the end of the chapter.

linked to a program that took him every year to the Middle East, where, as a lifelong pacifist, he sought a country shared by Israelis and Palestinians. Mendelsohn published several books in the Sociology of the Sciences Yearbook of Kluwer Academic Press (1968, 1988, 1993 among others), which were important sources of new knowledge in the international arena. International collaboration at a meeting in Paris resulted in another book in that collection, edited by the French scholar R. Waast, American expatriate in Paris T. Shinn, and J. Spaapen, from the Netherlands. Among other valuable people was Sandra Harding, an American researcher in the field of STS who, among other things would later contribute to the development of a valuable journal like *Tapuya: Latin American Science, Technology and Society*, published by Taylor & Francis from Puebla, Mexico. Thus, participation in experiences like these facilitated the expansion of inter-regional exchanges of ideas and experiences, in broadening chains of contacts and networks.

Alongside texts, scientific knowledge takes many other forms, including embodied knowledge and "know how". Information and knowledge circulating through samples, instruments, compounds, animals, patients have recently been put together under the umbrella concept of "intermediary objects" as research-enriching tools (Vinck & Pérez-Martelo, 2023, pp. 251–262). Artefacts appear as testimonies, symbols and indicators of economic and social orientation, dissemination of knowledge and stage in evolution. Several *Quipu* articles deal with those intermediary objects in case studies located in the various Latin American countries. We mention here a meeting at the Smithsonian Institution in Washington in the 1980s, where members of the *Quipu* network met, among other well-known figures in the history of science, someone like Nathan Reingold, senior scholar from that Institute. They also had the chance to spend a few exciting hours in some of the basements of the Smithsonian buildings, getting familiar with a huge diversity of intermediary objects, and became more aware of the complexities of conditioning such intermediary objects to ensure their stability, as well as the specific and always changing requirements of photography workshops and other technical activities related to the archival day-to-day of historians. Among other things, we learned about the chemical characteristics of the paper of books and manuscripts from different historical periods, the elaboration of blueprints in digital version of airplanes of different sizes and complexity that were housed in the Smithsonian Museum of Aeronautics, and about the machines that the Smithsonian was buying for this technical work from their position in the forefront of the age of archive digitalization. Given the perennial resource scarcity in most Latin American countries, establishing partnerships with institutions in more technically advanced countries was critical regarding knowledge circulation.

In Mexico City, in 2001, almost 20 years after Berkeley, the XXI International Congress on the History of Science and Technology was held. International participation had already diversified considerably, with some 50 countries represented and a significant Latin American presence. Juan José Saldaña was the event's organizer, closing with a flourish the period of institutionalization of the history of science and technology in Latin America.

17.3 *Quipu*'s Role in the Circulation of Academic History of Science Knowledge in Latin America and Venezuelan Participation

Quipu's role in the circulation, visibility and accessibility of socio-historical scientific knowledge was to provide a stable, dependable platform with guaranteed professional reliability for scientific publications. The scientific credibility of the journal rested on the names of the editorial board, which included José Babini (Argentina), Simâo Mathias (Brazil), Luís Carlos Arboleda (Colombia), José López Sánchez (Cuba), Rodrigo Fierro Benítez (Ecuador), José Luís Peset (Spain), Enrique Beltrán (Mexico), Arturo Alcalde Mongrut (Peru), Marcel Roche (Venezuela) and Dirk J. Struik (United States), and the associate country editors who acted as gatekeepers and guaranteed serious peer reviewing helping to identify possible authors and readers. It was published in Spanish. Being a social science and humanities journal devoted to history of science, an area that traditionally had books as the most prestigious types of publications, *Quipu* published research articles, book reviews, and reports three times a year. Most of the authors were Latin American, although some Northern authors also participated with articles.

At the time of *Quipu*'s birth, there was a community of historians in Venezuela, although the history of science was of interest to only a few of them. Germán Carrera Damas was one of the great contemporary Venezuelan historians, with an extensive and significant body of work. His life and academic work experience in other countries, including diplomacy, gave him a cosmopolitan flavor and made him a source of important ideas and criticisms. He was interested, among other things, in the history of historiography, concerned with understanding the foundations of the historical consciousness of Venezuelans. Other valuable historians were Elías Pino Iturrieta and Manuel Caballero. Good historians such as Ricardo Archila (1966) and Eduardo Arcila Farías (1961, 1979) had made their works directly related to specific disciplines, medicine, and economics and engineering, respectively. But more directly influential in the fledgling community of the social historians of science in the region, probably because of their contribution to the understanding of the scientific dynamic in a Latin American country, were the socio-historical works by Brazilian scholars like Simón Schwartzman (1979) and Mario Guimarães Ferri and Shozo Motoyama (1979–1980), Eli de Gortari (1963) and Elías Trabulse (1984) from Mexico and José Babini (1971, 1981) and Marcelo Montserrat (1993) from Argentina, as well as the young promising Peruvian historian of science Marcos Cueto (1989, 1994, 1995). A small group of hard scientists and social thinkers came strong with a new approach to the role of science and technology in Latin America, giving place to what came to be known as the "Pensamiento Latinoamericano en Ciencia y Tecnología para el Desarrollo" (PLACTD). Jorge Sábato (1975), Amilcar Herrera (1971, 1973), Edmundo Fuenzalida Faifovich (1971, 1997), Francisco Sagasti (1975), were some of them.

Among the Venezuelan social scientists were José Agustín Silva Michelena (1970), Heinz Sonntag (1975, 1988), Roberto Briceño León (1987), Maritza Montero (1984), Lourdes Yero et al. (1982), Moisés Naim and Ramón Piñango (1986).

Researchers interested in "modern" approaches to the social history of science and technology and new practitioners of what would come to be known as the local social studies of science and technology were Marcel Roche (1973, 1975, 1988), Ignacio Ávalos and Marcel Antonorsi (1980), Carlota Pérez (1983, 1985), Olga Gasparini (1969) whose untimely death prevented a larger contribution to the social study of science in Venezuela, Hebe Vessuri (1984, 1987a, b), Díaz et al. (1983), Yajaira Freites (1984, 1997), Yolanda Texera (1991), Arnoldo Pirela (1996), Rafael Rengifo (1986), Alexis Mercado (1995), Irene Plaz Power (1993), Ocarina Castillo (1990). Some of them were starting their research career practically simultaneously with the early years of *Quipu* and the LASHST.

Venezuelan participation in the journal and in the regional congresses was small but consistent, as revealed by a review of its 13 volumes. The presence of Venezuelan authors is already visible in the very first issue of the journal, with the review of a book by Marcel Roche, *Rafael Rangel, Ciencia y Política en la Venezuela de principios del siglo XX* "[Rafael Rangel, science and politics in Venezuela at the beginning of the twentieth century] (1973), by E. Yanes. Roche was one of the great men of science in democratic Venezuela. A researcher in the medical area, he was an institution builder creating the Luis Roche Foundation in 1952, the Venezuelan Institute of Scientific Research (IVIC) in 1959, the National Council of Scientific and Technical Research (CONICIT) in 1967, and the *Interciencia* journal in 1976. He collaborated with *Quipu* and accepted to be a member of its Editorial Board from the beginning of the journal's life. In 1985, volume 2, issue 2, two reviews appeared, the first one by M.L. Rodríguez was about the book edited by Hebe Vessuri on *La ciencia académica en la Venezuela moderna* [Academic science in modern Venezuela] (1984). This was the second book in which she and her co-authors dealt with the study of "peripheral science" and of "centers and peripheries", rapidly inducing responses like that of Marcos Cueto (1989), from Perú about "scientific excellence".

She had agreed to participate in *Quipu* as one of the region's associate editors with the commitment to bring Venezuelan and other Latin American authors and readers closer to it. She led a small group of people of several disciplinary origins who were interested in studying the process of implantation and consolidation of scientific activity in the country from the perspective of the social sciences. The other review, in number 3 of volume 2, was by J.A. Ramirez about the book by Waleska Lemoine and María Matilde Suárez, *Beauperthuy. De Cumaná a la Academia de Ciencias de París. Sobre miasmas, insectos y resistencia académica* [Beauperthuy. From Cumaná to the Paris Academy of Science. On miasmas, insects, and academic resistance] (1984). Waleska Lemoine was a promising young student of Marcel Roche at IVIC who soon would settle down in Denmark and never returned to the country. Maria Matilde Suárez is an anthropologist who wrote some history books unrelated to science.

In 1986, volume 3, issue 1, the first Venezuelan article by a Venezuelan author appeared in the journal. It was an article by Yolanda Texera Arnal, who was working on what would become both her master's and doctoral theses at CENDES, under Vessuri's supervision. Texera was part of the graduate group on the social study of science and technology that had developed there since the late seventies. She was perhaps the author that most systematically pursued the historiographical vein in that generation of scholars of science and technology. The paper in *Quipu* was "La Biología en un contexto periférico. La Escuela de Biología de la Universidad Central de Venezuela" [Biology in a peripheral context. The School of Biology of the Central University of Venezuela] (pp. 29–52). In issue 2 of that year, M. D. de Kury reviewed Canelón, Jesús Saturnino, *Odontología y Sociedad: Venezuela siglo XIX* [Dentistry and Society: Venezuela 19th century] (1981). In this issue is published the article by Waleska Lemoine, "Las mujeres y el conocimieneto científico en Venezuela a través de la historia" [Women and scientific knowledge throughout history] (pp. 189–211).

Volume 4, issue 2, 1987, gathers several works about Venezuela. It includes a work by Tulio Arends, "Bolívar y los científicos" [Bolívar and the scientists] (pp. 167–184), complemented in the same issue by the review of his book *Ciencia y tecnología en la época de Bolívar* [Science and Technology in the time of Bolivar] by G. Vargas. Arends was a physician, hematology researcher and university professor who also held positions as president of CONICIT and director of IVIC. Among his side interests were scientific publications and libraries, something that led him to deal with those subjects in different contexts. In the same issue 2 of 1987 there is another work by Texera, "Exploradores botánicos europeos en Venezuela durante el siglo XIX" [European botanical explorers in Venezuela during the nineteenth century] (185–211). A paper by Yajaira Freites, "La ciencia en la época del gomecismo" [Science in the age of gomecismo] (pp. 213–251). A fourth paper by Hebe Vessuri on "El proceso de profesionalización de la ciencia venezolana: la Facultad de Ciencias de la Universidad Central de Venezuela" [The process of professionalization of Venezuelan science: the Faculty of Sciences of the Central University of Venezuela] (pp. 253–281), showed, among other things, the importance of the contribution of foreign scientists in the set- up of contemporary social and intellectual landscape of Venezuelan society. Already in 1987, a Venezuelan Group of History and Sociology of Science figures among the National Societies Affiliated to the LASHST, organized by Humberto Ruiz Calderón and Yajaira Freites, which has lasted to the present.

In 1988 the journal did not receive collaborations from Venezuela but in 1989, volume 6, there is an article by Waleska Lemoine in issue 1, about "Lo cognitivo y lo social en la resistencia académica" [The cognitive and the social in academic resistance] (pp. 79–91), while in vol. 6, issue 2, there is a paper by Isabel Licha on "El impacto modernizador de la ingeniería sanitaria en Venezuela: el caso del INOS y de otras instituciones sanitarias" [The modernizing impact of health engineering in Venezuela: the case of INOS and other health institutions] (pp. 217–236). Licha was an industrial engineer who for a while was part of the CENDES group and then developed her career abroad in the field of social development. In 1990 (volume 7) there is a contributions by Claudio Bifano on "Vicente Marcano y la investigación química en Venezuela" [Vicente Marcano and chemical research in Venezuela] (pp. 61–71). Bifano is a distinguished Venezuelan inorganic chemist with extensive teaching and research activity and laterally interested in the history and social aspects of science. In issue 2 of that volume is an article by Hebe Vessuri, "El Sísifo sureño: las ciencias sociales en la Argentina" [Southern Sisyphus: the social sciences in Argentina] (pp. 149–187).

Volume 8, issue 1, 1991, was published with the institutional support of the National Council of Science and Technology of Venezuela and dedicated to the history of science and technology in that country. José Luis Ávila Bello is the author of the first article in this issue, called "El Instituto Nacional de Dermatología de Venezuela 1971-1985" [The National Institute of Dermatology of Venezuela 1971–1985] (pp. 135–170). Ávila Bello was a disciple of Jacinto Convit, a renowned researcher in the field of communicable diseases, particularly in the clinical treatment of leprosy. The second work in this issue is by Ana Teresa Gutiérrez called "La búsqueda de una ilusión: la investigación sobre malaria en Venezuela" [The Search for an Illusion: Malaria Research in Venezuela] (pp. 171–200). She was a CENDES researcher who wrote a doctoral thesis that was published (1987) about Arnoldo Gabaldón, the great malaria scientist who led Venezuela to be the first country in the world to eradicate the disease, although regrettably the disease later returned with force. Her paper is followed by Pablo Testa, "Antecedentes y orígenes de la Escuela de Estadística y Ciencias Actuariales de la Universidad Central de Venezuela" [Background and origins of the School of Statistics and Actuarial Sciences of the Universidad Central de Venezuela] (pp. 201–213). Testa is a statistician who made his career at CENDES in the technology innovation group, first with Arnoldo Pirela and then with Alexis Mercado and is currently retired.

Irene Plaz Power is the author of the third paper in that issue, "Informática en la sociedad venezolana: Una breve historia de una tecnología autónoma" [Computing in Venezuelan society: a brief history of an autonomous technology] (pp. 215–234). Irene Plaz also began her career as a student of the CENDES group, did her thesis with Marcel Roche at the IVIC, and developed a valuable career in Social Informatics, witnessing the process of academic telematization in the country in those years. She is currently retired from IVIC. Her paper is followed in the same issue by Miguel Szinetar Gabaldón with an article on "La primera estación experimental agrícola de plantas y semillas en Venezuela" [The first agricultural experimental station of plants and seeds in Venezuela] (pp. 235–253). Szinetar was doing his doctoral thesis with Vessuri on the recent history of science in the country, focusing on the study of Alberto Adriani, an interesting figure in the transition to the country's modernization. The thesis led him later to his publishing it as a book (1998) and separately he also published, among other works, the complete works of Adriani (1998). The last paper in that issue is an article by Hebe Vessuri "Universalismo y nacionalismo en la ciencia moderna. Un enfoque desde el caso venezolano" [Universalism and Nationalism in Modern Science. An approach from the Venezuelan case] (pp. 255–271), in which she develops some of her ideas about international asymmetries and inequalities in the scientific-technical field.

In 1992, volume 9, Issue 1 contains a contribution by Yajaira Freites, "De ilustrados a profesionales: los ingenieros venezolanos entre 1899 y 1933" [From Enlightened to Professional: Venezuelan Engineers Between 1899 and 1933] (pp.75–88) and a review of the book by Babini, N. *La informática en Argentina 1956–1966,* written by Irene Plaz Power (pp. 247–249). In issue 3 of this volume appears the article by

Carlos Schubert, "Emile Rod, la controversia sobre las grandes fallas de rumbo y el paradigma de la tectónica de placas en Venezuela: una visión personal" [Emile Rod, the controversy over strike-slip faults and the plate tectonics paradigm in Venezuela: a personal view] (pp.349–381). Schubert was a highly regarded geology researcher at the Ecology Center, IVIC.

In 1993, volume 10, there are no works from Venezuela but in 1994, volume 10, issue 1, Hebe Vessuri publishes "¿Estilos nacionales de ciencia?" [National styles in science?] (pp. 103–118). Here she introduces the notion of national science styles which she had already developed in a contribution to a previous book edited by Brazilian anthropologists Roberto Cardoso de Oliveira and Guillermo Rubén, to discuss the "sciences in national contexts", offering some heuristic guidelines. A discipline is seen as a non homogeneous, consensual community, consisting of diverse segments, often identified with competitive styles or programmes, adapted to different socio-institutional contexts and, very importantly, prescribing favoured relationships with other disciplines.

A series of financial difficulties led to the temporary suspension of the journal during the ensuing 4 years, between 1993 and 1998. It was re-activated in 1999 with volume 12, without the participation of Venezuelan authors. The last volume in which we found registers corresponds to the year 2000, and it includes two articles by Yajaira Freites. In issue 1 of volume 13 there appears "La visita de Humboldt (1799-1800) a las provincias de Nueva Andalucía, Caracas y Guayana en Venezuela y sus informantes" [Humboldt's (1799–1800) visit to Nueva Andalucía, Caracas and Guayana provinces and their informants] (pp. 35–52) and in issue 3, "El quehacer de la historia de la ciencia en Venezuela: entre la justificación y el conocimiento" [The work of the history of science in Venezuela: between justification and knowledge] (pp. 261–289). These last two publications complete the universe of works on Venezuelan topics published by *Quipu* both in its first period of 1984–1995 and in its short-lived second period: 1999–2000. Of a total of 279 articles published by the journal, 21 were Venezuelan contributions.

While it lasted, the journal served to make its participants think and grow as intellectuals, citizens, and human beings. The archive witnesses the ideas, themes and approaches common at the time and constitute a valuable repository.

References

Aceves Pastrana, P. (1996). Tradición y modernidad en la Nueva España: Estudios sobre aguas minerales (s.XVII-XVIII). Llul: *Revista de la Sociedad Española de Historia de las Ciencias y de las Técnicas*. Vol. 19, Nro. 37.

Adriani, A., & Gabaldón, M. S. (1998). *Textos escogidos*. Caracas.

Aguilera Ríos, S. (2000). Quipu: una revista latinoamericana de la historia de las ciencias y la tecnología, Biblos *3W. Revista bibliográfica de geografía y ciencias sociales*. Universidad Barcelona, Nr. 2012, 28 de febrero.

Arboleda, L. C. (1987). Acerca del problema de la difusión científica en la periferia: El caso de la física newtoniana en la Nueva Granada (1740-1820). *Quipu. Revista Latinoamericana de Historia de las Ciencias y la Tecnología, 4*, 7–30.

Arboleda, L. C. (2021). Ubiratam d'Ambrosio como historiador de las matemáticas y las Ciencias. Cua*dernos de Investigación y Formación en Educación Matemática.* Special issue.

Arboleda, L. C., Quevedo, E., & Hernández, M. (1991). ¿Epistemología versus Estudios Sociales de la Ciencia? *Interciencia. Revista de ciencia y tecnología de América, 16*, 266–268.

Archila, R. (1966). *Historia de la medicina en Venezuela*. Universidad de los Andes, Mérida: Ediciones del Rectorado.

Arcila Farías, E. (1961). *Historia de la ingeniería en Venezuela*. Caracas.

Arcila Farías, E. (1979). *El régimen de la encomienda en Venezuela*. Caracas. Universidad Central de Venezuela. Facultad de Ciencias Económicas y Sociales, Instituto de Investigaciones.

Ávalos Gutiérrez, I., & M. Antonorsi Blanco. (1980). *La planificación ilusoria: ensayo sobre la experiencia venezolana en política científica y tecnológica*. Caracas: CENDES/Editorial Ateneo de Caracas.

Babini, J. (1971). *El saber en la historia*. Buenos Aires: Centro Editor de América Latina. Colección enciclopédica Biblioteca fundamental del hombre moderno. Tomo 26.

Babini, J. (1981). *Historia de la medicina*. Buenos Aires.

Barros Da Silva, M. R. (2014). História e historiografia das ciencias latino-americanas: Quipu (1984-2000). *Revista Brasileira de História da Ciência, Rio de Janeiro, 7*(1), 47–57.

Briceño León, R., & M. Acosta. (1987). *Ciudad y capitalismo*. Caracas. Edic. de la Biblioteca, UCV.

Caballero, M. (1986). *Latin America in the World Revolution* (The Communist International 1919–1943). Cambridge University Press.

Cabral, R. (1986). *The interaction of Science and Diplomacy: The United States, Latin America and Nuclear Energy, 1945–1955*. Ph.D: Dissertation, Dept. of History, University of Chicago.

Capel, H. (1977). Institucionalización de la geografía y estrategias de la comunidad científica de los geógrafos, *Geo Crítica. Cuadernos Críticos de Geografía Humana*, Universidad de Barcelona, n° 8–9, pp. 1–58. http://www.ub.es/geocrit/geo8.htm; y http://www.ub.es/geocrit/geo9.htm

Carrera Damas, G. (2000). El siglo venezolano. Conversando con Germán Carrera Damas. A. Baptista (Ed.), *Venezuela siglo XX. Visiones y testimonios* (pp. 459–486). Fundación Polar, Caracas.

Castillo, O. (1990). *Los años del bulldozer: ideología y política, 1948–1958*. Caracas.

Cueto, M. (1989). *Excelencia científica en la periferia: actividades científicas e investigación biomédica en el Perú 1890–1950*. Lima.

Cueto, M. (1994). *Missionaries of science: The Rockefeller Foundation and Latin America*. Indiana University Press.

Cueto, M. (1995). *Saberes Andinos: Ciencia y Tecnología en Bolivia, Ecuador y Perú*. Instituto de Estudios Peruanos.

de Gortari, E. (1963). *La ciencia en la historia de México*. México.

Declaración de Bucarest. (1981). *Documento preparatorio de la fundación de la Sociedad Latinoamericana de Historia de la Ciencia y la Tecnología*. Adoptada en Bucarest, el 1ro. de septiembre de 1981, durante la celebración del XVI Congreso Internacional de Historia de la Ciencia, https://issuu.com/chcytal/docs/declaracion_de_bucarest http://www.rvistaccuba.cu/index.php/revacc/article/view/219

Díaz, E., Texera, Y., & Vessuri, H. (1983). *La ciencia periférica. Ciencia y sociedad en Venezuela*. Monte Ávila.

Freites, Y. (1984). La institucionalización del ethos de la ciencia: el caso del IVIC. H. Vessuri (compl.): *La ciencia académica en la Venezuela moderna: historia reciente y perspectivas de las disciplinas científicas*. Caracas. Fondo Editorial Acta Científica Venezolana.

Freites, Y. (1997). El problema del saber entre hacendados y comerciantes ilustrados de la provincia de Caracas-Venezuela (1793–1810), *Dynamis: Acta hispanica ad medicinae scientiarum que historiam illustrandam*, N° 17.

Fuenzalida Faifovich, E. (1971). *Subdesarrollo e investigación científica: una teoría sociológica. II Conferencia sobre el tema*. IEFF.

Fuenzalida Faifovich, E. (1997). "La investigación básica en derecho y sociedad. Un enfoque de sociología de la ciencia sobre el caso de Chile", *Anuario de Filosofía Jurídica y Social*, número 15.

Gasparini, O. (1969). *La investigación en Venezuela: condiciones de su desarrollo.* Publicaciones IVIC.

Guédon, J.-C. (1996). *La planète cyber. L'Internet et le cyberespace.* Éditions Gallimard.

Guimarães Ferri, M., & Motoyama, S. (1979). *História das Ciências no Brasil.* Editora da Universidade de São Paulo.

Gutiérrez, A. T. (1987). *Tiempos de guerra y paz: Arnoldo Gabaldón y la investigación sobre malaria en Venezuela (1936–1990).* Centro de Estudios del Desarrollo–Universidad Central de Venezuela.

Harding, S. (1986). *The science question in feminism.* Cornell University Press.

Herrera, A. O. (1971). *Ciencia y política en América Latina.* Siglo XXI.

Herrera, A. O. (1973). Los determinantes sociales de la política científica en América Latina: Política de Científica Explícita y Política Científica Implícita. *Revista de Ciencias Sociales, Venezuela, 13*(49), 113.

Keim, W., Rodríguez Medina, L., et al. (2023). *Routledge handbook of academic knowledge circulation.* Routledge.

Lafuente, A., & Mazuecos, A. (1987). *Los caballeros de punto fijo: ciencia, política y aventura en la expedición geodésica hispano-francesa al virreinato del Perú en el siglo XVIII.* Madrid. Consejo Superior de Investigaciones Científicas. Ediciones de El Serbal.

Lafuente, A., Mazuecos, A., Elena, A., & Ortega, M. L. (1993). *Mundialización de la ciencia y cultura nacional.* Doce Calles.

Lemoine, W., & Suárez, M. M. (1984). *Beauperthuy. De Cumaná a la Academia de Ciencias de París. Sobre Miasmas, Insectos y Resistencia Académica.* Universidad Católica Andrés Bello.

Livingstone, D. N. (2003). *Putting science in its place: Geographies of scientific knowledge.* University of Chicago Press.

Mendelsohn, E. (1968). *The politics of pessimism: Science and technology Circa 1968.* The Sociology of Sciences Yearbook 1968. Kluwer Academic Press. Dordrecht. Springer.

Mendelsohn, E., Roe Smith, M., & Weingart, P. (1988). *Science, technology, and the military.* The Sociology of Sciences Yearbook. Kluwer Academic Press. Dordrecht. Springer.

Mendelsohn, E., Ezrahi, Y., & Segal, H. (1993). *Technology, pessimism and postmodernism.* The Sociology of Sciences Yearbook. Kluwer Academic Press. Dordrecht. Springer.

Mercado, A. (1995). *Desarrollo Tecnológico en la Industria de Química Fina de Brasil: Clasificación Taxonómica y Determinación de una Secuencia Incremental de Capacitación.* Caracas.

Mercado, A, & Pirela, A. (1994). El papel de los Centros de Investigación en el Desarrollo Tecnológico de la Industria Química: Experiencias en Brasil y Venezuela. *Ciencia, Tecnología y Sociedad en América Latina.* H Vessuri (coord). Caracas.Edit Nueva Sociedad.

Montero, M. (1984). *Ideología, alienación e identidad nacional: una aproximación psicosocial al ser venezolano.* Universidad Central de Venezuela, Ediciones de la Biblioteca.

Montserrat, M. (1993). *Ciencia, historia y sociedad en la Argentina del siglo XIX.* Buenos Aires.

Naím, M., & Piñango, R. (1986). *El Caso Venezuela: una ilusión de armonía.* Caracas.

Pérez, C. (1983). Structural change and assimilation of new technologies in the economic and social systems. *Futures, 15*(5), 357–375.

Pérez, C. (1985). Microelectronics long waves and world structural change: New perspectives for developing countries. *World Development, 13*(3), 441–463.

Peset, J. L., & Lafuente, A. (1987). El conocimiento y el dominio de la naturaleza: la ciencia y la técnica. en Batllori Munné, M. (coord.). *La época de la Ilustración, El Estado y la cultura (1759–1808)* (Vol. 1, pp. 347–393). Madrid. Espasa-Calpe.

Peset, M., & Peset Reig, J. L. (1974). *La Universidad española (siglos XVIII y XIX).* Madrid Taurus.

Petitjean, P., Jami, C., & Moulin, A. M. (Eds.) (1992). *Science and empires.* Springer (Studies in the Philosophy of Science, vol. 136).

Pino Iturrieta, E. (1971). *La mentalidad venezolana de la Emancipación (1810–1812)*. México City.

Pirela, A., Rengifo, R., Arvanitis, R., & Mercado, A. (1991). *Conducta empresarial y cultura tecnológica: Empresas y centros de investigación en Venezuela*. Caracas.

Pirela, A., Rengifo, R., Arvanitis, R., Mercado, A., Rengifo, R., Mercado, A., & Arvanitis, R. (1993). Technological learning and entrepreneurial behavior: A taxonomy of the chemical industry in Venezuela. *Research Policy, 22*(5–6), 431–453.

Pirela, A., Rengifo, R., Arvanitis, R., & Mercado, A. (Eds.). (1996). *Cultura Empresarial en Venezuela. La Industria Química y Petroquímica*. Fundación Polar.

Plaz Power, I. (1993). *La informática en la sociedad venezolana: ¿símbolo de desarrollo?: viejas reflexiones sobre una nueva tecnología*. CDCH UCV.

Pyenson, L. (1975). "Who the guys were": Prosopography in the history of science. *History of Science, 15*(3), 155–188.

Pyenson, L. (1985). Cultural imperialism and exact sciences: German expansion overseas, 1900-1930. *The American Historical Review, 20*(1), 1–43. https://doi.org/10.2307/1866636

Quevedo, E. (1993). *Historia social de la ciencia en Colombia*. 10 vols. Proyecto COLCIDNCIAS OEA, 1983–1986. Bogotá. Tercer Mundo Editores-COLCIENCIAS.

Rengifo, M. R. (1986). Ciencia y política en Venezuela: del espejismo al simulacro. *Espacios, 6*(1).

Rengifo, R., Pirela, A., & Arvanitis, R. (1997). Science and production in Venezuela: Two "emergencies". In J. Gaillard, V. V. Krishna, & R. Waast (Eds.), *Scientific communities in the developing world*. Sage Publications.

Restrepo Forero, O. (1984). La Comisión Corográfica: un acercamiento a la Nueva Granada. En: México, *Quipu. Revista Latinoamericana de Historia de las Ciencias y Tecnología*, vol. 1, issue 3. Mexico City. UNAM.

Restrepo Forero, O. (1986). El tránsito de la Historia Natural a la Biología en Colombia, En: *Colombia Ciencia Tecnología y Desarrollo*. (Bogotá) 10(3/4): 181–275. Julio-Diciembre.

Roche, M. (1973). *Rafael Rangel: Ciencia y Política a principios de siglo*. Caracas.

Roche, M. (1975). *Descubriendo a Prometeo (Ensayos sobre ciencia y tecnología en Venezuela)*. Caracas.

Roche, M. (1988). *Mi compromiso con la ciencia*. Caracas.

Rodríguez Medina, L., & Vessuri, H. (2021). Personal bonds in the internationalization of the social sciences: A view from the periphery. *International Sociology, 36*, 2–21. https://doi.org/10.1177/0268580920962014

Sábato, J. (Ed.). (1975). *El pensamiento latinoamericano en la problemática ciencia-tecnología-desarrollo-dependencia*. Buenos Aires.

Sagasti, F. (1975). Hacia un nuevo enfoque para la planificación científica y tecnológica. In J. Sábato (Ed.), *El pensamiento latinoamericano en la problemática ciencia-tecnología-desarrollo-dependencia*. Paidós.

Sala Catalá, J. (1987). *Ideología y Ciencia Biológica en España entre 1860 y 1881. La difusión de un paradigma*. CSIC Press.

Saldaña, J. J. (1985). *El perfil de la ciencia en América*. XI Congreso Interamericano de Filosofía: actas del Simposio Historia y Filosofía de la Ciencia en América, Guadalajara, Jalisco, México, 12 a 14 de noviembre de 1985.

Saldaña, J. J. (Ed.) (1988). Cross cultural diffusion of science: Latin America. Vol. 5, *Acts of the 17th International Congress of History of Science*, Berkeley, 31 July-8 August 1985. Cuadernos *Quipu*, Mexico city, Nr.2.

Saldaña, J. J. (Ed.). (1996). *Historia social de las ciencias en América Latina*. Porrúa-Coordinación de Humanidades/UNAM.

Schwartzman, S. (1979). *Formação da comunidade científica no Brasil*. São Paulo. Ed.Nacional. https://archive.org/details/FormaoDaComunidadeCientficaNoBrasil/page/n3/mode/2up

Silva Michelena, J. A. (1970). *Crisis de la democracia*. Caracas.

Sonntag, H. (1975). *Teorías del subdesarrollo y el caso de Venezuela*. Centro de Investigaciones sobre América Latina de la Univ. de Bielefeld.

Sonntag, H. (1988). *Duda/certeza/crisis: la evolución de las ciencias sociales de América Latina*. Caracas.

Szinetar Gabaldón, M. (1998). *El proyecto de cambio social de Alberto Adriani, 1914–1936*. CENDES.

Texera, Y. (1991). *La exploración botánica en Venezuela 1754–1950*. Fondo Editorial Acta Científica Venezolana.

Trabulse, E. (1984). *Historia de la ciencia en México: estudios y textos, siglo XVI*. México.

Vessuri, H. (Ed.). (1984). *Ciencia académica en la Venezuela moderna*. Caracas.

Vessuri, H. (1987a). The social study of science in Latin America. *Social Studies of Science, vol.17*, 519–554.

Vessuri, H. (Ed.). (1987b). *Las instituciones científicas en la historia de la ciencia en Venezuela*. Fondo Editorial Acta Científica Venezolana.

Vessuri, H., Guédon, J.-C., & Cetto, A. M. (2013). Excellence or quality? Impact of the current competition regime on science and scientific publishing in Latin America and its implications for development. *Current Sociology, 62*, 647–665. https://doi.org/10.1177/0011392113512839

Vinck, S., & Pérez-Martelo, C. (2023). Circulating knowledge through intermediary objects in scientific cooperative networks. In W. Keim, L. R. Medina, et al. (Eds.), *Routledge handbook of academic knowledge circulation*. Routledge.

Yero, L., Aguilera, M., & Lemoine, V. R. (Eds.). (1982). *La participación de la comunidad científica frente a las alternativas del desarrollo*. Caracas.

Hebe Vessuri Argentine-Venezuelan social anthropologist who pioneered research and teaching in the social studies of science and technology in several American countries since 1966. She belonged to boards and research committees of several national, regional, and international organizations. Her focus of interest has been on science in the global peripheries, with emphasis on the Latin American region, the dilemmas of social participation/exclusion, the unequal globalization of science in the contemporary world, and expert knowledge at the interface between higher education, scientific research, and indigenous and traditional knowledge. She was awarded the Bernal 2017 Prize to Distinguished Contribution, from the Society for the Social Study of Science (4S) Boston, 2017; the Houssay Prize to the Trajectory in the field of Human Sciences from the Argentine Government in 2018, and the ESOCITE (Latin American Society for the Social Studies of Science)'s Oscar Varsavsky´s Prize to the Professional Trajectory in 2014.

Isabelle Sánchez-Rose Dr. Sc. IVIC in Social Studies of Science, M.Sc. CENDES-UCV in Policy and Management of Technological Innovation, anthropologist UCV, Diplome d'Études Approfondies en Études Latino-Américaines, option: Études de l'Amérique Latine, Université Paris III, France. Professor-Researcher (retired) of the Area of Scientific and Technological Development, CENDES, and currently research coordinator of CENDES.

Chapter 18
An Approach to the Institutionalization of the History of Science in Chile

Zenobio Saldivia Maldonado and Pablo Andrés Rojas Castro

Abstract The process of institutionalization of the discipline of the history of science in Chile is defined, described, and analyzed, and the phases of the process are determined. Consequently, the milestones of this process are outlined, highlighting their most relevant characteristics, predecessors, and practitioners in each phase, as well as the principal works of these participants. This process encompassed the involvement of researchers, journals, and institutions associated with the mobility of knowledge.

Keywords History of science · Institutionalization · Chile

18.1 Previous Background

When one thinks of a discipline like the History of Science, the first thing that catches our attention is its recent emergence within the Latin American academic and disciplinary universe. Indeed, in Europe, following the efforts of Jean Le Rond D'Alambert and Denis Diderot during the Enlightenment to establish and disseminate scientific knowledge in the Grande Encyclopédie, and subsequently with Auguste Comte's interest in the progress of science in the 1830s and 1840s as part of the theoretical foundation for the philosophical, sociological, and epistemic movement known as "positivism," there is not a noticeable systematic body of individuals, means, and institutions in our continent with a clear awareness of working within a new branch of historiography known as "history of science." This is understandable as the young republics of our countries were in the process of political

Z. Saldivia Maldonado (✉)
Universidad Tecnológica Metropolitana, Santiago, Chile
e-mail: zenobio@utem.cl

P. A. Rojas Castro
Universidad Metropolitana de Ciencias de la Educación, Santiago, Chile

M. de la P. Ramos-Lara, L. C. Arboleda (eds.), *History of Science in Latin America: The Construction of an Intellectual Field (20th century)*,
https://doi.org/10.1007/978-3-031-75536-1_18

consolidation, establishing educational institutions, and institutionalizing science itself. This was a priority in the nineteenth century. This doesn't mean that some scholars did not address topics of scientific dissemination. As we know in the case of Chile, individuals like Andrés Bello, Benjamín Vicuña Mackenna, José Victorino Lastarria, and others wrote articles and texts on scientific topics. The point is that the country had not yet seen a certain historiographical specialization with a focused and organized group of scholars who were clearly convinced of actively engaging in the history of science as a distinct field.

Therefore, it is possible to assert that the History of Science is a very young discipline that begins to gain prominence primarily from the mid-twentieth century, thanks to the work of the Belgian mathematician and chemist George Sarton (1884–1956), who resided in the United States, on one hand, and the Russian philosopher based in France, Alexandre Koyré (1892–1964), on the other. The former solidifies and disseminates this knowledge as a university chair in the 1950s in the aforementioned century in the United States, highlighting the triumphant progress of scientific reason and bringing together scientists and humanists on this subject, as already noted by Kragh (1989, p. 3). The latter manages to focus the attention of historians and university intellectual life on this discipline and its topics of interest in the European context, thanks to his books on the scientific work of scholars from the seventeenth and eighteenth centuries, published in the 1950s and early 1960s. However, the emergence of this form of knowledge exhibits a clear Eurocentric beginning, or at most, a noticeable concern for issues specific to the interests of the northern hemisphere. It becomes evident, then, that its proponents start from the analysis of scientific development and the quest for understanding cognitive apprehension processes specific to the northern hemisphere in general. Intellectuals in the Americas must wait for a few more decades to engage with this discipline and make their contributions to it.

In Latin America, the emergence of this discipline is much more recent. Thus, from the late 1960s onward, there are works related to this historiographical form that allude to the initial attempts to elucidate the processes of constructing episteme in our countries. These works also encompass research that sheds light on the political factors and cultural variables that influenced scientific consolidation. Indeed, one of the first countries in Latin America to show interest in the peculiarities of the genesis and development of episteme appears to be Mexico. Since the late 1960s, various titles in Latin American historiography have addressed the vicissitudes of science in this country. This is primarily due to the efforts of Dr. Juan José Saldaña, who, first through the creation of the journal "Latino América Anuario" in 1968, and later with the establishment of the journal "Quipú" in 1984—under his direction—began disseminating reflections and guiding doctoral theses and essays on the progress of science and technology in Mexico and other countries in America. Additionally, he authored works such as *The ideology of science in Mexico* in the nineteenth century, published in 1985, *Science and the State in Latin America* from the year 1988, *The influence of the French Revolution in the scientific modernization of Mexico* from 1989, to name just a few by this author. Subsequently, works of this nature are observed in Peru, Argentina, Colombia, Chile, and Nicaragua, among other countries.[1]

18.2 History of Science in Chile

On our part, we believe that the institutionalization of this discipline corresponds to a cultural, political-institutional process and social mobility of knowledge, which occurs mainly from the late sixties of the twentieth century to the first decade of the twenty-first century. This process would have taken place in two additional phases: the first, which we could call "articulation" and which would go from the late sixties to probably the mid-nineties of the twentieth century, and whose most relevant note is the fact that the majority of the authors do not clearly perceive themselves as members of a new historiographical discipline that is duly established in the community, with occasional exceptions. And the second, which we can call "consolidation", would be located approximately from the mid-nineties of the last century until the middle of the first decade of the twenty-first century; in which scholars, scientists, academics, researchers, journalists, magazines and various institutions participate. In this period, it is already observed that almost all of these authors and/or collaborating entities perceive themselves as working specifically on issues of the history of science in Chile. And, therefore, we estimate that already at this milestone, there is a high bibliographic, theoretical and analytical production and with agents and authors, who have a clear conviction of working expressly in the field of historiography that interests us.

Next, we present the main authors who, in our opinion, have contributed to the institutionalization process of the discipline of the history of science in Chile in its two complementary phases. We highlight some of their works in this field of knowledge in each case.

18.3 First Phase: Articulation

Before explicitly mentioning the practitioners of the history of science in Chile, it is worth noting that the institutionalization process does not indeed emerge at the mentioned historical milestone. As the epistemologist Jean Piaget points out, there

[1] As an illustration, for example, from Peru let us remember the Maritime History of Peru 17th and 18th Centuries (Lohmann Villena, 1981); or The Studies of the History of Science in Peru, 2 vol. (CONCYTEC, 1986); or The History of Chemistry in Peru (Guevara, 1993). From Argentina: The History of Engineering in Argentina (Centro Argentino de Ingenieros, 1981), or Ranch and Society in the Pampas 1740–1820 (Mayo, 1995). From Colombia let us bring to presence the text of Zenobio Saldivia (compiler) An Approach to the Development of Science in Colombia (Saldivia, 2018). From Nicaragua, An Approach to the Development of Science in Nicaragua (Saldivia, 2005).

is always a previous phase in intellectual development, "...since in no cognitive field is there an absolute beginning of development" (Piaget, 1981, p. 13). In this case, the identified previous moment corresponds to some forward-thinking authors who worked on these topics in isolation in the country. They would be the precursors.

Thus, one of the precursors of the history of science in Chile is Carlos Porter. Although he emerged a little before the sixties, he is known for his articles published in the Chilean Journal of Natural History (which he himself founded in 1897) and in the Annals of General and Systematic Zoology from 1928, as well as for his books on scientific topics and others about the development of some scientific disciplines, as Gazmuri (2006, p. 269) has already highlighted. He has left us numerous articles and various works in this line; for example, Materials for the history of natural sciences. Bibliographic notes, published in 1933.

Another scholar of interest is Ernesto Greve Schlege, who was born in 1873 in Santiago. His parents were German Greve Lenker and Ana Schlegel. He was a diplomat, ambassador of Chile in Peru, and a member of the Chilean Academy of Language. He died on January 5, 1959, in Santiago. Among his main works oriented to the history of science, technology, and industry in Chile, we highlight: History of Engineering in Chile (IV vol. 1938–1944), History of the Amalgamation of Silver, published in 1943, and Don Amado Pissis and his geographical and geological works in Chile, from the year 1946. Given the chronology in which his works are located, especially those related to the progress of technology and/or industry in the country, we can rightly include him in the universe of the precursors of the discipline that interests us.

Another precursor of the history of science in Chile, who is located in the first chronological stage of the process that interests us, is the Hungarian Desiderio Papp (1895–1993) who settled in Chile. Among his works, he has left us the text: Revolutionary Ideas of Science (Volumes I-II-III, from 1979). Although his works are focused on the evolution of science in Europe, the work drew attention, promoted reflection on these topics, and established the convenience of studying the development of science also in the country. So we can well consider him another precursor of the history of science in Chile.

Now, specifically among the authors already clearly located in the chronology indicated for the first phase that we have called "articulation", we can highlight Augusto Salinas Araya, Gabriel Guarda Geywitz, Juan Ricardo Couyoumdjian, Alamiro de Ávila Martel, Juan Benavides Courtois, Guillermo Guajardo Soto, Mario Berríos Caro and Zenobio Saldivia Maldonado, among many others, that we cannot cover in a modest chapter of text.

18.4 Augusto Salinas Araya

He is a distinguished academic and writer, specialized in topics of history and sociology of science, culture, and politics. He has been a professor at various universities in Chile, Argentina, and the United States and has published various books and articles on topics such as science in Chile and Latin America, or about the Spanish Renaissance, medicine, and health. Salinas Araya has also received several recognitions for his academic career and his contribution to culture. Among his works that allude to the progress of science in Chile, appeared in the milestone that interests us we can highlight: Science in Chile and in the United States of America: a comparative historical analysis (1776–1976), an essay on the history and sociology of science in both countries, highlighting the peculiarities of the development of the episteme in each case, published in 1976. And also—already in the milestone of "consolidation"—we can mention Science under Fire. Scientific research, University and Political Power in Chile. 1967–1973, edited 2012; which refers to a careful study on the development and challenges of science in Chile during the governments of the Christian Democracy and the Popular Unity.

18.5 Gabriel Guarda Geywitz

Fernando Guarda Geywitz was born in Valdivia in 1928. After completing his basic studies at the Salesian Institute of Valdivia, he was sent to Santiago to complete his education at the Internado Nacional Barros Arana. In 1947, he entered the School of Architecture of the Catholic University, becoming interested in history. And after a few years of travel in Europe, upon his return in 1958, he graduated as an architect and then entered the Benedictine Order, being ordained as a priest in 1968, at which point he changed his name to Gabriel Guarda. He passed away in 2020. One of his first works is titled History of Valdivia:1552–1952, edited in 1953. Later, he published books specialized in religious, urban, and regional history, focused on aspects of architecture and on the details and construction techniques of churches and other buildings; for example: The Chilean City in the eighteenth Century, published in 1968, Urban History of the Kingdom of Chile, edited in 1978, Historic-Architectural Urban Complexes. Valdivia S. XVIII XIX, published in 1980, Churches of Chiloé, edited in 1984, Flandes Indiano. The Fortifications of the Kingdom of Chile 1541–1826, published in 1990, The architect of the Moneda Joaquín Toesca, 1752–1799: an image of the Spanish empire in America, edited in 1997. For the above, it is not surprising that in 1984, he was awarded the National History Prize. Of course, Guarda has other publications that are not directly related to specific aspects that now interest us and also other texts that are outside the milestone here outlined.

18.6 Juan Benavides Courtois

Another architect very interested in studying the development of architecture, the use of materials, and construction techniques in Chile, is Juan Benavides Courtois. He was born in 1930 and passed away in 2019. Among his works, we can recall his thesis: Carpentry and Ironwork in Chilean Colonial Architecture, from the year 1956, or his texts Country Manor Houses: Rural Architectural Ensembles, 2 vol., edited in 1981, Colonial Carpentry and Ironwork, published in 1998; Altiplano Architecture: Hamlets and Villages of Arica, edited in 1977; Cities and Port Architecture: The Major Ports of the Chilean Coast, edited in 1994, in collaboration with Marcela Pizzi Kirschbaum and María Paz Valenzuela. Or also the meticulous essay "Architecture and Engineering in the Time of Carlos III. A Legacy of the Enlightenment to the Captaincy General of Chile" published in 1988. He has also left us other texts that are already outside the milestone we are considering here.

18.7 Juan Ricardo Couyoumdjian

The researcher Juan Ricardo Couyoumdjian Bergamali, was born in Santiago in 1939; he is a Doctor in History, professor at the Institute of History of the Catholic University of Chile, former director of the same entity and was also Dean of the Faculty of History, Geography and Political Science of the Catholic University of Chile. He has received numerous national and international recognitions, among them, the designation of Emeritus Professor of the aforementioned university (2016) and the Monsignor Carlos Casanueva Award (2009). He is the author and co-author of various books and articles on the history of Chile and America, specializing in topics of economic history, Chilean bibliography and some works on the history of science in our country. Among his studies most related to the topic that interests us, the article stands out: "Overview of public works in Chile. Context, sources and bibliography" (Rev. Latin America in Economic History, from the year 2000). And although already outside the milestones selected here, it is worth highlighting the careful and well-documented introductory work, very focused on the scientific contributions of its object of study: "Francisco Vidal Gormaz: his life, his professional career and the Nautical Geography of Chile", in Francisco Vidal Gormaz: Hydrographic Explorations and Recognitions, published in 2012.

18.8 Alamiro de Ávila Martel

Alamiro de Ávila Martel was born in Valdivia in 1918, he was a historian, lawyer, and bibliographer, interested in his studies of Roman law, history of law, and Indian law. He worked in teaching at the University of Chile, obtaining the recognition of emeritus professor. He was a member of the Chilean Academy of History. He passed away in Santiago in 1990. He has very meticulous works on what was understood and taught as scientific topics in universities during the Enlightenment period in Chile: Historical Review of the University of Chile (1622–1979), published in 1979, and a text entitled: Mora and Bello in Chile, edited in 1982, in which he carefully analyzes the documents related to the foundation of the Colegio de Santiago and El

Liceo Chile and the discussions of Bello and Mora, for the characteristics, teaching methodologies, and the science topics of the time in these establishments. Also the essay "The University and higher studies in Chile during the time of Carlos III", published in 1988. Precisely because of these works, and especially the last one mentioned, we allow ourselves to include him as another exponent of the history of science for this milestone, since this work is extremely careful in presenting all the scientific topics, the teaching methods, the teachers and their training, the years of study for the different professions, the form and demand of the exams, the study books, the topics of physics, philosophy, law, and other branches of knowledge, which were taught in the universities that existed in the eighteenth century in Chile: "The Conventual University and the Royal University of San Felipe" (Ávila, 1989).

18.9 Guillermo Guajardo Soto

He is a distinguished academic and researcher, with a doctorate in Latin American Studies from the National Autonomous University of Mexico (UNAM), and a historian from the University of Chile. He currently serves as a teacher in postgraduate courses in History and Latin American Studies at UNAM. He has also participated as a visiting professor in the master's degree in Economic History at USACH. His main interest is oriented towards studying the genesis and scientific and technological evolution of the FFCC in Chile and other countries; in addition, he has dedicated himself to studying topics of public companies and issues related to Latin American infrastructure. Among his works we can highlight: "The learning of railway technology in Chile, 1850-1920", published in the Quipú Journal, Vol. 9, N°1, 1992, Mexico; "Technological Changes and Economic Projects in the Armed Forces of Chile, 1860-1930", published in History Journal N° 41, vol. II, Santiago, 2008. And already outside the period of the consolidation of the history of science in Chile, he has published: Seven historical lessons for a new railway in Chile, in collaboration with Rodrigo del Río, in the year 2022.

18.10 The Duo Mario Berríos Caro and Zenobio Saldivia Maldonado

Now, specifically among the authors already clearly located in the chronology indicated for the first phase, we can highlight Mario Berríos Caro. His contributions to the history of science are mainly related to the study of the scientific and technological development of the nineteenth century in Chile and its link with the nation project. In the seventies, he went into exile in the German Democratic Republic, where he worked as a researcher and professor at the Institute of Latin American Studies of the Berlin Academy of Sciences and at Humboldt University in Berlin. Once back in Chile, he was one of the founders and directors of the Center for Latin American Studies, which was dedicated to the analysis and dissemination of Latin American culture, history, and thought. He also collaborated with various magazines and cultural and political organizations related to Latin American culture. Then, Berríos worked at the Professional Institute of Santiago—now the Metropolitan Technological University (UTEM) -, and later, when Professor Zenobio Saldivia Maldonado joined this corporation in 1989, they formed a working duo oriented towards the history of science in Chile; together they participated in Latin Americanist congresses in Chile and Argentina, published some texts and articles with contents about the evolution of science in Chile. Let's remember, for example, some of the peer-reviewed articles from these researchers: "The Context of Claudio Gay: Science and Education in the Chilean 19th Century", in Trilogía in 1992, or "The Epistemological Scope of Gay", published in Trilogía in 1994; "The Construction of a Concept of Science in Chile: Manuel de Salas and Claudio Gay" (Berríos & Saldivia, 1993, pp. 131–136), "Andrés Bello: Between Orality and Written Tradition" (Berríos & Saldivia, 1994, pp. 55–59), among others. And the peer-reviewed book: Claudio Gay and Science in Chile, published in 1995. Berríos had also previously published the essay: Identity-Origin-Models: Latin American Thought, edited in 1988. Let's remember that after the death of his friend in 1995, Saldivia continued to disseminate this line in various articles, but these are already located in what we have called "phase two" and will be mentioned in due course.

Of course, during this phase, a series of journals also participated, which, although they did not have the history of science in Chile as a central topic and systematically in their mission and vision; in many of their issues, some works related to the topics that interest us are observed. Among these we can mention indicating their year of inception: the Annals of the University of Chile in 1843, the Marine Journal in 1885, the Annals of the Institute of Engineers of Chile in 1889, the Chilean Journal of History and Geography in 1911, the Atenea Journal of the U. of Concepción in 1924, the Bulletin of the Society of Biology of Concepción (1927), the Bulletin of the Chilean Academy of History in 1933, or History, journal of the Pontifical U. Catholic of Chile in 1961; the Journal of Social Sciences, from the U. of Valparaíso in 1971; the Trilogy Journal, IPS in 1981, or the Universum Journal in 1986 (U. of Talca), the Bulletin of History and Geography, from the U. Catholic Blas Cañas in 1987; or the SOLAR Journal of Latin American Studies 1991–1998, among many others. And among the entities that collaborated indirectly or specifically in this period, with some study referring to what interests us, stand out: the Academy of Sciences in 1964, the Institute for Fisheries Development (IFOP) in 1964, the Society of Biology of Chile in 1976, or the National Commission for Scientific and Technological Research (CONICYT), in 1967, among others.

18.11 Second Stage: Consolidation

Among these exponents of what we have called the second phase of the institutionalization of the history of science in Chile, we can highlight: Rafael Sagredo Baeza, José Antonio González Pizarro, Francisco Díaz Céspedes, Augusto Millán U., Ian Thompson, Ricardo Rees Jones, Camilo Larraín Aguirre, Carlos Sanhueza Cerda, Patricio Leyton, Guillermo Latorre C., among others. And in the field of the history of technology in Chile, we have to mention for example Ian Thomson, Ricardo Rees Jones, Miguel Antonio Muñoz Asenjo, among others. And of course, some authors who had already been participating in the first stage, such as Zenobio Saldivia Maldonado, continue with their production in this phase.

Among the relevant characteristics of this milestone, there is a noticeable increase in articles, books, and conferences on the progress and challenges of various disciplines, but now considering much more an externalist approach to the sciences and their relationships with power, universities, the mobility of knowledge, and international entities. Of course, there is a greater awareness among writers, authors, and scholars in general, of delving into the history of science in Chile, properly speaking. Next, we can appreciate the contribution of the mentioned authors and researchers, whose production stands out notably in this period of time.

18.11.1 Ian Thomson Newman

Ian Thomson Newman was born in England in 1945. He is a transport economist and former head of the Transport Unit of ECLAC, a United Nations agency dedicated to promoting economic and social development in Latin America and the Caribbean. He studied at the University of Manchester, where he graduated in 1972, and then moved to Chile. In 2005, the State of Chile granted him Chilean nationality by special grace.

His historiographical interest focuses on the history of the railway in Chile, a topic he is passionate about and on which he has written various articles, short essays, and books. Among his articles, let's mention at least the following: "The first trains of the southern hemisphere" (EFE News, in 1987), "The hidden history of the Transandino Railway" (En Tren, 1995), and "Santiago suburban trains" (On Wheels, in 1996). Among his books, we can highlight: History of the Railway in Chile, published in 2000, in collaboration with Dietrich Angerstein; North Network: The History of the Chilean Northern Railways (2003). He has also researched the economic aspects of bioceanic corridors in South America through rail transport.

18.11.2 Ricardo Rees Jones

Ricardo Rees Jones is a lawyer, doctor of law, academic, and Chilean writer, who has been living in Argentina for many years. He has served as a professor at various universities, both in Chile and abroad, and has published several articles and books on topics of economics, politics, history, and culture.

He has also been recognized for his teaching work and his contribution to the academic development of various institutions. In 2016, he received a tribute from the IEEM (Institute of Business Studies of Montevideo) where he collaborated in the training of the first professors. His text most relevant to our current interests is: Carlos Lambert and technological innovation in the Chilean copper industry during the nineteenth century, published in 2016.

18.11.3 Augusto Millán Urzúa

Augusto Millán was born in 1925. He was a civil mining engineer (University of Chile) and obtained his Master of Sciences from the Royal School of Mines (University of London). In 2005, he received the National Award from the College of Engineers of Chile. He was president of the Institute of Mining Engineers of Chile (IIMCh). He passed away in 2014.

Among his books that focus on the development of technology and the mining industry in Chile, let's remember: History of Iron Mining in Chile, edited in 1999, History of Gold Mining in Chile, published 2001, Metallic Mining in Chile in the twentieth Century, edited in 2006.

18.11.4 Camilo Larraín Aguirre

Camilo Larraín was a distinguished doctor born in Santiago in 1918. He began his medical studies at the Pontifical Catholic University of Chile and continued them at the University of Chile, graduating as a doctor in 1943. He did an internship at Bellevue Hospital in New York, USA, to study internal medicine. He worked at the José Joaquín Aguirre Hospital. He passed away in 2007. One of his texts, which addresses the historical aspects of the Medical Society of Chile, is aptly titled: The Medical Society of Santiago and the historical development of internal medicine, edited in 2002.

18.11.5 *Rafael Sagredo Baeza*

He is a historian, academic, and researcher at the Pontifical Catholic University of Chile, in Santiago. He has specialized in the history of science in Chile and America, in cultural history, and other topics. He is the author of numerous books and articles, the responsible editor of the Library Fundamentals of the Construction of Chile, coordinator and organizer of more than a dozen works on the history of Chile and America. He has also researched and published on the history of economic ideas, mentalities, and political practices. He was the conductor and responsible editor of the publications of the Library Fundamentals of the Construction of Chile (2006–2014), re-editing a vast collection of texts by authors related to the scientific and technical development of Chile between 1830 and 1950, with an introduction and analysis from our contemporaneity. In 2000 he was appointed Conservator of the Medina Room of the National Library, a position he currently holds. In 2022 he was recognized with the National History Award. In relation to the cognitive universe that interests us, we highlight from this author, for example, the text The Malaspina Expedition in the Southern Frontier of the Spanish Empire, written in collaboration with José Antonio González (Sagredo y González, 2004). Sagredo also has many other works but they would already be outside the second milestone considered here.

18.11.6 *José Antonio González Pizarro*

For his part, researcher José Antonio González Pizarro is a professor at the Catholic University of the North, in Antofagasta, Chile. As a researcher, he has been interested in the history of his region, mainly in the era of the nitrate boom and the influence of European immigrants. He has received numerous recognitions for his historiographic work, including the Antofagasta Municipal Literature Prize in 1998, the Regional Culture Prize (2004), and the Academic Trajectory Prize in 2010. In addition, he is a member of several scientific associations, including the Chilean Academy of History, of which he has been a part since 1994. He has published various books, articles, and participated with presentations at congresses on historical topics, and his articles highlight various aspects of the past and identity of northern Chile, such as topics linked to the nitrate industry, popular culture, and historical memory, among others. González Pizarro has not dedicated himself exclusively to the history of science, but has addressed other aspects of regional history, such as literature, education, politics, and law. Therefore, his work is broader and more diverse than that of a traditional science historian. Among his texts linked to the history of science in Chile, we highlight: The Epic of Nitrate in the Atacama Desert, published in 2018, although it is already outside the milestone in question.

18.11.7 Carlos Sanhueza Cerda

He is a distinguished researcher with studies in Germany who currently works at the University of Chile, and has dedicated himself to researching the history of science in Chile, mainly focusing his attention on the themes of the mobility of scientific communication and the contexts and relationships that are generated from the instruments in certain disciplines. Among his latest works framed in the mentioned characteristics are: Geography in action. Hans Steffen's disciplinary practice in Chile (1889–1913), edited in 2014. Sanhueza also has a vast production in the line that interests us, but they are already outside the milestone of "consolidation". In any case, as a reference, let's highlight his most recent work: History of Science and Technology in Chile (4 Volumes), published in 2023, in which he acts as general editor, in collaboration with Lorena B. Valderrama.

18.11.8 Zenobio Saldivia M

This author continues to work alone after the departure to infinity of his friend Mario Berríos, in the consolidation phase and until now, but in relation to the "consolidation" phase (from the mid-90s to 2005) let's remember some works that he has left us, such as: "Claudio Gay, precursor of the Chilean agribusiness" (Saldivia, 1994), "Claudio Gay: from taxonomy to agribusiness in nineteenth-century Chile" (Saldivia, 1998, pp. 453–460), "The Institutionalization of science in Chile" (Saldivia, 1999, pp. 141–156), "Juan Ignacio Molina, first Chilean scientist" (Saldivia, 2004), to mention a few. And books such as: The vision of nature in Gay, Domeyko and Philippi (Saldivia, 2003), Science in nineteenth-century Chile (Saldivia, 2005b), among others that are located in this milestone.

18.11.9 Francisco Díaz Céspedes

He is a professor of Basic General Education with a mention in Technological Education and Language, Bachelor of Education (UTEM), Master in Political Science (UTEM), Master in Philosophy of Sciences (USACH), Creator of the Center for Pedagogical Research, in Maipú.

Since 2005, he has worked as a teacher of History, Geography and Social Sciences in various educational establishments in the Metropolitan Region. He has also been a teacher of Philosophy, Political Science and Research Methodology at UTEM and at the Central University, in Santiago. In addition, he has participated in several research, extension and dissemination projects on the history of science in Chile and Latin America, both in the academic and social fields.

His area of interest is the analysis of the relationship between science, politics and culture in the Latin American historical context, especially in the nineteenth and twentieth centuries. He has studied the role of scientists and intellectuals who contributed to the development and institutionalization of science in Chile, such as Claudio Gay, Ignacio Domeyko, Antonio María Vargas Reyes and Alejandro Venegas Carús. He has also examined the social, economic and cultural impact that natural sciences, medicine, engineering and education had on the formation of national identity and citizenship.

He has published several books and articles on the history of science in Chile and Latin America, although most are already outside the milestones of institutionalization; for this reason and because of the strength with which he is working on these topics, we will highlight: Claudio Gay and science in Chile, edited in 2015, The Institutionalization of Natural Sciences in Nineteenth-Century Chile, published 2016. He has also collaborated with Professor Saldivia in various research and collegial books.

18.11.10 Guillermo Latorre C

Full professor at the Institute of Letters of the Pontifical Catholic University of Chile until 1987, when he moved to the U.S. Ph.D. in Linguistics and Literature (Purdue University), U.S. His field of research interest ranges from English and Spanish syntax, Chilean toponymy, and the cultural impact of science and technology. Among his published texts -although outside the phases of articulation and consolidation- stand out: Chile and Darwin: The response to Evolutionism since 1869 in collaboration with Zenobio Saldivia (Latorre & Saldivia, 2014), Andrés Bello scientist. Published writings (1823–1843), in collaboration with Rodrigo Medel, published in 2018.

18.11.11 Patricio Leyton Alvarado

On the other hand, it is convenient to highlight the author Patricio Leyton Alvarado, who has a degree in History from the Pontifical Catholic University of Chile, a Diploma in Philosophy and Sciences from the Alberto Hurtado University. Although his work is very recent and escapes the milestone of consolidation, we allow ourselves to include him for the systematic nature of his work, which clearly leaves a trace of interest in the history of astronomy. He has published numerous articles on the history of astronomy and physics in Chile and America. Among these we can highlight: “Andrés Bello and Cosmography: scientific literature from the perspective of the cultural history of science”, edited in 2014, “The social role of physics in the scientific thought of Ignacio Domeyko in the republican Chile of the 19th century” (2014), “Physics in the Malaspina expedition (1789–1794): Ideology, practice and experimentation", edited in 2015, the latter in collaboration with Professor Saldivia. Or also "The government of Manuel Montt and the financing of astronomy: science and politics in the conservative republic (1852–1861)", published in 2016, "Renaissance astronomy in the theological work of the Chilean Jesuit Manuel Lacunza in the 18th century", edited in 2017, "The reception of X-rays in Santiago de Chile, circulation and appropriation of experimental physics at the end of the 19th century", published in 2017. And even more recently, the reader can look for other works of this researcher in collaboration with Verónica Ramírez Errázuriz, such as: "The Trail of Halley's comet: Circulation and legitimization of astronomical knowledge in Chile (1910)", published in 2021.

18.11.12 Miguel Antonio Muñoz Asenjo

He holds a degree in international trade from the Metropolitan Technological University in Chile, a Master's degree in International Studies from the Institute of Advanced Studies at the University of Santiago de Chile in Santiago, and a doctorate in American Studies from the same institution. He is currently a faculty member of the Faculty of Economics at UTEM, Santiago. His intellectual trajectory, as can be deduced, has been interdisciplinary, combining works that allude to the history of science, technology, and international trade.

Dr. Muñoz has examined the characteristics of the scientific links sustained between Chile and the United States in the early years of the twentieth century (Muñoz, 2017). In this sense, he has highlighted the way in which the Chilean scientific and engineering community became involved with its American counterparts, in a context in which the North American country was experiencing high scientific, economic, and technological development. Parallel to the increase in economic and commercial ties that the United States was establishing with Latin American countries, scientific ties also expanded within the framework of Pan-American ideas.

Additionally, the author has advanced in the study of the arrival of specific technologies in Chile, such as the sanitation system (Muñoz, 2019a), or information technologies for the management of foreign trade (Muñoz, 2019b), from the point of view of interdisciplinary studies of technological change. And although his production is already outside the milestone of consolidation, we have mentioned it because of the novelty of these analytical approaches that are also giving life to the history of science and technology in Chile.

In addition to the researchers mentioned here, who participated in the consolidation phase of the institutionalization of the history of science in Chile, some entities are added, such as the Association of Fishing Industrialists from the Bío Bío Region (ASIPES), which funded and published the text: History of the Fishing Industry in the Bío Bío Region, published in 2000, or the National Mining Society (SONAMI), which funded the book: History of Gold Mining in Chile, edited in 2001, or the

Institute of Engineers of Chile, which published the text North Network: The history of the railways of Northern Chile, published in 2003, among other entities.

Among the universe of journals interested in social, historical, and cultural topics, which at this milestone contributed insights on topics of the history of science in Chile, many of the already mentioned stand out, which continue with their cognitive heritage such as the Annals of the University of Chile, the Chilean Journal of History and Geography, the Bulletin of the Chilean Academy of History, the Atenea magazine of the U. of Concepción; or the Medical Journal of Chile (1872), History, journal of the Pontifical U. Catholic of Chile (1961); the Journal of Social Sciences, from the U. of Valparaíso (1971); the Trilogy Journal, IPS (1981), the Universum Journal (U. of Talca). And among other more recent ones, it is possible to name the defunct Cultural Heritage Journal (1995–2009), from the Directorate of Libraries, Archives and Museums (DIBAM), or Thélos from the Metropolitan Technological University (UTEM), an electronic journal that started in 2005 and continues to date. And more recently, other academic journals that are already outside the milestone that interests us.

18.12 Towards a Conclusion

Apparently, in this process of institutionalization of the history of science in Chile, we did not have the strong presence of an author or academic who would gather and guide various doctoral theses on specific topics of the genesis and evolution of national science, in the manner of Dr. Juan José Saldaña, in Mexico, who managed to articulate from UNAM a working model and a School with numerous researchers on the development of science and technology in Latin America. A relevant characteristic of this institutionalization process that we have been analyzing in Chile, is the fact that the concern to account for the universe of science, had the participation of numerous precursors, as mentioned at the time, but despite the fact that they did not have a collective and integrating roof, they managed to leave their contribution, motivated by their own passion for investigating these topics. And within the universe of exponents of authors who contributed to the consolidation of the institutionalization of the discipline that we have been working on, it is observed that they have been and/or are still, university professors, botanists, entomologists, philosophers, historians, doctors, journalists and communicators and researchers on the presence of women in the sciences.

Bibliography

Ávila, M. (1989). *La Universidad Conventual y la Real Universidad de San Felipe, Estudios de Carlos III en el reino de Chile*. Edic. U de Chile, Stgo.

Berríos, M., & Saldivia, Z. (1993). La construcción de un concepto de ciencia en Chile: Manuel de Salas y Claudio Gay. *Revista de Sociología, 8*, 131–136.

Berríos, M., & Saldivia, Z. (1994). La construcción de un concepto de ciencia en Chile: Manuel de Salas y Claudio Gay. *Revista de Sociología, 9*, 1–6.

Centro Argentino de Ingenieros. (1981). *La Historia de la Ingeniería Argentina*. Buenos Aires.

CONCYTEC. (1986). *Los Estudios de Historia de la Ciencia en el Perú* (Vol. 2 vol).

de Guevara, J. D. (1993). *La Historia de la Química en el Perú*. CONCYTEC.

Gazmuri, R. C. (2006). *La Historiografía Chilena (1842–1970)*, T. I, Aguilar Chilena de Ediciones, Santiago.

Kragh, H. (1989). *Introducción a la Historia de la ciencia*. Ed. Crítica, Barcelona.

Latorre, G., & Saldivia, Z. (2014). Chile y Darwin: La respuesta al evolucionismo desde 1869. *RIL editores*.

Lohmann Villena, G. (1981). *Historia Marítima del Perú Siglos XVII y XVIII*. Lima.

Mayo, C. A. (1995). *Estancia y Sociedad en la Pampa 1740–1820*. Ed. Biblos, Bs. Aires.

Muñoz, M. (2017). Dos formas de acercamiento del progreso estadounidense a la comunidad científica chilena, a inicios del siglo XX (1908-1916). *Bicentenario, Revista de Historia de Chile y América, 16*(2).

Muñoz, M. (2019a). Imaginarios y transición socio-técnica: Estudio de los imaginarios de actores políticos sobre la implementación de un sistema de saneamiento en Santiago de Chile (1869-1931). *Revista de Estudios Políticos y Estratégicos, 7*(1).

Muñoz, M. (2019b). Establecimiento de un Régimen de Facilitación del Comercio en Chile: Estudio del Comercio Internacional desde la Perspectiva de las Transiciones Socio-técnicas. *Información Tecnológica, 30*(6), 233–242.

Piaget, J. (1981). *Psicología y Epistemología*. Ed. Ariel, Barcelona.

Sagredo, R., & González, J. A. (2004). *La Expedición Malaspina en la Frontera Austral del Imperio Español*. Editorial Universitaria, Stgo.

Saldivia, Z. (1994). Claudio Gay, precursor de la agroindustria chilena. *Cuadernos Docentes, 1*(1), 1–10.

Saldivia, Z. (1998). Claudio Gay: de la taxonomía a la agroindustria en el Chile decimonónico. In *Inter-American review of bibliography*. OEA.

Saldivia, Z. (1999). La Institucionalización de la ciencia en Chile. In *Inter-American review of bibliography*. OEA.

Saldivia, Z. (2003). La visión de la naturaleza en Gay, Domeyko y Philippi. Edic. Universidad de Santiago de Chile. Santiago.

Saldivia, Z. (2004). Juan Ignacio Molina, primer científico chileno. *Rev. En Foco*, (1). Brasil.

Saldivia, Z. (2005a). *Una aproximación al desarrollo de la Ciencia en Nicaragua*. Bravo y Allende Editores, Stgo.

Saldivia, Z. (2005b). *La ciencia en el Chile decimonónico*. Santiago: Ediciones Universidad Tecnológica Metropolitana, 34.

Saldivia, Z. (Ed.). (2018). *Una aproximación al desarrollo de la ciencia en Colombia*. Bravo y Allende Editores.

Dr. Zenobio Saldivia Maldonado is a Professor of Philosophy at the University of Chile, Santiago, holds a Master's degree in Philosophy of Sciences from the University of Santiago de Chile (USACH), and a Doctorate in American Thought with a mention in History of Sciences from USACH, Santiago. He is currently an Emeritus Professor at the Metropolitan Technological University and Director of the Department of History of the same corporation. He has been awarded Doctor Honoris Causa by the Ada Byron University, Peru, and by the Academic Conclave Fiat Lux, Mexico. He has numerous national and international recognitions and to date has published 27 books related to philosophy, epistemology, education, and history of sciences.

Dr. Pablo Andrés Rojas Castro holds a Bachelor's degree and is a Professor of Philosophy from the Pontifical Catholic University of Valparaíso, holds a Master's degree in Educational Informatics from the University of Chile, a Master's degree in Education and a Doctorate in Education from the Pontifical Catholic University of Chile. He is an Associate Professor at the Metropolitan University of Educational Sciences, performing functions as a researcher and teacher in undergraduate and postgraduate programs, mainly in teacher training programs.

Chapter 19
The Development of Studies on the History of Sciences, Medicine, and Engineering in Perú in the Twentieth Century

Leticia Quiñones, Jorge Lossio, and Lizardo Seiner

Abstract Until the 1980s, studies on the history of science, engineering, architecture, and health in Peru were developed mainly by professionals in these areas. However, from this decade onwards, historians became more interested in these lines of research. This paper offers a historiographic balance of the research in these branches of history in the last 40 years through their characters and contributions, the institutions to which they belonged, and the activities they developed relevant to Peru's progress.

Keywords Peru · History of science · History of engineering · History of health · Bibliography

19.1 Four Decades of Development of the History of Science in Peru, 1986–2024

A little less than 40 years ago, in November 1984, the First National Congress on Historical Research was held in Lima, an event at which several round tables were organized, one of them devoted to the History of Science, convened by the *Sociedad Peruana de Historia de la Ciencia y la Tecnología.* As a result of those sessions, two compilation volumes were published, the first volume covering Basic and Technological Sciences [sic], while the second volume included contributions in the area of Social Sciences (Yepes, 1986).

Forty years later, the present study seeks to present a synthesis of the main contributions made in the History of Science, although only in this field, without considering those made in the History of Technology, a discipline that has its own

L. Quiñones
Centro Historia, UNI, Rímac, Peru

J. Lossio · L. Seiner (✉)
Pontificia Universidad Católica del Perú, Lima, Peru

M. de la P. Ramos-Lara, L. C. Arboleda (eds.), *History of Science in Latin America: The Construction of an Intellectual Field (20th century)*,
https://doi.org/10.1007/978-3-031-75536-1_19

autonomous expansion. We will also focus on the field of Basic Sciences, in which we contemplate Biology, Astronomy, Chemistry and related branches and not on the Social Sciences, as was the modality used in the 1984 event. We chose this period (1986–2024) because the publication of an early and important article by Marcos Cueto in 1986 had the merit of grouping more than 400 bibliographic references that appeared in the twentieth century, although including three areas, Science, Technology and Medicine. For the author, the guide was conceived with the purpose of facilitating the consultation work of those researchers who wanted to venture into the field of the *Social* History of Science, observing also that the studies properly directed to the knowledge of the History of Science were noticeably scarcer than those that were elucidating the History of Peruvian Medicine (Cueto, 1986: 147), field in which he has been one of the referents in Latin America (Cueto, 1997, 2020a, b).

It should not be overlooked that before the celebration of the 1984 event, very isolated works dedicated to the History of Science had already been developed, such as the one offered by the famous botanist Fortunato Herrera (1943), a native of Cusco[1] although we did not find the celebration of any event alluding to the historical theme. From the institutional point of view, it should be noted that the sciences in Peru were beginning to be seen as activities evidently centered on research, but it was understood that they could provide direct technical advice to the State. Such was the case of the *Academia de Ciencias Exactas, Físicas y Naturales de Lima* founded in 1938, later called the *Academia Peruana de Ciencias* since 1966 and still in force today, which publishes a Bulletin containing, among others, articles dedicated to reviewing the careers of outstanding scientists[2] or institutions such as the *Observatorio Magnético de Huancayo* since its foundation in 1922 (Alberto Giesecke & Mateo Casaverde, 2007).

Historical incursions undertaken by scientists themselves in their fields of study are not frequent in Peru. Among the few that can be found after searching in different journals, there is, for example, an article that evoked the centenary of the beginning of macromolecular science in Germany, also called polymer science, in the early 1920s (Ayarza, 2021).

We believe it is relevant to give an account of those contributions focused on highlighting **institutional commemorations**. Of the many that can be found, the one that dates to 1992 stands out, when Hugo Sarabia Swett, Rector of the Pontificia Universidad Católica del Perú, gave a speech at the inauguration of the Scientific-Technological Seminar on Chemistry, organized as part of the celebrations for the 75th anniversary of the founding of the PUCP, and the 25th anniversary of the Chemistry Section. Similarly, on another scale, in 2013, the 75th Anniversary of *the Academia Nacional de Ciencias* was commemorated, an entity that brings together various scientific disciplines grouped into six sections: Agricultural, Earth, Life,

[1] Academia Nacional de Ciencias. https://ancperu.org/ancperu/historia-anc/

[2] Biographies such as those of Santiago Antúnez de Mayolo or Godofredo García are included in the respective bulletins for the years 2021 and 2018.

Materials, Exact Sciences and Social Sciences Sections. This contrasted, nominally with the original denominations of the seven sections originally contemplated in 1938 at the time of its creation: Section of Exact Sciences, Physical Sciences, Chemical Sciences, Botany, Archaeology, Biology and Anthropology and Geology (Boletín de la ANC, 2013: 10, 18). It will be the Sciences themselves that through some of their members will elucidate the reasons behind the change of denominations or else History so that it can better define contexts and discourses around such changes.

The commemorations offer a valuable input to identify institutionalization processes, not only by contextualizing the emergence of the scientific entity, but also because sometimes they offer their own documentation, expressed for example in the minutes of the board of directors' meetings or similar, irreplaceable to appreciate the dynamics of the entity and to see the influence of individuals or ideas in the institutional progress.

Within the institutional range, the *Consejo Nacional de Ciencia, Tecnología e Innovación Tecnológica (*Concytec), the leading institution of the Peruvian State in these three áreas, promoted the dissemination of the History of Science appealing to the use of contemporary technological tools and hand in hand with the *Asociación Peruana de Historia de la Ciencia, la Tecnología y la Salud*, founded in 2018, inaugurated a cycle of talks from 2021, naming it explicitly as "Cycle of online talks "History of Science"".[3]

Peruvian scientific institutions are not numerous, but some have accumulated decades of institutional life. The balance sheets on their historical trajectory are valuable and allow us to glimpse the circumstances in which they were born and also to appreciate the contributions of the men of science who launched these projects. In 2006, the Entomological Society of Peru celebrated its 50th Anniversary, although with a different name, since in its beginnings it was called the Agricultural Entomological Society of Peru because it wanted to highlight the participation of agricultural entomologists, both agronomists and biologists (Aguilar, 2006: 1). The institutional paths are an essential part of the history of science. The identification of its milestones shows the different ways in which it was linked to the demands of society or the economy itself. The Entomological Society took on the challenge of finding a solution to the pests affecting crops that at the time were the protagonists of exports, namely sugar cane and cotton (Dale, 2006: 8). At the same time, these same entities recognized themselves as useful to the Nation and the State with a clear sense of service, affirming that ".... The Society is a spectacular institution. It should be a source of pride to belong to it. The State should be flattered by its performance [...] The years have shown that good and successful organizations endure, and this will be the case with the Entomological Society of Peru..." (Dale, 2006: 10)

[3]Consejo Nacional de Ciencia, Tecnología e Innovación Tecnológica [2021] https://www.facebook.com/concytec/videos/ciclo-de-charlas-online-historia-de-la-ciencia/783321735647906/

The development of Entomology in Peru in the long term, for example, throughout the twentieth century, can be seen in valuable contributions such as that offered by Beingolea (1991) in which he identifies Peruvian scientists, many linked to the former School of Agronomy, now the *Universidad Nacional Agraria La Molina*, who faced the development of pests that affected agricultural production by appealing to strictly natural methods based on the introduction and propagation of insect enemies of those that formed pests, a method called "biological control".

From another scientific field, Geology is an essential knowledge in a country with such characteristics for mining as Peru. Preceded by scientific institutions such as the *Sociedad Geográfica de Lima* (1888) o the *Sociedad de Ingenieros* (1898), the *Sociedad Geológica del Perú* was founded in Lima in June 2024, por consiguiente, próxima a celebrar su centenario, y reuniendo en sus inicios a ingenieros, de minas y civiles, así como a doctores en Ciencias que ejercían como catedráticos en varios centros de educación superior de la república, sea universidades o centros de formación militar (*Sociedad Geológica del Perú*, 1925: ii–vi). Even though neither the original statutes (1924) nor the current ones contain references to the need to develop historical studies as part of the institutional purposes, the authorities of the Society recognized the enormous contribution of those who preceded them in the cultivation of their discipline, evoking them at the moment when their institution was born, affirming that "... our Society is born with the legacy of the scholars who since the last century have been concerned with Peruvian geology. The Humboldts [...], D' Orbigny, Castelnau, Darwin [...] Depositaries of this heritage, we must claim their property rights, both on observations and geological theories, as well as on the denomination of fossils based on Peruvian [paleontological] material..." (Lissón, 1924: 11).

Although we have not been able to undertake an exhaustive review of the contents of the successive issues of the Bulletin published by the Society since the year after its foundation, we have found exciting contributions to the history of Geology. In 2005, two researchers developed a comparative study on three maps made in the eighteenth century of the famous Santa Barbara mercury mine located in the department of Huancavelica in the central highlands of Peru, considered the fourth largest mercury mine in the world and the first in America (Wise & Féraud, 2005: 23) and at the time the fundamental input on which silver production in the viceroyalty of Peru rested. By superimposing the three maps, they were able to identify the different galleries that made up a complex network and the progressive advance shown by the exploitation works carried out in the deposit, determining which mineshafts from the viceroyalty period are still working today and in which areas the landslides occurred that paralyzed production for a time at the end of the eighteenth century (Wise & Féraud, 2005: 31).

Among the institutions, the *Sociedad Geográfica de Lima*, which has just celebrated 136 years of institutional existence, provides a space through events and the publication of its centenary Bulletin, to offer studies dedicated to approaches that address geographic problems in the long term. It is significant, for example, that the XV version of the National Congress of Geography, held in 2023, will be named after a Peruvian scientist of deep impact in the aerospace field, Pedro Paulet Mostajo, to whom a recent biography has also been dedicated, although highlighting his work as an official of the Peruvian Foreign Ministry (Fernández Prada, 2015) or rather venturing into his multidisciplinary work (Mejía, 2017).

Regarding the reprinting of geographical works of high cartographic relevance, an essential contribution was the publication of the Geographical Atlas of Peru by Mariano Felipe Paz Soldán, the first official map of the republic and a true bibliographic jewel (Chaumeil & Delgado, 2012), apart from being a work of difficult consultation because there are not many copies in good condition. Paz Soldán's own work is enriched by the works dedicated by the editors, illuminating the development of geographic science in Peru in the mid-nineteenth century, a central element in the laborious construction of the material identity of the Nation. In the same line, the publication of Raimondi's Map of Peru, published in 1877 and which has not yet merited the due interest to undertake a reprint with which to compare the specific contribution of Paz Soldán, image of the erudite and studious bibliographer and statesman, although without the capacity of constant travel undertaken by Raimondi, is missed.

In the field of Astronomy is where a greater number of academic contributions can be found. The work of Roberto Takayama, a philosopher by training, identifies the figure of the physician and cosmographer Cosme Bueno as the one in whose time the slow transition from the classical paradigm to the modern one began, clarifying that it had not yet reached a totally modern character, also showing an eclectic position since he was ascribing part of the theories of Tycho Brahe as well as some Newtonian ones (Katayama, 2014: 114, 116). He will conclude by affirming the existence of three astronomical imaginaries in Peru between the seventeenth and nineteenth centuries, Ptolemaic, Tyconian and Keplerian-Newtonian (Katayama, 2014: 118). Expanding his research, Katayama continues with the study of the original works of several scientists who lived in the viceroyalty of Peru such as Francisco Ruiz Lozano, José Eusebio Llano Zapata or Pedro de Peralta Barnuevo, analyzing to which astronomical affiliation they ascribed. Physics is the second of the knowledge that Katayama seeks to trace in the viceregal scientific works by reviewing three authors, who were also at the head of the Cosmografiato at different times (Katayama, 2018). One of the conclusions deals with the role of safeguarding the classical paradigm played by the authors studied and not to understand them as thinkers who wished to annihilate scholastic remnants in thought.

It is worth mentioning that in Katayama's work he uses the expression Natural Philosophy, understanding it as what will later represent the future fields of physics and chemistry, and which will be understood generically as Science. Likewise, in a critical effort, he alludes to authors such as Fernández de Piérola as someone ignored by the "official canon", an expression of clear critical nuance, although without being clearly defined.

In conclusion to a long academic project initiated 20 years earlier, Margarita Suárez published a study on the astronomical ideas in force in the viceroyalty of Peru in the seventeenth century, based on the reproduction of three texts of the period, corresponding to Juan Jerónimo Navarro, Juan de Figueroa and Francisco Ruiz Lozano, one of the first to occupy the position of Cosmographer of Peru.

Following the suggestive proposal of identifying phases that explain the passage from astrology to astronomy in America, initiated with the so-called "Astrología Judiciaria" until reaching the questioning of these predictions due to the influence of enlightened thought (Suárez, 2019: 40–41)).

Studies on the Cosmografiato, a viceregal institution that extended its functions well into the republic, have followed the path laid out by Federico Schwab's pioneering work published in 1948. To the initial contribution by Dargent (1992) followed by a second one (1995), and to the several that Ortiz Sotelo dedicated to the subject, focusing on the first officials to occupy this position in the seventeenth century (1993, 1997), we can add a more recent one that proposes a long-term view specifying their contribution in the establishment of meteorological observations between the centuries (Seiner, 2004). The knowledge of cosmographers has been enriched with biographies dedicated to physicians who successively occupied the position of Cosmographer of the Kingdom, to later give way to the name Cosmographer of the Republic, Dr. Cosme Bueno (Cama & Cama, 2010), Gabriel Moreno (Huaraj, 2014) and José Gregorio Paredes, the latter, who also extended his services to the Nation with the cultivation of Astronomy, Mathematics, apart from dabbling in politics and representing the country in his capacity as Peruvian diplomat in Europe (Varillas, 2019: 113 y ss.).

The development of the History of Science has also followed paths centered on the work of a particular scientist, and in the last decades there has been notable progress, for example, in the study of Antonio Raimondi's contribution to Natural History, considering him one of the outstanding figures of Peruvian science in the nineteenth century, even though he was born in Italy. To the initial contributions provided by researchers associated with the Raimondi Museum (La Torre Silva,) was added the publication of the epistolary of the Italian scholar (Inglesi et al., 2005), being substantially enriched by the republication of several of the valuable works that Raimondi published during his lifetime and that were almost impossible to find in modern editions, all of them accompanied by valuable introductory studies highlighting the value of each of the works and their scientific or economic significance (Villacorta, 2003, 2004, 2006a, b, 2007, 2009, 2010, 2012).

Inquiries about Raimondi's exciting life have focused on clarifying commonplaces about phrases attributed to the sage within a broad reconstruction of his life (Bonfiglio, 2004: 265 y ss.) or determining how he was establishing his extensive networks of contact with the outside world (Seiner, 2010). A last contribution has allowed the reproduction of the graphic information, mainly watercolors, contained in the dozens of notebooks that Raimondi used throughout his 20 years of travel through Peru and that are currently kept in the *Archivo General de la Nación del Perú* (AGN, 2020).

Another Peruvian scientist of international stature was Mariano de Rivero y Ustariz, a character with links to European scientific institutions, scientific publications in the field of mineralogy, in addition to exercising regional political power in Peru and being the country's diplomatic representative in Europe. Diverse works have focused for example on his relations with Alexander von Humboldt (Alaperrine, 1999) or his valuable studies contributions in the field of mineralogy (Deustua,

2017) or the reprinting of works focused on the field of Archaeology (Coloma Porcari, 1994) or the study and facsimile reproduction of the *Memorial de Ciencias Naturales y de Industria Nacional y Extranjera*, jointly edited by Rivero and Nicolás Fernández de Piérola, considered the first scientific journal of republican Peru (Málaga Núñez-Zeballos, 1998, 2000, 2013).

Despite his short stay in Peru, the influence of Alexander von Humboldt's thoughts and work in the country was important. Not only are the observations and ideas that he devoted to the cold sea that bathes the Peruvian coasts (Vegas Vélez, 1987) but also the observations of various kinds that he undertook, be they astronomical, meteorological, technological, and even ethnographic. Although Humboldt did not dedicate a specific work to a country as he did to Mexico or Cuba, the lines he dedicated to Peru in his larger works are relevant because they show one of the earliest and most meritorious attempts to observe nature in the southern hemisphere. Therefore, the interest of publishing the extracts alluding Peru in the major works published personally by him (Vegas Vélez, 1987; Núnez, 2002).

With the work of Hipólito Unanue and in his condition as a physician, he should be classified within a History of Medicine, which is understandable and has been done for decades, thanks to the disciples he left behind. An outstanding contribution identified Unanue's work as part of the organicist tradition on which, in turn, the modern scientific tradition in Peru has been built (Ballón, 2006: 41). To this was added a doctoral work of deep reflection on the main work published by Unanue (Salaverry, 2016), *Observaciones sobre el clima de Lima y su influencia sobre los seres organizados* (Lima, 1806; Madrid, 1814, 2nd edition) in which the dimensions of the scholar's medical and scientific thought are differentiated, the latter focusing on his studies on geography and botany. Soon after, a critical edition of *Observaciones* appeared, filling a gap of almost half a century without a new edition where the inventory of Unanue's library was also inserted, the basis for new scientific studies that allow us to see the composition of the different knowledge by identifying works and authors (Seiner, 2018a, b).

The final part of the eighteenth century offers Peru the appearance of an exceptional figure: Santiago de Cárdenas, designer of a flying machine in Lima, who made converge the interest of two researchers, one physicist and the other biologist, as well as brothers, who published an exciting study that dealt with the knowledge of Cárdenas in Mechanics and Physics, apart from the ornithological ones, aimed at knowing and studying exhaustively the conditions that would allow to reproduce the flight of the condor (Stucchi & Stucchi, 2019).

In the field of seismology, scientists themselves have made important contributions. Since its origins, the work of the Geophysical Institute of Peru has given greater relevance to seismic studies and often directed efforts to collect valuable information on seismic occurrences in the past. Eng. Enrique Silgado fulfilled this task in a highly satisfactory way, highlighting for this publication only two works (1978, 1992), within a greater scientific production. It was the same Institute, in coordination with the Regional Center of Seismology for South America (CERESIS), an autonomous inter-governmental organization, which encouraged the preparation of historical seismic catalogs (SISRA, 1985; Huaco, 1986).[4]

Noting that these catalogs reproduced, understandably, only seismic parameters of almost exclusive use for geophysicists, the preparation of historical catalogs of earthquakes occurred in Peru was undertaken (Seiner, 2009, 2011) following the path opened by Silgado (1978) but especially by the contributions made for Mexico by researchers of the Center for Research and Higher Studies in Social Anthropology (García Acosta & Suárez Reynoso, 1996).[5] The Peruvian catalogs emphasized establishing traceability of information and identifying which testimony provided the most reliable information on a specific seismic occurrence.

In the field of natural phenomena that occurred in the past, scientists and historians have made efforts to elucidate, with effort, the manifestations of such events with the purpose of approaching and defining a typical behavior. Such is the effort aimed at understanding the El Niño phenomenon that occurred in 1925, one of the great events of that type in the twentieth century (Takahashi & Martinez, 2019), or the study dedicated to the great event of 1877–1878 (Aceituno et al., 2009). The desire to understand the value of the sources and the need to critically evaluate them, highlighting the documentary contribution of former engineers, became necessary when analyzing events associated with high rainfall and for which there are reasonable indications to consider them El Niño-type phenomena. However, we are still far from having a certainty supported by evidence that shows the typical manifestations of a contemporary El Niño (Seiner, 2003, 2018a, b).

As part of the research, texts have been found that, in spite of bearing the name History of Science, they do not offer any outstanding contributions because in the eagerness to synthesize, for example, the 200 years of Peruvian Republican history, they have tried to identify, although in a very superficial way, ".... the economic, political, cultural and social conditions in which the scientific products were achieved; thus, the scientists, their institutions and their most representative products [sic] appear in each of the moments of the development of scientific research in the last 200 years ...", a range as wide as it is incomprehensible in an academic product with a specific objective such as an article (Larico et al., 2022).

At the conclusion of this article, we present some intriguing theses that have been published in relation to our field of study. The evolution of IMARPE, the technical entity that established fishing quotas in times of the buoyant abundance of anchoveta in the Peruvian sea in the early 1960s is approached through the consultation of first-hand documentation (Osorio, 2017). Theses dedicated to the study of the History of Science in Arequipa and Cusco are *rara avis*. A few delve into the

[4] Centro Regional de Sismología para América del Sur. *Catálogo de Terremotos para América del Sur*. Lima, CERESIS. Programa para la Mitigación de los Efectos de los Terremotos en la Región Andina. Lima: CERESIS, 1985. Of the nine volumes published, the seventh was dedicated to Peru, covering three volumes. HUACO Oviedo, Daniel (ed.): *Catálogo Sísmico del Perú 1500–1982*. Lima: Instituto Geofísico del Perú, 1986.

[5] García Acosta, Virginia & Suárez Reynoso, Gerardo: *Los sismos en la historia de México*. Vol. 1. México D.F. Fondo de Cultura Económica, 1996.

analysis of disasters of seismic origin in Arequipa (Chire, 2003; Amanqui, 2006), but none touch upon the subject in Cusco.[6]

19.2 Studies on Engineering and Architecture

Engineering and architecture are forms of art and human expression -without exaggeration- millennia old. The need for individuals to adapt to their environment, forced them (and forces them) to adapt and develop various types of instruments, spaces and forms of production that improve the quality of life. Therefore, there is an intrinsic relationship between engineering, architecture with technology and the material progress of a society; therefore, its lack or excess affects the balance of this, so it has been necessary the introduction and progressive specialization of these arts.

Studies[7] in Peru on engineering and architecture have focused, above all, on activities such as mining in terms of the search for mineral deposits and forms of exploitation; public works such as roads, railroads, irrigation systems and terraces; the construction of buildings and public spaces, the urban sense of cities as well as the manifestation of modernity in those buildings; in agriculture due to the need to take advantage of the richness of the land; the development of education and the expansion of its specialties; the interest in establishing a close relationship between the State and the business sector to promote modernization with the help of these professions; and the way of thinking of the professionals in these specialties with respect to the Peruvian nation.

Research on these arts compared to other branches of Peruvian history is not abundant. The *Centro de Historia UNI* is still the place that is especially dedicated to this type of studies. However, professors and researchers from higher centers such as the *Universidad Nacional de Ingeniería*, the *Universidad Agraria La Molina*, the Universidad de San Marcos, the *Pontificia Universidad Católica del Perú* and the *Universidad de Lima*, have been able to dedicate themselves to research in these areas. Journals such as *Informaciones y Memorias. Boletín de Minas, Industrias y Construcciones*, *Boletín de la Sociedad de Ingenieros*; el *Arquitecto Peruano*; *½ de Construcción*; *U[r]bes. Revista de ciudad, urbanismo y paisajes; Revista de Arquitectura* (of the *Universidad Femenina del Sagrado Corazón*); *Arquitextos* (of the *Universidad Ricardo Palma*), y *Puente* (of the *Colegio de*

[6] Ragas, José: Catálogo de tesis de Historia de la Universidad Nacional San Agustín de Arequipa, 1985–2009 [2012]. See also by the same author and year: Catálogo de tesis de Historia de la Universidad Nacional San Antonio de Abad (Cusco), 1968–2006.

[7] There have been works, we could say, fundamental, because they intend to offer a comprehensive view of the topics and allow readers to be introduced to the issues that revolve around engineering and architecture. We can mention the publications of García Bryce (1986), López Soria (2004), Ludeña (2004), Gallegos (2006), Meza (2008) and the *Colegio de Ingenieros* (2021).

Ingenieros del Perú), to mention a few, contain interesting papers that -in certain cases- can be the starting point for more in-depth research.

In this intervention, I intend to expand on the papers published by Marcos Cueto (1986) and José Ignacio López Soria (2005). It should be noted that only the bibliography of the republican stage will be examined in depth, since the vision of engineering, architecture and technology are found within the ideas of modernity and western modernization that openly entered Peru from this period onwards.[8]

In general terms, we could affirm that the development of engineering and architecture, until the end of the eighteenth century, was one of contrasts, with notable advances in the pre-Hispanic period and a contraction in the viceregal period. However, the consolidation of Independence and the establishment of a republican system in Peru, in principle under liberal principles, did not mean the opening and access to the theoretical and practical knowledge of these arts. Moreover, it can even be said that this lack hindered the construction of the nation-state.

During the nineteenth century,[9] the development of engineering and architecture had ups and downs. The first republican decades were complicated, political struggles and the lack of experience in the administrative and institutional management of the Peruvian State did not allow the development of adequate policies to articulate the territory and economic activities as we can observe in the experience of Mariano Eduardo de Rivero in Peru. Miguel Rosas, states in an apt phrase to the lack of space for the development of professionals at this time: Rivero was an "official without a state and a scientist without an academy in the nascent republic" (2019, 12).[10] And the fact is that in general there was no labor supply and demand for engineers and architects. This does not mean that the Peruvian state did not generate directives: Rivero had been entrusted with the task of exploring mining deposits and in fact founded a Escuela de Minería in Huánuco; General José de San Martín had decreed, in 1822, that military engineers should assume the tasks of public works; in the 1840s a ministerial resolution had been promulgated for the cosmographer to evaluate and approve those who sought to develop the activities of "bricklayers, architects and builders" (Oviedo 1862, IX, 68). But these efforts did not determine the decision to forge a state policy regarding who would be the professionals in charge of public works and the articulation of the territory to promote economic activities.

It is from the second half of the nineteenth century that the professions of engineering and architecture will take on greater importance, and the works that were projected and executed allowed the generation of incentives for greater national articulation, not only at the territorial level, but also at the level of political and social discourse, as explained by López Soria (2021) when dealing with the link

[8] In the specific case of mining, it is worth mentioning the bibliographic compilation prepared by Samamé (1974).

[9] We can see a vision of engineering in the nineteenth century in Gallegos (2001) and López Soria (2021).

[10] On the trajectory of Mariano Eduardo de Rivero we can mention the works of Deustua (2017) and Rosas (2019).

between modernity, rationality and engineering. By the beginning of the twentieth century, engineers and architects were properly a modernizing elite. It was the Peruvian state that encouraged the arrival of foreign engineers and architects. This can be seen in research on the School of Engineers or in the biographies of Ernesto Malinowski (López Soria & Cazorla, 2018) and Eduardo de Habich (López Soria 1999a). But not only foreigners were hired, there were also Peruvian engineers and architects who managed to become accredited in the Cuerpo de Ingenieros del Estado, created in 1860 and reformed in 1872, as stated in the biographies of Felipe Arancivia (Vidal, 2001), Teodoro Elmore (2014), and Joaquín Capelo (Laya, 2020; Rodríguez, 2011). It was from 1876, with the founding of the Escuela de Ingenieros, that Peruvian engineers and architects such as Pedro Remy (Ueda, 2007), Federico Villarreal (Altamirano, 2013; Basadre, 2009; Watanabe, 2017), José Balta Paz (Quiñones, 2014b), José J. Bravo (Consejo, 1975), Carlos I. Lissón (Ueda, 2005) and members of the Grieve family (Ueda, 2008) and Basadre (Ueda, 2009b), among others. A common feature in all of them is that at some point they worked for the Peruvian State and were engaged in teaching. Not all Peruvian engineers were trained in Peru, there were those who managed to get funding and studied abroad, that was the case of Felipe Arancivia, Eduardo López de Romaña, Santiago Antúnez de Mayolo (Ueda, 2010) and Pedro Paulet (Madueño, 2002; Martuccelli, 2019; Mejía, 2019).[11]

But not only professionals who stood out in the 19th and early 20th centuries have been studied. For the twentieth century we can mention biographies on engineers Luis Jochamowitz (Araujo, 2009), Mario Samamé (Huamaní, 1999) and Orlando Olcece (Cabrera, 2014). In the case of architects the production has been more prolific, works have been published on Emilio Harth Terré (Mattos-Cárdenas, 2016), Rafael Marquina (Jiménez & Santiváñez, 2005), Adolfo Córdova (Martuccelli, 2012; Núñez, 2022), José García Bryce (Cabrera, 2012; Beingolea, 2020), Héctor Velarde (Cisneros, 2015; Universidad Lima, 2013), Walter Weberhofer (Universidad de Lima, 2016), Mario Bianco (Universidad de Lima, 2017), Luis Miró Quesada (Utia, 2001) and Fernando Belaunde Terry (Dorich, 2001; Ortiz de Zevallos, 2001). At the level of studies on women architects we can mention two works, "Las primeras mujeres en la Escuela de Ingenieros" (Ueda, 2009a) and Arquitectos / Arquitectas. Pioneers / Pioneras (Molina & Ludeña, 2011).

Until the 1930s, two centers of higher education had the prerogative of teaching engineering and architecture: the Escuela de Ingenieros (today the Universidad Nacional de Ingeniería) and the Escuela de Agricultura (today the Universidad Nacional Agraria La Molina). However, engineering and architecture programs were progressively developed in other Peruvian universities, both national and private.[12]

---

[11] En cuanto a publicación de biografías de ingenieros podemos considerar los trabajos de Bolaños (1929) y Tamayo (2015).

[12] On the *Escuela de Ingenieros*, see the collection of the History of the *Universidad Nacional de Ingeniería* (Cazorla, 1999; López Soria, 1999b; Rodríguez, 1999; López Soria et al. 2012). On the *Escuela de Agricultura* see Olcese (2002). On the professionalization of engineering see Lizarme

Education was not only limited to higher education, there was an incentive to develop technical professions, such as the Escuela Naval, the Escuela de Artes y Oficios, and there was even an attempt to found the Escuela Nacional de Electricidad. Engineering and architecture were definitely becoming the representative professions of the modern discourse.[13]

During the nineteenth century there was a strong relationship between engineering and architecture professionals and the Peruvian State. The Cuerpo de Ingenieros y Arquitectos depended on the Ministerio de Gobierno, and after 1896, when the Ministerio de Fomento was created (Lara, 1936; Quiñones, 2014a), relations became closer and more specialized work teams were created, such as the Cuerpo de Ingenieros Civiles (or Caminos) and the Cuerpo de Ingenieros de Minas. In the twentieth century, as the State developed more complex state policies due to social demand, new ministries were created, such as those of Transporte y Comunicaciones, Vivienda, Agricultura, Comercio, Minería, etc.[14]

In the nineteenth century, public works such as roads, railroads, bridges, docks, and the construction of public buildings were under the responsibility of the Ministerio de Gobierno (Regal, 1967). In this sense, the works with the greatest projection were railroads (Kemp, 2002; Galessio, 2007; Ribas, 2011), public buildings such as the panopticon (Michilot, 2021), irrigation works and roads. As already mentioned, the objective was the articulation of the territory (Contreras & Cueto, 2008; Contreras, 2010), not only for nationalist and economic purposes, but also in the hope of increasing foreign migration, fulfilling -in part- the modernizing vision (Sala i Vila, 2006). Although the Peruvian State was the one who projected the works with the reports prepared by engineers and architects, many times, those who executed them were businessmen (Wong et al., 2018) who achieved a contract with the State, that was the case of the businessman Henri Meiggs (Hampe, 2010) and Gustavo Eiffel's company (Gutiérrez, 2013).

For the twentieth century, priority was given to road works (Diez Canseco, 1929; Dagicour, 2019) and irrigation (Glave, 1986; Arroyo, 2022), construction of public buildings such as the Hospital Larco Herrera (Amaya, 2018) and urbanization projects. Only in Lima, the construction of both popular and residential neighborhoods

(2022). On the profile of professionals in engineering and agriculture for the decade of 1939 the speeches of Rómulo Ferrero (1939) and Cristóbal de Losada y Puga (1940) should be considered. On the teaching of architecture see Álvarez (2006), Córdova (2011) and Espejo (2023). On industrial engineering and metallurgical engineering programs see Garfias & Vega (2015) and Tantaleán (2012) respectively. On architecture and urban planning and engineering in the Pontificia Universidad Católica del Perú. see the publication that both faculties presented in 2013.

[13] On technical training in Peru, for the case of the late eighteenth century and the first half of the nineteenth century in useful the work prepared by Luque (2014) and from the second half of the nineteenth century and the first decade of the twentieth century see Guizado (2022). In the case of the National School of Electricity project, the paper prepared by Guizado & Ragas (2021) is recommended.

[14] In general, there is still little research on the study of the history of the Peruvian state administration. In the specific case of the link between the Peruvian State and engineering and engineers, we can mention the works of Cornejo (2010) and Velázquez & La Serna (2021).

was planned. Thus, in the context of the 1950s, the "Agrupación Espacio" (Space Group) arose within the academic sphere, at the initiative of students and professors of the School of Engineering, which proposed a modern, renovating perspective in the conception of the art of architecture, taking into account the architects' own cultural influences (Córdova, 2007; Itabashi, 2009; López Soria, 2017 [1997]; Kahatt, 2019). Later, around the 1970s, "Brutalism" developed, a trend that represented the political regime of that time (Ferrer, 2011).

Mining activity has been one of the most studied from an economic perspective, since the arrival of the Spaniards it became the activity with the greatest impetus (Samamé, 1979; Contreras, 1994; Tumialán, 2003; Morales et al., 2021). In spite of this, during the nineteenth century there were no major incentives. It was not until the 1890s and especially since the creation of the Ministerio de Fomento that it became a state policy. Within this formalization process, the Cuerpo de Ingenieros de Minas was created in 1902, which was not only dedicated to the exploration of new deposits, but also to the elaboration of production statistics, reporting on the conditions of the mines and labor relations. Subsequently, in 1923 the Dirección de Minas was created and in 1968 it was elevated to a ministry.[15]

Since the arrival of the first foreign engineers, there was a need to be linked to spaces of sociability. The Cuerpo de Ingenieros y Arquitectos itself allowed a dynamic interaction among them and, in addition, they joined other civil societies. After the shock of the war against Chile, these spaces of interaction were reorganized and new ones were created, where engineers and architects were incorporated. The Ateneo de Lima, the Sociedad Geográfica de Lima, and the Sociedad de Ingenieros (López Soria, 2003) were some of the spaces for social and intellectual interaction. Associations with a professional and business profile were the Sociedad Nacional de Minería, the Sociedad Nacional de Agricultura and the Sociedad Nacional de Industrias. In addition, several of these professionals assumed a political role and joined parties such as the Civil, Democratic and Liberal parties. In the twentieth century, their associative spirit was strengthened, and associations were created according to their specialties, such as the Instituto Geológico del Perú (1924), the Instituto de Ingenieros de Minas (1943), the Instituto de Electrotécnica Peruana (1943), to mention a few. It is pertinent to value how from the classrooms, both engineers and architects generated spaces for dialogue between them and their authorities, forming a student movement with continuity (Huamaní, 2015, 2017).

Finally, we must consider that the professional activity of engineers and architects, both in the state and private spheres, allowed and allows them to know Peru both in its territorial and social extension, being able to observe Peru in an integral way (López Soria & Lizarme, 2021, Huamaní, 2022; López Soria, 2022). In that sense, thanks to this experience they are able to formulate a rational and intellectual vision of Peru as a problem and as a possibility, where their art -in the modern

[15]On the mining activity linked to the professional activity of engineering we can mention the works of the Ministerio de Fomento (1924), Sotillo (1976) and Jochamowitz (1992). For the study of the *Cuerpo de Ingenieros de Minas* we should mention the research of Samamé (1984a, b), Seiner (2000) and Ueda (2002).

sense- is fundamental to adapt the environment to the needs of Peruvians (Martuccelli, 2017; Ludeña, 2017; Castañeda, 2022a, b; Jaimes, 2023).

19.3 The History of Health in Peru: Objectives, Assumptions, and Actors

The history of medicine is a discipline that, in the Peruvian case, can be traced institutionally to the third decade of the twentieth century, when a group of physicians and psychiatrists founded the *Sociedad Peruana de Historia de la Medicina.* Its first president was Carlos Enrique Paz Soldán, a renowned sanitarist promoter of social medicine and eugenics, and had renowned physicians among its most active members such as Juan B. Lastres, Honorio Delgado, Jorge Arias Schreiber, among many others. In January 1941, the publication of *Anales de la Sociedad de Historia de la Medicina* began. It was the first Peruvian journal dedicated to the history of medicine (Cueto, 2022).

New individual and institutional efforts were added throughout the twentieth century and in the twenty-first century, and new societies were created to study the history of medicine and health in Peru. An important change occurred in the 1980s, when historians, sociologists, and anthropologists ventured into this area and began to use the term 'history of health', instead of 'history of medicine'. Scholars of health history understood that beyond physicians and viruses, their work should include the view of patients, the role of healers, midwives, activists, incorporate different worldvies and a analyze social responses to epidemics. Another important change has occurred in the first decades of the twenty-first century, when the history of health has been analyzed from the perspectives of race, gender, and class with a global approach.

In this paper, we will analyze the history of health in Peru and its main actors, assumptions, and objectives. *This research hypothesizes* that in the Peruvian case different historiographical traditions have taken root that coexist to this day.

19.4 Beginnings of the Discipline: The *Sociedad Peruana de Historia de la Medicina*

In September 1939, a group of physicians founded the *Sociedad Peruana de la Historia de la Medicina.* Its first president was Carlos Enrique Paz Soldán, and among its most prominent members were Juan B. Lastres, Jorge Arias Schreiber, Honorio Delgado, Sergio Bernales, Constantino Carvallo, Pedro Weiss, Ángel Maldonado, among others. It also incorporated as members outstanding historians of that time such as Jorge Basadre, Víctor Andrés Belaúnde, Carlos Romero, José de la Riva Agüero, Raúl Porras Barrenechea, Horacio Urteaga and Rubén Vargas

Ugarte. In January 1941, the Peruvian Society of the History of Medicine began publishing the *Anales de la Sociedad de Historia de la Medicina*. Ten volumes were published until December 1949, when the Society ceased functioning (Cueto, 2022).

The first volume of the journal *Anales de la Sociedad Peruana de Historia de la Medicina* reproduced the minutes of the first sessions of the Society, including the founding act, and also collected some original works of its members such as Carlos Enrique Paz Soldán, Juan B. Lastres, Guillermo Fernández Dávila, Lizardo Velez Lopez, Jorge Basadre, Luis A. Eguiguren, Tomás Escajadillo, Maxime Kuczynski-Godard, among others.[16] At the founding session (September 21, 1939) Dr. Carlos Enrique Paz Soldán acted as President, and Dr. Juan B. Lastres as Secretary. A commission was appointed to draft the regulations of the Society, composed of Dr. Honorio Delgado, Dr. Guillermo Fernández Dávila, Dr. Ángel Maldonado, Dr. Carlos Enrique Paz Soldán, and Dr. Juan B. Lastres. The Society was sponsored by the government of Manuel Prado Ugarteche and his Minister of Education, Pedro M. Oliviera, and followed the parameters of the nascent societies of the history of medicine that appeared at the beginning of the twentieth century in European countries. As Cueto points out: "In July 1940, it was declared an official institution by Pedro de Olivera, Minister of Public Education in the government of Manuel Prado Ugarteche (1939-1945). Paz Soldán made sure that both Olivera and Prado were honorary members of the Society. As an official institution, it could receive some subsidy from the State to complement the membership dues" (Cueto, 2022).

It is worth mentioning that it was a Society with that was relevant not only in the local academic environment, but also in international networks, particularly with medical history societies of other countries. Likewise, foreign physicians were incorporated as honorary members of the society, both European, American, and Latin American (Paz Soldán, 1943). According to Marcos Cueto, both Carlos Enrique Paz Soldán and Juan B. Lastres can be read as "scientific intermediaries", in the sense that "they do not passively transfer knowledge from abroad, but seek to adapt knowledge, build new scientific fields and make their experience and production relevant in international knowledge networks. The exchanges between Peruvian medical historians, as well as European and North American specialists, allowed ancient medical practices–especially Andean pre-Columbian medicine– to be consecrated as an important reference in the unfinished project of a universal history whose central axis was to consider European medical culture as the valid paradigm for all cultures" (Cueto, 2022).

The objectives of the Society were to provide a space for support, discussion, and dissemination of the research of its members; to rescue the contribution of Peruvian pre-Hispanic and colonial societies to universal medical knowledge; to maintain active links with journals in the same field throughout the world; and to find a museum of the history of medicine. In other words, the Society sought to make the practice of the history of medicine an institutional rather than an individual task; it

[16] https://fuenteshistoricasdelperu.com/2023/08/09/anales-de-la-sociedad-peruana-de-historia-de-la-medicina-lima-1939-1949/

sought to rescue, with a certain nationalistic desire, the contributions of pre-Hispanic cultures to medicine; it saw internationalization as a key element of validation; and it assumed that the history of medicine contributed to the progress of the nation, offering lessons from the past that should serve as a guide for physicians today.

Among the characteristics of the historiography produced by members of the Society, or what is known as 'traditional history of medicine', we find, for example, an idea of historical continuity and evolution from the Incas to the Republic. (See for example the major work of Dr. Juan B. Lastres, *Historia de la medicina peruana (1951),* for the 40th anniversary of the foundation of the *Universidad Nacional Mayor de San Marcos*, by the then rector Luis Antonio Eguiguren).[17] We also found that the study of pre-Hispanic medicine was based on an anthropological desire to understand non-Western health worldviews that were considered current in the country. This is to be able to treat patients with non-Western views of health from a practical point of view (Andean, Amazonian). Although Lastres recognized some specific contributions of pre-Columbian medicine, he considered it inferior to Western medicine. According to Cueto: "Juan B. Lastres' perspective of Peruvian pre-Columbian medicine was most easily inscribed in a nascent narrative of a unique universal history in which the Andean Indians appeared as precursors of Western medicine, but primitive practitioners; not as possessors of an alternative knowledge of the official" (Cueto, 2022).

There was also an eagerness to identify the history of health institutions that emerged during the colonial and republican era, and to highlight with laudatory biographies the 'Peruvian' contributions to universal medicine, such as the case of quinine. Among the most studied characters we have the cases of Hipólito Unanue and Daniel A. Carrión, considered 'founding fathers' of medicine in Peru. The justification of the discipline was based on the importance attributed to the humanities as a fundamental part of medical education. In other words, the history of medicine was considered useful for training better physicians, for linking medicine with the humanities, and for learning about the culture, social contexts and expectations of their patients. Likewise, it was justified by the need to know the founding fathers of the different medical specialties and to show the evolution of universal medicine. In other words, factors of professional identity and eagerness to demonstrate the superiority of Western medicine over other actors and knowledge in the way of treating diseases (for example, shamans or midwives) converged. Their main audience was medical students or health professionals.

The development of a professional history of medicine responded to a series of factors, such as the important scientific advances of the late 19th and early 20th centuries, such as the germ theory of disease, advances in surgery, the discovery of antibiotics, and the development of vaccines, among others. These advances transformed medicine and generated interest in its study from a historical approach.

[17] On the work of Juan B. Lastres see: https://fuenteshistoricasdelperu.com/2022/04/06/juan-b-lastres-1902-1960/ y Cueto, M. (2022). La historia de la medicina, Henry E. Sigerist y los intermediarios médicos en el Perú de mediados del siglo XX. *Historia*, *1*(55), 329–350. Retrieved from https://revistahistoria.uc.cl/index.php/rhis/article/view/33765

There was also an important process of institutionalizing health in the first decades of the twentieth century, which led to the creation of medical agencies and institutions that began to be interested in building their histories.

19.5 The Social History of Health

In the 1980s, a different way of approaching the history of medicine appeared when historians, anthropologists, and sociologists began to get involved in the area. The arrival of professional historians in the field implied new assumptions and questions. By 'social history of health', we refer to an approach that included in its research the social responses to epidemics, the view of patients, and the non-medical actors who were part of the world of health (healers, herbalists, midwives). It was an approach that analyzed how cultural beliefs, values, and practices have affected the perception and treatment of health and disease. Its practitioners were very emphatic about including the impact of socioeconomic conditions on health and access to medical care. Another topic of interest was developing and implementing public health policies over time. Similarly, unlike traditional history, it was an approach that questioned and negatively viewed some aspects of the medicalization of society and showed that science and public health could serve as instruments of social control. That is, there was, more in the 1990s, a clear influence of Foucault. The very fact of using the term health, rather than medicine, indicates an intention to broaden explanations beyond the purely biological realm (Armus, 2002).

As Diego Armus points out: "Traditionally, the subject of the disease has been a kind of controlled preserve by medical historians. They wrote not only a history of changes in treatments but also biographies of famous doctors. Beyond their specific contributions, these stories seem to have endeavored to reconstruct the "inevitable progress" generated by graduated medicine, to unify the past of an increasingly specialized profession and to highlight a certain ethics and moral philosophy that is intended to be distinctive, unaltered and symbolic of the medical practice throughout the time. The new history of medicine, on the other hand, tends to highlight the uncertain developments of medical knowledge, dialogs with the history of science, and discusses not only the context—social, cultural, and political—in which some doctors, institutions, and treatments "triumphed," making a place in history, but also those who were lost in oblivion. It is a narrative that strives to tension the natural history of the disease and some dimensions of its social impact" (Armus, 2002).

The emergence of this new approach was a consequence of several factors. Among them was the expansion of social history, i.e., a greater interest in the everyday lives of 'ordinary' people, in contrast to traditional history, which was more focused on political events and major figures. Dependency theory, with a focus on the relationship between industrialized and 'developing' countries and the inequalities between the two, and the demographic changes of the twentieth century that made the living and working conditions of the less developed sectors, including the poverty-health relationship, more visible in the cities and to the academic communities as well.

In Peru, the leading promoter of this discipline has been Marcos Cueto, whose works have inspired generations of historians. In particular, among the works that aroused interest among the community of Peruvian historians in the social history of health, we can highlight *Excelencia científica en la periferia: actividades científicas e investigación biomédica en el Perú*, 1890–1950 (1989) and *El regreso de las epidemias* (1997). Marcos Cueto Caballero earned a bachelor's degree in Humanities from the *Pontificia Universidad Católica del Perú* and obtained his Ph.D. from the Department of History at Columbia University. He then completed a postdoctoral fellowship at MIT (Lossio, 2022).

Cueto advocated a social history of health, where the biological is one more component within a much more complex framework that should not neglect the political, economic or social. Thus Cueto inaugurated a new perspective in the history of Peruvian science and medicine, which left aside praise and laudatory biography to concentrate on the social and cultural aspects of health policies and epidemics. In the case of *El Regreso de las Epidemias*, he showed how epidemics, being moments of crisis, allowed us to see often hidden aspects of society (such as racial prejudice). He also focused not only on official responses and medical explanations, but also studied social responses that could be religious, spiritual, or based on non-Western traditional knowledge. In other words, the study of the history of medicine and health as a way of making social history, of approaching society.

Among the topics of interest were, among others, the relationship between the centers producing science and medicine with the 'peripheral' scientific communities (a notion that was discussed), as well as the local reception of medical theories and advances developed in Europe (such as eugenics, Darwinism or bacteriology). The current sought to give visibility to local scientific communities and their contributions to scientific theories and medical discoveries. For example, in the case of quinine, history would not be centered so much on the role of the Jesuits or the British doctors who synthesized it but rather on the indigenous knowledge that first allowed Europeans to "discover" its value against malaria. Or, for example, the interest may lie not only in describing eugenics but in understanding how the European eugenic discourse was reinterpreted in Latin America, which actors participated in the Latin American eugenic movement, and to what extent ideas coming from Latin America may have contributed to forge global eugenic ideas. Other topics of interest were scientific and health institutions, hospitals, and the state's role in public health.

19.6 The Re-launching of the *Sociedad Peruana de Historia de la Medicina* in the 1990s

In 1981, when Dr. Uriel Garcia was Minister of Health, a commission was created to pay tribute to distinguished Peruvian physicians, which led to the creation of the "Asociación Peruana de Parques de la Medicina". A little more than a decade later, in 1995, Dr. Gustavo Delgado Matallana proposed a merger between this Association and the "Sociedad de Historia de la Medicina Peruana". In the same year of 1995, Dr. Fernando Cabieses proposed the change of name to 'Asociación de Historia de la Medicina Peruana y Parques conmemorativos'. Thus, the Asociación de Historia de la Medicina Peruana y Parques conmemorativos was founded to bring together in a single institution the Sociedad Peruana de Historia de la Medicina (1939–1949) and the *Asociación Peruana de Parques de la Medicina* (1982–1995) (See the institutional history at https://asociaciondehistoriadelamedicinaperuana.com/). That is to say, at the same time that a new current of health history was emerging, the Society for the History of Medicine, which was mentioned in the first pages, re-emerged, driven by humanist physicians.

Although most of its members are physicians, there are some historians and sociologists among the associates. The members of the Association meet periodically (virtually or in person) to discuss topics related to the history of medicine and health in Peru. Regarding the assumptions, objectives, topics, and ways of approaching the history of medicine of those who guide this institution, a fusion between the traditional history of medicine and the contributions of social history is discovered. An example is that among the most recurrent themes are the biographies of the 'fathers of Peruvian medicine', but from perspectives that are not necessarily laudatory. For example, Uriel García developed an research on Daniel A. Carrión. Still, he is focused not on his sacrifice (self-inoculation with Peruvian wart) but on how his image was modified in the portraits that were made of him to Europeanize his physical features (García Cáceres Uriel, Daniel Alcides Carrión. Acta Médica Peruana, 2006).

On the other hand, more classical objectives of traditional history survive. For example, the desire to write comprehensive and encyclopedic works and the history of medical subspecialties. As for the institutional purposes indicated by the Association itself, they include, among others, to bring together physicians and other researchers dedicated to the history of national or universal medicine, to pay homage to illustrious physicians, to collaborate with teaching institutions to institute courses on the history of medicine, to carry out cultural activities on the subject of the history of medicine, to establish links with similar institutions inside and outside the country, among others.

Although its members question the uncritical use of historical practice in the past, there is a certain continuity with traditional history in terms of rescuing important figures and major events. In other words, there is a fusion between the aims of the history of health (broadening the spectrum beyond physicians and viruses, understanding the institutional development of public health, questioning aspects of the medicalization of society) and the history of traditional medicine (paying homage to illustrious physicians, a mostly medical public).

19.7 The Twenty-First Century and the *Asociación Peruana de Historia y Estudios Sociales de la Ciencia, la Tecnología y la Salud*

In 2018, a group of academics (historians, sociologists, anthropologists) founded the *Asociación Peruana de Historia y Estudios Sociales de la Ciencia, la Tecnología y la Salud*, with the aim of "seeking collaborative ways to overcome the fragmentation of studies referring to these topics both in Peru and abroad. Our main purpose is to research, teaching and dissemination of works related to Peru's history of science, technology, and health. We also seek to contribute to the preservation of the monumental heritage and historical sources related to the topics that bring us together" https://aphcts.wordpress.com/. Among the institutional objectives are: to seek collaborative ways to overcome the fragmentation of studies in science, technology and health; and to promote research, teaching and dissemination of works related to the history of science, technology and health in Peru. https://aphcts.wordpress.com

The works of its members, although varied in focus and interests, bring back a greater emphasis on gender (for example, Dr. Raul Necochea's studies on cancer and family planning or Ruth Iguiñiz's work on childbirth), race (for example, Patricia Palma's work on Chinese herbalists), indigenous medicine (especially Amazonian), history of pandemics (obvious effect of Covid-19), analysis of policies and programs conducted by the State, and a look at the local and the global (obvious effect of Covid-19), indigenous medicine (especially Amazonian), history of pandemics (obvious effect of Covid-19), analysis of health policies and programs conducted by the State, and a look at the local and the global (for example, the work of Marcos Cueto on global health), among others. However, there is a clear line with the social history of health in that it is based on the premise that to understand the history of health, it is necessary to understand and address the broader social and political contexts. Unfortunately, in the Peruvian case, they are societies that converse little, that each one attends to its public and works from its assumptions and interests. At the same time, each holds its first history conferences or congresses.

According to Cueto, "In the last fifteen years, a third historiographic tradition has been taking shape, focused not only on events of the second half of the 20th century, and concerned with the reception, negotiation and local reworking of European and North American influences, but also with the circulation of ideas, institutions, and people. Part of this tradition is also the contextualized study of key figures in the international agencies that interacted with the Cold War, such as the Canadian George Brock Chisholm, the first director of the WHO. It is important to emphasize the return of the biographical genre to the history of health -something that never ceased to be important in the history of science and the history of medicine- which was harshly criticized in the seventies of the 20th century. Thus, a social turn in the history of medicine occurred. Biography was seen as a style of a traditional history obsessed with the contributions of "great doctors" (generally white European or North American men) and ignoring contextual factors such as race, gender, and economic interests, as well as the views and practices of patients, nurses, and healers. It is clear that the re-emergence of biography in health history—often understood as a critical study that can be of little-known characters and that complements social history—is the result of the influence of other currents such as microhistory, labor history, gender history and postcolonial history" (Cueto, 2020a, b).

References

Aceituno, P., Prieto, M. D. R., Solari, M. E., Martínez, A., Poveda, G., & Falvey, M. (2009). The 1877-1878 el niño episode: Associated impacts in South America. *Climatic Change, 92*(3–4), 389–416.

Alaperrine-Bouyer, M. (1999). *Mariano Eduardo de Rivero en algunas de sus cartas al barón Alexander von Humboldt.* Arequipa: Universidad Nacional de San Agustín de Arequipa: Centros de Estudios Arequipeños.

Alberto Giesecke, M., & Mateo Casaverde, R. (2007). Historia del Observatorio Magnético de Huancayo y el Ecuador magnético. *Boletín de la Academia de Ciencias, 1*, 1–11.

Altamirano, A. (2013). El ingeniero Federico Villarreal y el encomio al conocimiento. *Tipshe: Revista de Humanidades, 13*(11), 39–54.

Álvarez, S. (2006). *La formación en arquitectura en el Perú. Antecedentes, inicios y desarrollo hasta 1955.* Lima: Instituto de Investigación de la Facultad de Arquitectura Urbanismo y Artes. Proyecto Historia UNI. Universidad Nacional de Ingeniería.

Amaya, E. (2018). Ingenieros, arquitectos, médicos y locura. La construcción del asilo colonia de la Magdalena (1896-1918). *Devenir, 5*(10), 65–90.

Araujo, K. (2009). *Dignos de su arte. Sujeto y lazo social en el Perú de las primeras décadas del siglo XX.* Iberoamericana–Vervuert. Universidad de Santiago de Chile.

Armus, D. (2002). La enfermedad en la historiografía de América latina moderna. *Asclepio, 54*(2), 41–60. https://doi.org/10.3989/asclepio.2002.v54.i2.140

Ayarza, J. (2021). 100 años de la Ciencia Macromolecular: Orígenes y perspectivas de futuro. *Revista de Química, 35*(2), 36–43.

Ballón, J. C. (2006). Hipólito Unanue visto por Augusto Salazar Bondy: la tradición organicista de la ciencia en el Perú. In *Aproximación a Unanue y la Ilustración peruana* (pp. 17–41), Augusto Salazar Bondy. UNMSM: COFIDE.

Basadre, A. J. (2009). *Vida y Obra del sabio Federico Villarreal.* Universidad Nacional Mayor de San Marcos Fondo Editorial.

Beingolea, J. (2020). José García Bryce. *Devenir, 17*(14), 5–8.

Bolaños, F. (Ed.). (1929). *La Obra de los ingenieros en el Progreso del Perú* (Vol. 4 vols). Editorial Perú Moderno.

Cabrera, J. (2012). Un arquitecto con historia. *Puente. Ingeniería. Sociedad. Cultura, VII*(27), 19–25.

Cabrera, J. (2014). El embajador de la ciencia. *Puente. Ingeniería. Sociedad. Cultura, IX*(32), 18–25.

Cama, J. M. M., & Cama, M. A. M. (2010). *La ilustración en Lima: Vida y Obra del doctor Cosme Bueno y Alegre (1711–1798).* UNMSM.

Castañeda, C. (2022a). Discurso arquitectónico e identidad cultural. La vigencia ideológica de seis autores peruanos en la elaboración de un universo simbólico contemporáneo. *Limaq, 10*(dic), 131–159.

Castañeda, C. (2022b). Arquitectura e identidad cultural en el Perú. El universo simbólico en el encuentro de lo auténtico y lo apropiado durante el siglo XX. *Area, 28*(2), 1–17.

Cazorla, I. (1999). *Historia de la Universidad Nacional de Ingeniería. Tomo II. El crecimiento y la modernización (1909–1930).* Proyecto Historia UNI, Universidad Nacional de Ingeniería.

Chaumeil, J. P., Delgado Estrada, J. M., & Lomné, G. (2011). Proyecto de publicación del Atlas geográfico del Perú de Mariano Felipe Paz Soldán, París 1865. *Bulletin de l'Institut français d'études andines, 40*(3), 599–601. https://doi.org/10.4000/bifea.1346

Cisneros, M. (2015). *Héctor Velarde. Equilibrio y proporción de tiempo y espacio entre lo clásico, la tradición y la modernidad.* Tesis para optar el grado de Magíster en Arte Peruano con mención en Historia del Arte. Lima: UNMSM, Unidad de Posgrado.

Colegio de Ingenieros del Perú. (2021). *50 siglos de ingeniería en el Perú. Conmemorando el Bicentenario de la Independencia del Perú.* Argos Productos editoriales.

Coloma Porcari, C. (1994). *Los inicios de la Arqueología en el Perú o "Antigüedades Peruanas" De Mariano Eduardo de Rivero.* Instituto Latinoamericano de Cultura y Desarrollo.

Consejo Nacional de la Universidad Peruana. (1975). *Homenaje a José J. Bravo, 1874–1927.* Lima.

Contreras, C. (1994). La minería en la historia del Perú. *América Latina en la Historia Económica, 1*(1), 33–40.

Contreras, C., & Cueto, M. (2008). Caminos, ciencia y estado en el Perú, 1850-1930. *História, Ciências, Saúde—Manguinhos, 15*(3), 635–655.

Contreras, C. (2010). La economía del transporte en el Perú, 1800-1914. *Apuntes. Revista De Ciencias Sociales, 66*, 59–81.

Córdova, A. (2007). La Agrupación Espacio y la generación del 50. *Puente: Ingeniería, Sociedad, Cultura, 2*(7), 12–17.

Córdova, A. (2011). Cien años de la enseñanza de la Arquitectura. *Puente. Ingeniería. Sociedad. Cultura, V*(22), 12–15.

Cornejo, E. (2010). *Caminos y caminantes en el Perú. Infraestructura Para el crecimiento económico y la inclusión social.* Editorial San Marcos.

Cueto, M. (1986). Bibliografía sobre la historia de la ciencia y la tecnología en el Perú. *Apuntes. Revista De Ciencias Sociales, 19*, 147–162.

Cueto, M. (2020a). *El regreso de las epidemias: Salud y sociedad en el Perú del siglo XX.* IEP.

Cueto, M. (2020b). La salud global, la salud planetaria y los historiadores. *Quinto Sol. Revista de Historia, 24*(3), 1–21. https://doi.org/10.19137/qs.v24i3.4834

Cueto, M. (2022). La historia de la medicina, Henry E. Sigerist y los intermediarios médicos en el Perú de mediados del siglo XX. *Historia, 1*(55), 329–350.

Dagicour, O. (2019). Régénérer la Patrie, construire l'État. Savoirs géographiques et production du territoire, Pérou (1900-1930). *Bulletin de l'Institut Pierre Renouvin, 49*, 173–178.

Dargent Chamot, E. (1992). El cosmógrafo mayor Juan Ramón Coninck. In *Actas del Primer Simposio de Historia Marítima y Naval Iberoamericana.* Lima: Dirección de Intereses Marítimos: Instituto de Estudios Histórico-Marítimos.

Dargent Chamot, E. (1995). El observatorio astronómico de Lima. *Derroteros de la Mar del Sur, 3*, 29–36.

de Ribas, N. (2011). El tren de Lima a la Oroya: construcción e idea de progreso en el proyecto ferroviario transandino del ingeniero polaco Ernesto Malinowski (1818-1898). *Itinerarios: revista de estudios lingüísticos, literarios, históricos y antropológicos, 14*, 251–261.

Deustua, J. (2017). Sociedad, ciencia y tecnología: Mariano de Rivero, la minería y el nacimiento del Perú como República, 1820-1850. *Apuntes. Revista de ciencias sociales, 80*, 51–77.

Diez Canseco, E. (1929). *La red nacional de carreteras* (2nd ed.). Imp. Torres Aguirre.

Dorich, L. (2001). Fernando Belaunde Terry y la planificación en el Perú. In *En Construyendo el Perú II. Aportes de ingenieros y arquitectos* (pp. 233–244). Proyecto Historia UNI, Universidad Nacional de Ingeniería.

Elmore, T. (2014). *Lecciones de Arquitectura.* Wiley Ludeña (ed.). Lima: Facultad de Arquitectura y Urbanismo, PUCP.

Espejo, Á. (2023). *La educación del arquitecto peruano. Reflexiones históricas Para comprender su situación actual. Tesis Para optar el Grado académico de Magíster en Educación con mención en docencia universitaria.* UNMSM, Unidad de Posgrado.

Espinoza, C., & Arroyo, E. (2022). Irrigaciones, tierra y sociedad en la costa peruana: 1890-1930. *Investigaciones Sociales, 47*, 199–212.

Facultad de Arquitectura y Urbanismo. (2013). *Bases. De la materia a la metrópolis*. Facultad de Arquitectura y Urbanismo PUCP.

Facultad de Ciencias e Ingeniería. (2013). *Ciencias e Ingeniería: 80 años*. PUCP.

Fernández Prada Franco, Alberto. (2015). *Pedro Paulet Mostajo (1874–1945)*. Universidad Ricardo Palama.

Ferrer, F. (2011). El Brutalismo. Expresión arquitectónica de una época de la historia del país. *Moneda, 148*, 47–51.

Ferrero, R. (1939). La agricultura y la profesión agronómica. *Revista de la Universidad Católica, VII*(8–9), 494–507.

Galessio, E. (2007). *Ferrocarriles del Perú. Un viaje a través de su historia*. Lima.

Gallegos, H. (2001). La ingeniería en el Perú del siglo XIX. In *Construyendo el Perú II. Aportes de ingenieros y arquitectos* (pp. 33–53). Proyecto Historia UNI, Universidad Nacional de Ingeniería.

Gallegos, H. (2006). *La ingeniería*. GCAQ Fondo Editorial.

García Bryce, J. (1986). La arquitectura en el Virreinato y la República. En *Historia del Perú*. Tomo IX: Procesos e instituciones (pp. 11–89). Lima: Editorial Juan Mejía Baca.

Garfias, M., & Vega, J. (2015). *La ingeniería industrial en San Marcos. Cincuenta años forjando una nueva tradición universitaria*. UNMSM, Facultad de Ingeniería Industrial.

Glave, M. (1986). El agrarismo "suttoniano" y el conflicto de aguas de 1928–1930 en el departamento de Lambayeque. En *La historia agraria* (pp. 64–100). Chiclayo: Centro de Estudios Sociales "Solidaridad".

Guizado, Y., & Ragas, J. (2021). La Escuela Nacional de Electricidad y la educación técnica en Perú. *Diseña, 18*, 3–14.

Guizado, Y. (2022). De la palabra a la mano: La formación de la Educación Técnica en el Perú, 1864-1915. *ISHRA, Revista del Instituto Seminario de Historia Rural Andina, 9*, 71–84.

Gutiérrez, D. (2013). *Gustave Eiffel en el Perú: 1870–1890. Proyectos y obras*. INIFAUA, Universidad Nacional de Ingeniería.

Hampe, T. (2010). Meiggs, Grace y la obra del Ferrocarril Central Andino (1870-1885). *Historia: Revista de la Escuela Profesional de Historia, 9*, 83–94.

Huamaní, E. (1999). *Mario Samamé Boggio*. Proyecto Historia UNI, Universidad Nacional de Ingeniería.

Huamaní, E. (2015). *La Escuela de Ingenieros y el movimiento estudiantil. Tomo I: 1900–1932*. EDUNI.

Huamaní, E. (2017). *La Escuela de Ingenieros y el movimiento estudiantil. Tomo II: 1933–1955*. EDUNI.

Huamaní, E. (2022). *Ingeniería e ingenieros en el Perú (1900–1950). 1. Ubicación sociopolítica*. Centro Historia UNI y Bibliometría, EDUNI.

Huaraj, J. C. (2014). *El doctor Gabriel Moreno y Espinosa. Apuntes Para la historia de la ilustración borbónica tardía en el Perú*. UNMSM.

Inglesi, P., Inglesi, S., & La Torre, R. (2005). *Antonio Raimondi. Mirada íntima del Perú. Epistolario 1849–1890*. Fondo Editorial del Congreso del Perú.

Itabashi, E. (2009). *La Agrupación Espacio. Edificio residencial de la FAP. 1959, 1961. Chiclayo-Perú*. Universidad de Cuenca.

Jaimes, F. (2023). *Huellas y voces en el Espacio en el Tiempo. La arquitectura moderna de Miró Quesada: Influencias e interpretación*. Escuela Técnica Superior de Arquitectura.

Jiménez, L., & Santiváñez, M. (2005). *Rafael Marquina. Arquitecto*. INIFAUA, Universidad Nacional de Ingeniería.

Jochamowitz, L. (1992). *Hombres, minas y pozos 1896–1996*. Instituto Andino de Estudios Arqueológicos.

Kahatt, S. (2019). El Perú como proyecto. *A&P Continuidad, 6*(11), 122–133.

Katayama Omura, R. (2018). La transdisciplinariedad como método de investigación en el pensamiento filosófico natural en el virreinato peruano. In C. Monteagudo & P. Quintanilla (Eds.),

Los caminos de la filosofía: diálogo y método (pp. 403–424). Pontificia Universidad Católica del Perú.

Kemp, K. (2002). *El desarrollo de los ferrocarriles en el Perú*. Proyecto Historia UNI, Universidad Nacional de Ingeniería.

Lara, E. (1936). *Historia del Ministerio de Fomento y Obras Públicas: 1896–1936*. Tomo I.

Larico, A. Y., et al. (2022). Análisis del desarrollo de la investigación científica en la historia republicana en el Perú. *Revista Internacional de Derecho, 3*(1), 33–44.

Laya, A. E. (2020). Joaquín Capelo: comisionado especial del Supremo Gobierno y prefecto del departamento de Loreto (1899-1900). *ISHRA, Revista del Instituto Seminario de Historia Rural Andina, 4*, 23–34.

Lizarme, N. (2022). La profesionalización de la ingeniería y la construcción del Perú moderno (1899-1930). *Apuntes. Revista de Ciencias Sociales, 49*(91), 97–131.

López Soria, J. I. (1999a). *Eduardo J. de Habich (1835–1909). Homenaje a los 90 años de su muerte*. Proyecto Historia UNI, Universidad Nacional de Ingeniería.

López Soria, J. I. (1999b). *Historia de la Universidad Nacional de Ingeniería. Tomo I. Los años fundacionales (1876–1909)*. Proyecto Historia UNI, Universidad Nacional de Ingeniería.

López Soria, J. I. (2003). *La Sociedad de Ingenieros del Perú. Primera década (1898–1908)*. Proyecto Historia UNI, Universidad Nacional de Ingeniería.

López Soria, J. I. (2005). Historia de la ingeniería y la arquitectura en el Perú. Bibliografía comentada. In M. Guerra, M. Guerra, & D. Rouillon (Eds.), *Historias paralelas: Actas del primer encuentro de historia Perú-México* (pp. 331–348). Fondo Editorial PUCP, El Colegio de Michoacán, A. C.

López Soria, JI. (2017 [1997]). Aproximaciones a la "Agrupación Espacio". In *Filosofía, arquitectura y ciudad* (pp. 24–30). Lima: Centro Historia UNI, EDIFAUA, Universidad Nacional de Ingeniería.

López Soria, J. I. (2021). Modernidad, racionalidad e ingeniería en el Perú. *Puente. Ingeniería. Sociedad. Cultura, XVI*(63), 2–9.

López Soria, J. I. (2022). Los ingenieros y "la nacionalidad en construcción". *Summa Humanitatis, 12*(1), 26–60.

López Soria, J. I., & Cazorla, I. (2018). Malinowski y la guerra con Chile. *Summa Humanitatis, 10*(2), 58–115.

López Soria, JI, & Lizarme, Nashely. (2021). *El pensamiento de los ingenieros sobre el Perú (1850–1930)*. 2 tomos. Lima: Centro Historia UNI, EDUNI.

López Soria, J. I., et al. (2004). *Enciclopedia Temática del Perú. Tomo XI. Ciencia y Tecnología*. Empresa Editora El Comercio.

López Soria, J. I., Quiñones, L., & Ueda, M. (2012). *Historia de la UNI. Volumen IV. Institucionalización Como universidad frente a los retos del desarrollo (1955–1984)*. Centro Historia UNI, Universidad Nacional de Ingeniería.

López-Ocón, L. (2001). La Sociedad Geográfica de Lima y la formación de una ciencia nacional en el Perú Republicano. *Terra Brasilis (Revista da Rede Brasileira de História da Geografia e Geografia Histórica), 3*, 1–24. https://doi.org/10.4000/terrabrasilis.330

Losada y Puga, Cristóbal de. (1940). La profesión de ingeniero. *Revista de la Universidad Católica, VIII*(1), 63–75.

Lossio, J. (2022). Salud en emergencia. Historia de las batallas contra las epidemias y la COVID-19 by Marcos Cuetos. *Argumentos, 3*(1), 61–63.

Ludeña, W. (2004). *Lima: historia y urbanismo*, 4 tomos Lima: Ministerio de Vivienda, Construcción y Saneamiento; Facultad de Arquitectura, Urbanismo y Artes, Universidad Nacional de Ingeniería.

Ludeña, W. (2017). *Ideas y arquitectura en el Perú del siglo XX*. SEMSA.

Luque, J. (2014). La formación técnica minera en el Perú: 1779-1876. *Nueva Corónica, 3*, 295–317.

Málaga Núñez-Zeballos, A. (1998). *Memorial de Ciencias Naturales y de Industria Nacional y Extranjera/Redactado Por M. de Rivero y N. De Piérola (Edición facsimilar)*. Cámara de Comercio e Industria de Arequipa.

Málaga Núñez-Zeballos, A. (2000). Mariano E. de Rivero, Nicolás de Piérola y el primer periódico científico del Perú. In S. L. Febres & G. L. Villena (Eds.), *Homenaje a Félix Denegri Luna* (pp. 417–433). Pontificia Universidad Católica del Perú.
Málaga Núñez-Zeballos, A. (2013). *La modernización de la República. La prensa científica del Perú (1827–1829)*. Asamblea Nacional de Rectores.
Maqueira, Á. (comp.). (2021). *José García Bryce. Los goces de la memoria.* Lima: Universidad de Lima Fondo Editorial.
Martuccelli, E. (2012). *Conversaciones con Adolfo Córdova*. INIFAUA, Universidad Nacional de Ingeniería.
Martuccelli, E. (2017). *Arquitectura Para una ciudad fragmentada. Ideas, proyectos y edificios en la Lima del siglo XX*. Universidad Ricardo Palma. Editorial Universitaria.
Martuccelli, E. (2019). Pedro Paulet. *En Anales del IAA, 49*(1), 27–42.
Mattos-Cárdenas, L. (2016). *Lima y la Plaza de Armas. Historia y aportes de Emilio Harth-Terré*. Colegio de Arquitectos del Perú.
Mejía, Á. (2019). Pedro Paulet, sabio multidisciplinario. *Persona & Cultura, 14*(14), 95–122.
Mejía Salvatierra, Á. (2017). Pedro Paulet, sabio multidisciplinario. *Persona y Cultura, 14*, 95–122.
Meza, L. (2008). *La ingeniería en el Perú. Evolución y perspectivas*. Lima*: Colegio de Ingenieros del Perú*. C. P. S. Ingeniería S. A. C.
Michilot, M. (2021). Heterotopías panópticas. Análisis tipológicos de tres instituciones de encierro de la ciudad de Lima (siglos XIX y XX). In S. Kahatt, E. Martuccelli, & V. Mejía (Eds.), *Arquitectura & Investigación. Arte, tipología, política* (pp. 159–178). Arquitectura PUCP.
Ministerio de Fomento. (1924). *Síntesis de la minería peruana en el Centenario de Ayacucho*. Imprenta Torres Aguirre.
Molina, A., & Ludeña, W. (Eds.). (2011). *Arquitectos/Arquitectas. Pioneros/Pioneras*. Facultad de Arquitectura y Urbanismo, PUCP.
Morales, B., et al. (Eds.). (2021). *160 años de historia e investigación geológica, minera y metalúrgica en el Perú*. INGEMMET.
Núñez, E. (2002). *Alexander von Humboldt en el Perú: Diario de viaje y otros escritos*. Banco Central de Reserva del Perú.
Núñez, O. (2022). Semblanza y recuerdo de Adolfo Córdova. *Puente. Ingeniería. Sociedad. Cultura, XVII*(67), 4–9.
Olcese, O. (2002). *"Enfrentando la adversidad. Camino a la Gloria". Historia de la Universidad Nacional Agraria La Molina. 1902–2002*. de la Universidad Nacional Agraria La Molina.
Ortiz de Zevallos, A. (2001). Fernando Belaunde Terry, pensador y arquitecto: aportes a la definición del oficio y de la responsabilidad del arquitecto en el Perú. In *Construyendo el Perú II. Aportes de ingenieros y arquitectos* (pp. 227–232). Proyecto Historia UNI, Universidad Nacional de Ingeniería.
Ortiz Sotelo, J. (1993). *El Contralmirante Eduardo Carrasco (1779–1865)*. Dirección de Intereses Marítimos. Fondo de Publicaciones.
Ortiz Sotelo, J. (1997). Los cosmógrafos mayores del Perú en el siglo XVI. *Boletín del Instituto Riva-Agüero, 24*, 369–389.
Osorio Tarazona, A. (2017). *Pesca, ciencia y nacionalismo: El IMARPE Como centro de Excelencia Científica (1964–1982)*. Pontificia Universidad Católica del Perú.
Oviedo, J. (1862). *Colección de leyes, decretos y órdenes publicadas en el Perú desde el año de 1821 hasta 31 de diciembre de 1859. Reimpresa por orden de materias por el Dr. Dn… Tomo nono. Ministerio de Beneficencia, Instrucción Pública y Justicia*. Felipe Bailly.
Paz Soldán, C. E. (1943). *Sociedad de Historia de la Medicina: Memoria de sus trabajos, años 1942–1943*. Imprenta América.
Quiñones, L. (2014a). *Construir y modernizar: el Ministerio de Fomento (1896–1930)*. Centro Historia UNI, EDUNI.
Quiñones, L. (2014b). *José Balta Paz. Ingeniero y político*. Centro Historia UNI, Universidad Nacional de Ingeniería.

Regal, A. (1967). *Castilla Constructor. Las obras de ingeniería de Castilla*. Instituto Libertador Ramón Castilla.

Rodríguez, K. (1999). *Historia de la Universidad Nacional de Ingeniería. Tomo III. La apertura a espacios nuevos (1930–1955)*. Proyecto Historia UNI, Universidad Nacional de Ingeniería.

Rodríguez, K. (2011). *Joaquín Capelo: la obra de un ingeniero, sociólogo y luchador social*. https://www.academia.edu/37398427/Joaquín_Capelo_La_obra_de_un_ingeniero_sociólogo_y_luchador_social. Accessed 15 december 2023.

Rosas, M. (2019). *Mariano Eduardo de Rivero y un diálogo tecnológico con el mundo andino. Tesis Para optar el Grado de doctor en Filosofía*. Brown University.

Sala i Vila, Nuria. (2006). Ingenieros y colonización amazónica en el Perú, 1821-1930. *Anuario IEHS, 21*, 441–466.

Salaverry, G. O. (2016). *Las "Observaciones sobre el clima de Lima" (1806) de Hipólito Unanue Como Topografía Ilustrada*. Universidad de Salamanca.

Samamé, M. (1974). *Minería peruana. Bibliografía*. Editorial Jurídica.

Samamé, M. (1979). *El Perú minero* (Vol. 14 vols). INCITEMI.

Samamé, M. (1984a). Del Cuerpo de Ingenieros de Minas al INGEMMET. *Re-Metallica* 2 (may-jun): 25-32. Sala i Vila, Nuria. 2006. Ingenieros y colonización amazónica en el Perú, 1821-1930. *Anuario IEHS, 21*, 441–466.

Samamé, M. (1984b). Del Cuerpo de Ingenieros de Minas al INGEMMET. *Re Metallica, 2*(May-Jun), 25–32.

Schwab, F. (1948). Los almanaques peruanos y guías de forasteros, ¿1680?-1874. *Boletín Bibliográfico de San Marcos, 19*, 1–2.

Seiner, L. (2000). Orígenes y funcionamiento inicial del Cuerpo de Ingeniero de Minas del Perú. In *Construyendo el Perú I. Aportes de ingenieros y arquitectos* (pp. 29–34). Proyecto Historia UNI, Universidad Nacional de ingeniería.

Seiner, L. (2001). El Fenómeno El Niño en el Perú: reflexiones desde la Historia. *Debate Agrario, 33*, 1–18.

Seiner, L. (2003). Antonio Raimondi y sus vinculaciones con la ciencia europea, 1851-1890. *Bulletin de l'Institut Français d'Études Andines, 32*(3), 517–537.

Seiner, L. (2004). Los Inicios de la meteorología en el Perú y la labor del Cosmografiato: 1753-1856. *International Comission for the History of Meteorology, 1*, 14–27.

Seiner, L. (2009). *Historia de los sismos en el Perú, siglos XV-XVII*. Universidad de Lima.

Seiner, L. (2010). Antonio Raimondi en el Perú: viajes, obra científica y redes de influencia en la periferia, 1851-1890. In R. S. Baeza (Ed.), *Ciencia-Mundo. Orden republicano, arte y nación en América* (pp. 255–280). Editorial Universitaria: Centro de Investigaciones Diego Barras Arana.

Seiner, L. (2011). *Historia de los sismos en el Perú, siglos XVIII-XIX*. Universidad de Lima.

Seiner, L. (2018a). Estudio introductorio, edición e índices. In H. Unanue (Ed.), *Observaciones sobre el clima de Lima y su influencia sobre los seres organizados*. Fondo Editorial del Congreso: Instituto Riva Agüero: Pontificia Universidad Católica del Perú.

Seiner, L. (2018b). *El Niño y sus diversas manifestaciones en la historia*. Correlación con el cambio climático. https://portal.concytec.gob.pe/images/noticias/Art%C3%ADculo_-_Lizardo_Seiner_-_PUCP.pdf

Seiner, L. (2020). Rasgos del discurso sismológico en el Perú, siglos XVI-XVII. *Nuevo Mundo Mundos Nuevos*. https://doi.org/10.4000/nuevomundo.79677

Silgado Ferro, E. (1992). *Investigaciones de Sismicidad Histórica en la América del Sur en los siglos XVI, XVII, XVIIII Y XIX*. CONCYTEC: Centro Regional de Sismología para América del Sur.

Sotillo, F. (1976). Cien años de ingenieros de minas en el Perú. *Minería y petróleo, 168*(May-Jun), 37–52.

Stucchi, L., & Stucchi, M. (2019). *Navegar por los aires. Análisis físico y biológico del sistema propuesto por Santiago de Cárdenas en el siglo XVIII*. Fondo Editorial de la Universidad del Pacífico.

Takashi, K., & Martínez, A. (2019). The very strong coastal El Niño in 1925 in the far-eastern Pacific. *Climate Dynamics, 52*, 7389–7415.

Tamayo, A. (2015). *Hombres de ciencia y tecnología en el Perú. Cinco siglos de investigación científica en territorio peruano*. Argos productos editoriales.

Tantaleán, G. (2012). *Ingeniería metalúrgica en la República*. UNMSM, E. A. P. de Ingeniería Metalúrgica.

Tumialán, P. (2003). *Compendio de Yacimientos minerales*. Instituto Geológico, Minero y Metalúrgico.

Ueda, M. (2002). *Historia del Cuerpo de Ingenieros de Minas del Perú. 1902–1950*. Proyecto Historia UNI, Universidad Nacional de Ingeniería.

Ueda, M. (2005). *Carlos I. Lissón. Ingeniero, geólogo y paleontólogo*. Proyecto Historia UNI, Universidad Nacional de Ingeniería.

Ueda, M. (2007). Remy. El inicio de la ingeniería. *Puente. Ingeniería. Sociedad. Cultura, II*(7), 2–7.

Ueda, M. (2008). Los Grieve: el ingeniero al servicio del Perú. *Puente. Ingeniería. Sociedad. Cultura, III*(10), 2–7.

Ueda, M. (2009a). Las primeras mujeres en la Escuela de Ingenieros. *Puente. Ingeniería. Sociedad. Cultura, IV*(12), 4–9.

Ueda, M. (2009b). Los Basadre, la Escuela de Ingenieros y la ingeniería nacional. *Puente. Ingeniería. Sociedad. Cultura, IV*(13), 2–7.

Ueda, M. (2010). Santiago Antúnez de Mayolo. *Puente. Ingeniería. Sociedad. Cultura, V*(19), 2–7.

Universidad de Lima. (2013). *Héctor Velarde. Arquitecto y humanista*. Universidad de Lima Fondo Editorial.

Universidad de Lima. (2016). *Walter Weberhofer. El proyecto moderno en el Perú*. Universidad de Lima Fondo Editorial.

Universidad de Lima. (2017). *Mario Bianco. El espacio moderno en el Perú*. Universidad de Lima Fondo Editorial.

Utia, F. (2001). Luis Miró Quesada Garland y la modernidad en arquitectura. In *Construyendo el Perú II. Aportes de ingenieros y arquitectos* (pp. 209–224). Proyecto Historia UNI, Universidad Nacional de Ingeniería.

Vegas Vélez, M. (1987). *La Investigación oceanográfica en el Perú: Desde Humboldt hasta nuestros días*. Presidencia de la República.

Vegas Vélez, M. (1991). *Humboldt en el Perú: Diario de Alejandro de Humboldt durante su permanencia en el Perú, agosto a diciembre de 1802*. Centro de Investigación y Promoción del Campesinado.

Velázquez, D., & La Serna, J. C. (2021). *200 años de historia, 52 conectando vidas*. Ministerio de Transportes y Comunicaciones.

Vidal, J. (2001). Felipe Arancivia Bercolme (1840-1915). Un ingeniero al servicio del estado. In *Construyendo el Perú II. Aportes de ingenieros y arquitectos* (pp. 193–203). Proyecto Historia UNI, Universidad Nacional de Ingeniería.

Villacorta, L. F. (2003). *Informes y polémicas sobre el guano y el Salitre (Perú: 1854–1877)*. UNMSM: COFIDE.

Villacorta, L. F. (2004). *Oro del Perú*. UNMSM: Asociación Educacional Antonio Raimondi: Compañía de Minas Buenaventura: COFIDE.

Villacorta, L. F. (2006a). *El Departamento de Ancash [sic] y sus riquezas minerales (1873)*. UNMSM: Golden Associates: Asociación Educacional Antonio Raimondi.

Villacorta, L. F. (2006b). *La sierra y Selva central: Morococha, Cerro de Pasco y Chanchamayo*. UNMSM: Golden Associates: Asociación Educacional Antonio Raimondi.

Villacorta, L. F. (2007). *Minerales del Perú y la búsqueda de una imagen republicana*. UNMSM: Golden Associates: Asociación Educacional Antonio Raimondi.

Villacorta, L. F. (2009). *Aguas minerales y potables Del Perú: Autoridad científica y nuevos espacios de consagración republicanos*. UNMSM: Golden Associates: Asociación Educacional Antonio Raimondi.

Villacorta, L. F. (2010). *Flora Perpetua: Arte y ciencia botánica de Antonio Raimondi*. Asociación Educacional Antonio Raimondi.

Villacorta, L. F. (2012). *Terra nostra: Antonio Raimondi y la Cartografía del rumbo de la república*. Asociación Educacional Antonio Raimondi.

Watanabe, L. (2017). *Federico Villarreal. Senador por Lambayeque*. Fondo Editorial del Congreso del Perú.

Wong, D., Hernández, H., & Carrasco, J. M. (2018). Gringo entrepreneurship in Latin America. The Thorndikes of Peru, 1901-1938. *Journal of Evolutionary Studies in Business, 4*(1), 180–207.

Yepes, E. (1986). *Estudios de historia de la ciencia en el Perú*.

Leticia Quiñones is a doctoral candidate in the Pontificia Universidad Católica del Perú Doctoral Program. She lectures in the Faculty of Humanities of the University of Piura. She is the author of *Construir y Modernizar: El Ministerio de Fomento (1896–1930)* (UNI, 2014); and *José Balta Paz: Ingeniero y Político* (UNI, 2014). His areas of interest are the administrative history of the Peruvian state, modernity and material progress, and the political activity of engineers.

Jorge Lossio is a senior lecturer in the Pontificia Universidad Católica del Perú in the Department of Humanities, where he is also director of the Instituto Riva-Ag*üero*. He is a research member of the Asociación Peruana de Historia y Estudios Sociales de la Ciencia, la Tecnología y la Salud. He is the author of several papers and books on the history of Peruvian health. Among them: *¿Qué hicimos mal: la tragedia de la covid-19 en el Perú* (IEP, 2022) and *Salud pública en el Perú del siglo XX: paradigmas, discursos y políticas* (IRA, 2017).

Lizardo Seiner holds a Ph.D. in History from the University of Grenoble (France), a Master's and Bachelor's degree from the Pontificia Universidad Católica del Perú, and lectures in its Department of Humanities. He specializes in Environmental History, a field in which he has published several works, including two seismic catalogs from the 16th to the nineteenth century and the historical recurrence of the El Niño phenomenon, research in the History of Science on the development of meteorology in Peru and a study on Antonio Raimondi.

Chapter 20
The History of Science and Technology in the Cuban Context: Institutionalization and Development

Diego de Jesús Alamino Ortega and Yenile Aguilar Rodríguez

Abstract International knowledge and transparency of the process of foundation and development of science in Cuba has not received enough visibility due to historical conditions; three centuries of Spanish colonization and then controlled by the United States, delayed Cuba's full participation in the concert of Latin American nations and the world. In this essay, without attempting to present an exhaustive study or a strictly chronological account, the authors, using documentary sources, have wanted to reflect the emergence of genuinely Cuban scientific activity. Scientists, professionals and intellectuals, from different professions, at the end of the eighteenth century and the beginning of the twentieth, felt urged to reason on the future of science in Cuba. The main purpose of this essay is to show, in bold strokes, the process of organization, association and institutionalization of the history of science and the emergence of historians of science and technology in the Cuban context as well as their main contributions and works.

Keywords Science in Cuba · Historians · Institutionalization · Associations · Publications

20.1 Introduction

It is not possible to know the history of science and technology and the role of historians of science and technology in Cuba if we do not go down to the origins of its development as a social practice, to the beginnings of scientific and philosophical thought aimed at response to the problems of the Cuban context.

If we intend to address how the history of science and technology have been developed and institutionalized in the Cuban context, it is necessary to begin by

D. de J. Alamino Ortega (✉) · Y. Aguilar Rodríguez
Centro Universitario "Enrique Rodríguez-Loeches Fernández", Universidad de Matanzas, Matanzas, Cuba
e-mail: dalaminortega@gmail.com

M. de la P. Ramos-Lara, L. C. Arboleda (eds.), *History of Science in Latin America: The Construction of an Intellectual Field (20th century)*,
https://doi.org/10.1007/978-3-031-75536-1_20

providing in these essay arguments that allow the reader to assess the peculiar historical, political, economic and social development. Likewise, it is necessary going up to the emergence of science, technology and historians of science and technology in the constitution of Cuba as a nation, as a manifestation of what the mode of production of an era relate to the superstructure of society. This essay also serves to recognize Cuban historians of science, although it is impossible to mention all, in the references and the text, you can find the vast majority of them.

20.2 Origins of Science and Technology on the Island of Cuba

Christopher Columbus, traveling in search of new routes, stumbled upon Cuban land in 1492, a place where he found human beings with his image and likeness, living in a state of primitive community, while Spain, which had not yet overcome the feudal stage, had Universities. In their early trips, the colonizers, in order to survive in these lands unknown to them, had to learn from the knowledge and experiences of the native settlers whom the Admiral called Indians. To the cultivation of cassava and its processing, tobacco and cotton, the construction of single-trunk canoes and the use of palm trees for the construction of houses, precarious technologies that the aborigines mastered, the colonizers contributed large and minor livestock, poultry and sugar cane. As the years went by, the European life experience adapted to the conditions of the tropics, stone was used in construction, clothing adapted to the climate, household goods began to be used and logically it became present as the dominant ideology the Catholic religion, thus starting a process of evangelization, not strictly educational.

The brutal exploitation native inhabitant, were submitted to by the colonizers, forced them to work as slaves; diseases imported from Europe made them sick and in 1540 they were close to extermination. Nevertheless, this weak population showed signs of rebellion that were unsuccessful due to the military superiority of the colonizers. In this way, Africans brought to Cuba under subhuman conditions of slavery, painfully burst into the Cuban context, also showing signs of rebellion. These three major ethnic groups: aborigines,[1] Africans and the Spanish colonizers later led to the formation of the Cuban nationality, as a process of transculturation (Ortiz, 1940). Without failing to take into account: Chinese, Portuguese, Anglo-Saxons, French, Indians from the American continent, Jews, Arabs, West Indians and others, who arrived in search of settling in a promising territory of well-being given its virgin status (García Galló, 1978).

For historians to emerge and persist, and in particular historians of science and technology in the Cuban context, this work deals with a *sine qua non* condition is having a favorable socioeconomic, intellectual and educational environment in the

[1] The native inhabitant of Cuba have been called: *aborígenes*.

population. This latter was not the interest of the colonizers, who limited knowledge to the Spanish language and the Catholic religion. Until 1840, the Colony did not establish a Law of Public Instruction on the Island, leaving elementary education under the protection of particular initiatives and institutions such as the *Sociedad Económica de Amigos del País*[2] (SEAP), which emerged almost three centuries after colonization, in 1793 (Batista et al., 2018). In contrast, higher and secondary education was favored by the purpose of enlightening the clergy and representatives of the hegemonic power, so education was largely restricted to white Christians.

One of the most distinctive institutions of the Cuban colonial period that extended until 1898, when the North American intervention occurred, turned out to be the San Basilio Magno Seminary, founded in 1722 in Santiago de Cuba. In this institution, an excessively rigid discipline was practiced and blacks, mulattoes, descendants of Jews or Moors could not attend. In the same way, different educational institutions emerged that made teaching an ecclesiastical responsibility, in which glimpses of science were taught, but mostly unrelated to the scientific advances that had been taking place in Europe, linked to names such as Copernicus, Vesalius, Galileo, Descartes, Newton and Leibniz (Clark, 2018).

The studies developed at the beginning of the eighteenth century in most educational institutions followed scholastic thought, as were the texts that were interpreted in light of this philosophical theological current, including the *Real y Pontificia Universidad de San Jerónimo de La Habana,*[3] founded in 1728 by the Dominicans (Order of Preachers). The Royal and Pontifical University of Havana, almost two centuries after colonization was established, prepared for the first time on the Island: theologians, jurists and doctors. Despite its delay and shortcomings,[4] the mere existence of this official institution opened new possibilities for the country's culture and for those born in Cuba, who still did not stop traveling to study in Europe or Mexico. The university was secularized in 1842 and was renamed *Real y Literaria Universidad de La Habana,*[5] now under the aegis of the Spanish government (LeRoy, 1979).

The first demonstrations against scholasticism in Cuba originate in the *Conciliar Colegio Seminario de San Carlos y San Ambrosio,*[6] which received authorization for its foundation in 1773[7] and began with the priest José Agustín Caballero (1771–1835), a man of an outstanding performance in the transformation of education in Cuba. Caballero supported by elective thought, highlighted in his work *Philosophia Electiva*, which circulated in Havana in 1797, was a determined representative of

[2] Economic Society of Country's Friends or Patriotic Society.

[3] Royal and Pontifical University of San Jerónimo of Havana.

[4] The Havana University was subject to scholasticism, keep in mind that the first public defense in Cuba of the Copernican theses that were published in 1543, was made at the Royal and Pontifical University of Havana almost at the end of the eighteenth century, in 1797.

[5] Royal and Literary University of Havana.

[6] Conciliar Seminary College of San Carlos and San Ambrosio.

[7] In 1767, the Seminary began to operate, in the facilities that were formerly the Colegio San José, when the Jesuits were expelled from the Spanish empire.

Creole enlightenment, which earned him the title Father of Cuban Philosophy, as stated by García Galló (1978). Another representative of the anti-scholastic struggle in the Seminary was the priest and *patriota entero,*[8] Félix Varela Morales (1788–1853), a disciple of José Agustín Caballero, and according to Torres-Cuevas et al. (1997) *el que nos enseño primero en pensar.*[9] It will be necessary to elaborate more on Varela when this work enters the nineteenth century and it will not only be because of his anti-scholastic position but also because of the harmonious conjunction in his personality of scientific, religious and independence thought (Alamino, 2001).

The books by which the instruction was followed in the aforementioned educational institutions and those that circulated in Cuba, were not writing by Cuban authors, were not in Spanish, nor were they produced on the Island. The printing press arrived in Cuba in 1723, although when Havana became a major port for the transit of the *Flota de Indias,*[10] came from Europe much current information. The Sevillian doctor living in Havana Lázaro de Flores wrote the first book of Cuban origin published in Spain: *El Arte de Navegar,*[11] from a scientific point of view presented the first explanations of geoastronomical natural phenomena that serve to guide navigation. Flores' book contains a calculation of the geographical longitude of Havana with an error of around 1%, the first reference to an astronomical instrument manufactured in the city, and the first reference to Copernicus in a work written on the Island. The book did not circulate in Cuba (López Sánchez, 1998).

The first interests in science on the Island are given by the conjunction between medical sciences and natural sciences. It is another doctor, Antonio Riaño de Gamboa (1672–1729) born in Havana, who is recognized as having carried out in 1714 astronomical observations to determine the latitude and longitude of Havana and other cities, for which he has been considered the first Cuban astronomer. The doctors established on the Island, who were the first professionals, were inclined towards the study of natural sciences and mathematics, perhaps recognizing for to practice their profession needed physics, botany, chemistry and other scientific disciplines.

In order to legitimize the history of science and technology, documents must exist. Of the *Hispanic period*[12] of science in Cuba, named after the prestigious Cuban historian of medicine and science José López Sánchez (1998), one can almost only know about science from the notes of the council, notarial protocols and the funds preserved in the *Archive of the Indies*. The first printed document known

---

[8] Complete patriot, qualified as such by the Cuban National Hero José Martí.

[9] Qualified as like the one who taught Cubans to think first by José de la Luz Caballero.

[10] Flota de Indias*:* The group of ships that met in Havana to travel together to Spain loaded with the riches of America.

[11] The Art of Navigation.

[12] According to López Sánchez's periodization of science in Cuba, the *Hispanic period* extends until 1790, the *Cuban period* occurs between 1790 and 1861, the so-called *Academic period* begins in 1861 and ends in 1902, the *Republican period* covers from 1902 to 1998, and *The recovery and advancement of science* period, begins in 1959.

in Cuba turns out to be a *Price List for medicine* (Espinosa, 2007). According to Rosell et al. (2006) since 1711, the *Royal Protomedical Tribunal* had been created to regulate the practice of medicine and health care.

It was in 1787 that the first science book was printed in Cuba and wrote it a Portuguese who became a naturalist and lived in Havana, Antonio Parra. The book name was: *Descripción de diferentes piezas de historia natural, las más del ramo marítimo, representadas en setenta y cinco láminas,*[13] it was the first scientific approach to Cuban nature (Clark, 2018).

Before 1790, some newspapers circulated but were limited to transmitting notices. Around this time was published the *Papel Periódico de La Havana,*[14] a publication of economic and literary interest that in its first issue announced its purposes: agriculture, commerce, news of scientific knowledge with application to economics, in addition to literature, anecdotes and news about science and art. Such was the significance of this publication that the Cuban scholar Tomás Romay Chacón[15] (1764–1849), who established the scientific bases of medicine in Cuba, expressed with reference to this newspaper: *It was the first route that our spirit was drawn, directing us to the Sanctuary of Science* (Batard, 2010).

With the use and consultation of the aforementioned publications, historians can now refer to printed records, and a space appears for the dissemination of science and technology, both from abroad and that originating on the Island, required by the needs of the country. The *Papel Periódico de La Havana* would reflect the growing cultural and philosophical movement aimed at disseminating updated knowledge on topics of interest to the ruling class, which would also have impacts on the main ways of instruction of the social strata with access to them.

The publishing boom to disseminate the scientific knowledge of those established on the Island and those born here (Creoles), and the interest in history, are progressively manifesting themselves in Cuba. The policy of Enlightened Despotism, applied in the second half of the eighteenth century, favored the appearance of the Creole sugar bourgeoisie, which gave a notable boost to science and technology, linked to its class objective. The result of this historical event is the *Sociedad Económica de Amigos del País* (SEAP) created in 1793, by the interest of distinguished personalities and intellectuals, with the aim of promoting the economic and sociocultural development of the Island. It is characteristic of this stage, the need to establish the historical facts, which makes the first historians appear and together with the signs of the existence of science and technology in Cuba, the possibility opens for historians not to stop attending to this important area of the future of the nascent Cuban nation.

The year 1797, described by López Sánchez (1998) as the year of the *Cuban scientific emergence*, had significant relevance. This year circulated seven

[13] Description of different pieces of natural history. The most from the maritime branch, represented in seventy-five plates.

[14] Newspaper of Havana.

[15] It will be necessary to elaborate on this Cuban Academician of the Royal Academy of Medical, Physical and Natural Sciences of Havana in other moments of the essay.

pamphlets containing manifestations of native Cuban science, they were: Dissertation on some Cuban plants, Inaugural Oration in Praise of surgery, Dissertation on yellow fever commonly called black vomit, by Tomás Romay Chacón, Discourse on the good properties of red earth for the cultivation of sugar cane, Report on the breeding of bees, Report on the best way to manufacture sugar. These works are linked to the knowledge prevailing in the eighteenth century, used for the good of the country's economy and culture. The inclination towards the application of the modern scientific method, observation, experience and reasoning, contrasted with scholasticism and in opposition to the obscurantist and backward conceptions, limiting knowledge, which the metropolis suffocated Cuba with (Clark, 2018).

The significance of the port of Havana with all the assurances needed to maintain the supply of ships, made the shipyard *Real Arsenal de La Habana*[16] become the main industry of the capital of the metropolis on the Island. At certain times, the Arsenal employed up to 2000 workers and in 1784, it had a large saw moved by water from the *Zanja Real*, an aqueduct completed in 1592 and the only one that Havana had until 1853. The rise of boat manufacturing and production sugar companies competed in deforestation and threatened the environment (Funes, 2002). The first launching of a ship built in Havana, the *San Andrés,* of 350 tons, occurred in 1551 and in 1769 a ship of more than 20, 000 tons of displacement, with 140 cannons, was the largest warship that sailed through the seas in the eighteenth century, named the *Holy Trinity* (Rodríguez, 2020).

The sugar industry was developing vigorously on the Island to the detriment of tobacco production (Ortiz, 1940), under the concern of Cuban landowners about the increase in productivity and the threat of slave revolts as had occurred in Haiti. In 1792, around 6000 slaves worked in sugarcane plantations. This is the reason for the impetus for the introduction of the steam engine, which arrived in Cuba in 1796, before the metropolis. The first experiment with the steam engine was not successful, but the sugar planters were not discouraged and the new technology began to operate a few years later[17] as a substitute for slave and animal motive power, with the consequent increase in productivity (Gouzevitch & Gouzevitch, 2022).

20.3 The Fruitful Cuban Nineteenth Century for Science

With the economic boom of the last third of the eighteenth century, the fruitful nineteenth century broke out for the nascent Cuban intellectuality, research in favor of economic development, science and technology continued on its good course. Félix Varela Morales who, with genuinely Cuban training at the Seminary of San Carlos

[16] Royal Arsenal of Havana shipyard.

[17] In 1797, the Count of Mopox installed the first steam engine to arrive in Cuba in one of his mills in Matanzas, which would be coupled to the cane mills 20 years later.

and San Ambrosio and the University of Havana, unique the last one authorized to grant degrees promoted the teaching of science. Varela is recognized by the publication in 1814 of the first physics text written by a Cuban. This is the book: *Volume IV of Institutions of Eclectic Philosophy for the use of youth*, which is a compendium of elements of mathematics and physics. Breaking with scholasticism, Varela embraces elective philosophy, without adhering to any sect or current, upholding that experience and reason are the sources of knowledge (Altshuler & Baracca, 2004).

In the first half of the nineteenth century, medicine, pharmacy and sugar production were of priority interest and since these were related to chemistry, Varela also taught it in the Seminar. In the case of physics, he turned out to be pioneer in the introduction of experimental work in teaching and makes attempts at research and verification through experiment. In particular, in his significant work *Lecciones de Filosofía*, Varela presents laboratory instruments, describes a mill for grinding sugar cane and the steam engine (Varela, 1941).

In relation to chemistry, it is recognized that in 1814, the first Cuban chemist, José Estévez y Cantal (1771–1841), carried out chemical analysis of the composition of medicines and medicinal waters. It is up to the Spanish chemist and industrial technologist José Luis Casaseca y Silván (1800–1869), Graduate of Sciences at the University of Paris, to systematically begun the experimental teaching of Chemistry in the Chair of Chemistry of the *Real Junta de Fomento*[18] for what he has been considered the Father of Cuban chemistry.

Once the *Protomedicato* was established, which brought order to the rules on the practice of medicine and health care, in 1817 the *Jardín Botánico de La Habana*[19] was created, it was the first scientific research institution. Its second director, the Spanish Ramón de la Sagra (1798–1871), taught botany courses, collected plants and animals that were later described, especially by French naturalists, who collaborated with him in his monumental work *Physical, Political and Natural History of the Island of Cuba*. It was considered the most significant carried out in the first half of the nineteenth century, a work consisting of 12 volumes and published between 1837 and 1857. Even before, in 1827, the Political Essay on the Island of Cuba, by Alejandro de Humboldt, had been published in Europe, but it did not have at that time much impacts on the Island, in particular, the Havana City Council forbid its circulation due to the unfavorable criteria contained regarding slave work (Clark, 2018).

A significant event occurred when Ramón de la Sagra founded and directed the magazine *Anales de Ciencia, Agricultura, Comercio y Arte de La Habana,*[20] which was the first Cuban serial publication that preferably dedicated to the dissemination of scientific topics. This journal began to be published in 1827 and lasted until 1831.

In 1823, Tomás Romay, introducer of vaccination against smallpox in Cuba and a prominent hygienist beyond Cuban borders, promoted the continuation of the

[18] Royal Development Board.

[19] Havana Botanical Garden.

[20] Annals of Science, Agriculture, Commerce and Art.

practical teaching of medicine that had been interrupted for some years. The event takes place at the Royal Military Hospital of Havana, where an Anatomy Museum was opened in 1823, an institution that like all museums, due to its scientific and cultural nature, contributes to human development. This year marked the first attempt to establish a medical society, but it did not come true. In 1840, the first Cuban medical journal, the *Repertorio Médico Habanero,*[21] was founded on the initiative of the prominent surgeon Nicolás José Gutiérrez Hernández (López Sánchez, 1998).

The leading institution of applied science in Cuba was the *Instituto de Investigaciones Químicas de La Habana,*[22] which was created in 1848 and was directed by the afore-mentioned Spanish chemist José Luis Casaseca Silván. When Casaseca retired, his student, the Cuban Álvaro Reynoso Valdés (1829–1888) took over the management of the Institute, who stood out in the branches of chemistry, physiology, agronomy and industrial technology and is considered The Father of Cuban Scientific Agriculture. Reynoso's masterpiece was the *Essay on the Cultivation of Sugar Cane,* published in Havana in 1862 (LeRoy, 1971).

There are many intellectuals who at this stage of the nineteenth century strived to introduce science, technology, philosophical, social and natural studies in Cuba: Esteban Pichardo, geographer (1779–1879), Felipe Poey (1799–1891), naturalist, and ichthyologist, Andrés Poey (1825–1919), meteorologist, Tranquilino Sandalio de Noda (1808–1856), learned man of science, Desiderio Herrera (1792–1856), mathematician, surveyor, and Álvaro Reynoso (1829–1888), all named by Clark (2018).

The authors consider back up the previous ones, with names such as José Antonio Saco (1797–1879), one of the founders of the Cuban nationality. He persistently opposed annexation and trafficking, and was a polemicist, sociologist, historian and jurist, explained Physics and Chemistry topics to the students of the Seminary of San Carlos and San Ambrosio. José de la Luz Caballero (1800–1862), educator and philosopher, student and follower of Varela's ideas regarding the teaching of science, proponent of feelings for the Cuban nationality, of progressive thought inspired by philosophers such as Descartes, Bacon, Newton, Locke, Condillac and the French Enlightenment. His conceptions led him to become involved in sharp philosophical controversies, in the Cuban context. It is also worth including Enrique José Varona y Pera (1849–1933), member of the Academy of Medical, Physical and Natural Sciences of Havana, who directed the newspaper *Patria* upon the death of José Martí. Although his political and pedagogical performance in the first third of the twentieth was very significant, in 1880 he gave, at the Academy of Sciences, a series of lectures on psychology, logics and ethics that greatly influenced the intellectuality of the time.

Due to his influence on Cuban thought of all times, particularly for his scientific-philosophical projection, it is strictly necessary to bring to these pages the Cuban

[21] Havana Medical Repertoire.

[22] Havana Chemical Research Institute.

National Hero José Martí Pérez (1853–1895). From a reading of José Martí's Complete Works, his accurate appreciations emerge about the role of practice and the scientific method, the importance of knowledge not being the privilege of elites. The *Cuban Apostle* promotes popular scientific education, supported by his own philosophy of education, with itinerant teachers who go teaching the best ways to cultivate the land so that men can be good and prosperous with the results of their work. He also remarked, in different publications, the work of prominent men of science with axiological nuances and with the interest of disseminating scientific and technological advances in what he called *Nuestra América.*

With the return of absolutism to Spain, the most notable event in Cuban cultural and political life in the first half of the nineteen century took place on the Island: the so-called Philosophical Controversy, when the interests, of the cultured and proud Creoles of their country, were truncated. The progress being made then was hindered, a debate arose between 1838 and 1840 in the press of that time, that went beyond the strictly philosophical conception. It was a controversy about what the Creole minority had been thinking and feeling in the face of the barriers established by the colonial reality, in the face of what cultured Cubans and lovers of their homeland had assimilated through travels and books; of those progressive ideas that European enlightened thought had been deploying. The main contender in the Philosophical Controversy on the side of lovers of sovereign thought turned out to be José de la Luz Caballero, who championed the educational ideas and independence thought of Varela (Piñera, 1960).

20.4 Interest in Studies in the History of Science: Inkling of Association, Institutionalization and Publications

On May 19, 1861, the *Real Academia de Ciencias Médicas Físicas y Naturales de La Habana*[23] (RACMFNH) was founded, which constitutes a milestone in the process of institutionalization of science and culture, as it integrates everything that best and shines of the Cuban intellectuality, with an inclination towards the research and promotion of science. It was integrated by academics, with their corresponding admission exercises, as occurred in the Economic Society of Country's Friends (SEAP) that preceded as a forum for promotion of development plans in all branches of Cuban colonial society, not strictly created for scientific and research purposes like the Academy.

In an article related to the eminent surgeon and scientist Nicolás José Gutiérrez (1800–1890), president and founder of the Academy of Havana (RACMFNH), Mesa Trujillo et al. (2019) quoted his words:

[23] Royal Academy of Medical, Physical and Natural Sciences of Havana.

> Siquiera no fuese más que por orgullo nacional, debiera hacerle entender a los forasteros y extranjeros principalmente, que no nos ocupamos sólo en hacer azúcar y cosechar tabaco, sino que cultivamos también las ciencias.[24]

This reasoning by Gutiérrez were premonitory, since on so many occasions and scenarios Cuba has been identified with rum, rumba and mulattas, ignoring the essences of its culture. In this way, the stage that López Sánchez (1998) has described as the *Academic period* of science in Cuba (1861–1902) begins, and extends until the constitution of the Republic that emerged threatened by a new colonizer, the USA.

The Royal Academy of Medical, Physical and Natural Sciences of Havana was created after a long gestation process that began in 1826. This process had its beginnings in an extraordinary assembly composed of Havana doctors, chemists and naturalists, expressly convened for that purpose and to being established turned out to be the first institution of this type in America (Editorial Rev Cubana Med Trop, 1996). The Academy had as an organ for the dissemination of its activities and as a scientific publication, the journal *Anales de la Academia*, which, as Valero and Jiménez (2022) pointed out, is a mandatory reference for any study of the history of science in Cuba, because although published science can lose its validity any material that allows us to rebuild the history of scientific institutions and thought, makes it valuable. The *Anales de la Academia* journal has had several stages since its first publication in 1864, not being published for a long period until the current one, starting again in 2011. It is released now in digital format. This journal is a reference for the study of the history of science in Cuba, although it is not characterized, at least in its initial stage, in the publication of articles on the history of science.

As has been said before, science had to appear first in Cuba so that someone would dedicate themselves to reason on this form of social and cultural awareness; in the rest of the world this occurred much earlier than in Cuba, so an attempt has been made to bring the reader closer to the development of science in the Cuban context so that it can serve as a frame of reference. A particularity in this essay is that it must be taken into account that the historians of science came before its organized forms and that the anecdotal about the course of science and the personalities involved in its development should not be confused with the history of science and technology. In this regard, Capote (2015) quotes very illuminating words from the prominent Cuban science historian José López Sánchez (1911–2004) expressed in 1971:

> La literatura histórico-científica cubana aparecida hasta ahora está constituida, en su mayoría, por artículos o ensayos breves o incluso libros que, salvo contadas excepciones, no son más que reseñas cronológicas o conjuntos de notas biográficas y selección de trabajos de los autores y que adolecen, por consiguiente, de rigor metodológico.[25]

---

[24] If only for national pride, it should make strangers and foreigners understand mainly that we are not only busy making sugar and harvesting tobacco, but we also cultivate sciences.

[25] The Cuban historical-scientific literature that has appeared so far is made up, for the most or short essays or even books that, with few exceptions, are nothing more than chronological reviews

However, the authors believe that we should not fail to take into account how guiding the published reviews or notes may be, due to data they provide and that allow us to go deeper into the essence of the development of the history of science and technology in Cuba. Another aspect of this same issue is the omissions or oversights that may arise regarding the inclination or not of historians for the history of science and technology, which Capote (2015) comments, since in a review of the thematic social historiography, carried out in 1985 by the Cuban historian, academic and university professor Oscar Zanetti (Havana 1946), does not identify the history of science as such.

To have an idea of the early interests in the history of science, one can consider works such as that of the professor at the University of Havana, Antonio Caro y Cerecio (1826–1891), who published in 1865 the history of the Institute of Chemical Research of Havana, this institution, as has been said, was created in 1848. Another work is The History of Natural History in Cuba by Manuel Presas from Matanzas (1845–1874), which was also published in 1865, both marking the advent of historians of science in Cuba (LeRoy, 1979; Pruna, 2015).

One of the Cuban science historians with a notable scientific production in the second half of the twentieth century and the beginning of the twenty-first, Dr. in Biological Sciences, Pedro Marino Pruna Goodgall (1943–2017), has considered as a continuity of the attempts that given in 1865, the mining works of Antonio Calvache, in 1944, the history of chemistry by Luis Felipe Le Roy in 1944, and the five volumes of the history of natural history, published by José Álvarez Conde between 1956 and 1961. Pruna also mentions the studies on the history of agronomy published by Rafael Martínez Viera and Francisco Díaz Barreiro, which confirms not only the interest in the history of technology but the establishment of genuine historians of science and technology in Cuba (Pruna, 2015).

According to Capote (2015), of more recent invoice and located in what has been called the social history of science and technology, they are the rigorous works: Finlay Man and scientific truth¸ by José López Sánchez; Science and scientists in colonial Cuba: the Royal Academy of Sciences of Havana 1861–1898, by Pedro Marino Pruna, A light that is here to stay, by José Altshuler Gutwert and Miguel González Royo and a masterpiece: the Albear Aqueduct of Havana, by Rolando García Blanco and collaborators.

At one point in this essay, it was highlighted that medicine was one of the first professions that were established and taught in Cuba with a scientific nature, so it is not surprising the preeminence of the historiography of medicine and the dedication of doctors to the history of medicine science in relation to other scientific disciplines, as well as the intention to associate and teach the history of medicine in the training of doctors, which contrasts with other careers in the Cuban context, where the history of particular sciences has been and is almost absent, which has motivated the 2005 National Literature Prize winner, Graciela Pogolotti (2015), to demand

---

or sets of biographical notes and selections of works by the authors and that They therefore lack methodological rigor.

that knowledge of the particular history of each science be included in the study plans of all majors.

Based on the above, it is justified that to continue with the course of the historiography of science in Cuba and its institutionalization, can serve as a guide the work of the prominent Doctor José López Sánchez (1911–2004) who turned out to be one of the most important historians Cubans of science, crossing the weak border of the particular history of medicine and entering into the history of science in general. He established methodological guidelines for the historiography study of science, given the limited possibility of pursuing these studies in Cuba, at least until the last third of the twentieth century.

The lack of training in the history of science means that López Sánchez, like the majority of Cuban science historians, has resorted to self-taught preparation, starting from his particular discipline. With this approach for his first studies he took the work of the hygienist doctor Tomás Romay Chacón (1764–1849), introducer of vaccination in Cuba and founder of the Cuban scientific movement, who played a leading role in the moments of the awakening of national consciousness. The extensive work that López Sánchez published about Romay had its first book in 1950: Tomás Romay. Biographical notes and speeches, and continued with other publications in which almost all of Tomás Romay's works and documents are found, which has earned López Sánchez international recognition and awards.

Following a laborious and erudite research carried out in the 1940s and 1950s in the Capitular Acts of the Havana City Council, López Sánchez published his results in the monographic collection *Cuadernos de Historia de la Salud Pública.*[26] A new research endeavor led him to a biography about Carlos Juan Finlay (1833–1915), a considerable contribution to the interpretation of the thought of the most eminent Cuban scientist of all time, nominated five times for the Nobel Prize, for the discovery of a new mode of disease transmission through a vector (metaxenic), which has sometimes been distorted or minimized, voluntarily or involuntarily, perhaps a victim of Eurocentrism.

With the social change brought about by the advent of the Revolution, in 1962 a deep reform of university studies was carried out in Cuba with Latin American roots in the Córdova's Reform of 1918, and in the student movement of the University Student Federation (FEU), founded by Julio Antonio Mella (1903–1929) in 1922. The Reform purpose of stripping university education of the burden of rote and reproductive systems, but even more so, promoting the involvement of the University in the scientific-technical development of society.

Many distinguished intellectuals, professors and university students were involved in the University Reform. Dr. López Sánchez took advantage of the opportunity to reopen the Chair of History of Medicine at the Faculty of Medicine of the University of Havana, for which he wrote the textbook that would serve as support: History of Medicine. Medicine. Vol. 1 (From Primitive Times to the Renaissance). This new moment in the history of medicine in the university environment has the

[26] Public Health History Notebooks.

antecedent that medical historians had established the Cuban Society of History of Medicine (SCHM) in 1956, the first association of this type that emerged in Cuba, chaired by the prominent medical historian Horacio Abascal Vera (1899–1964) and the academic César Rodríguez Expósito (1904–1972) as secretary. López Sánchez was the founder of this society, who among others was a member of the International Society for the History of Medicine. As editorial support for the SCHM, the Journal of the Cuban Society of History of Medicine appeared in 1958.

20.5 From the Royal Academy of Physical and Natural Medical Sciences to the Academy of Sciences of Cuba (ACC)

In the process of institutionalization of science, the Royal Academy of Medical, Physical and Natural Sciences of Havana, founded in 1861, has played a leading role in the development of science in Cuba, by bringing together the largest sum of knowledge that the Island in the nineteenth century, keep in mind that at the RACMFNH the Academic Carlos Juan Finlay submitted his memorable work on the transmission of yellow fever in 1882. The Havana Academy has gone through several stages, in terms of name, structure and operation, it lost its Royal status in 1898, when Spain lost its control over Cuba and the Republic was established, remaining as Academy of Physical and Natural Medical Sciences of Havana (ACMFNH).

Starting in 1898 and until the end of the first half of the twentieth century, the ACMFNH suffered from a period of decline due to little action in the promotion of a science of a national nature, resting the efforts of the development of science fundamentally on personalities and institutions on an individual basis. Such is the situation that, during a visit to Cuba in 1950, a commission from the Inter-American Development Bank ruled in a report that it did not find any adequate applied research laboratory and that the existing ones were dedicated to quality control or adjustment of the most basic characteristics external and superficial aspects of the products (Clark, 2018).

In 1962, the National Commission for the Academy of Sciences of Cuba was created, the Academy of Medical, Physical and Natural Sciences, extinguished. This commission reached a national level, with managing and planning functions, promoting the founding of a significant number of institutes and research centers (González & Arduengo, 2021). The National Commission for the Academy of Sciences of Cuba was identified as the Academy of Sciences of Cuba and in 1976 acquired the status of National Institute. In 1980, when the functions of the governing body of science and technology at the national level were transferred to the Academy, it acquired ministerial rank. In 1994, with the process of improvement of the Central Administration of the Cuban State, the administrative structures and governing functions of the Academy were merged together with those of the

National Commission of the Environment and Natural Resources and the Executive Secretariat of Nuclear Affairs, in the current Ministry of Science, Technology and Environment (CITMA). After this entire institutionalization process in 1996, the Academy was reestablished in its current capacity as the Academy of Sciences of Cuba (ACC), made up of scientists of relevant merits, representatives in a personal capacity, on an honorary basis and as Academicians of the national scientific community. Historians of science have been and are members of the ACC.

The process of institutionalization of science and scientific research that occurs after the arrival of the Revolutionary Government in 1959 brings with it the founding of institutions and research centers brings with it the founding of institutions and research centers. Only between 1962 and 1973, 53 Research Development (I + D) entities were organized in Cuba. To meet the purpose of this work and remark the process of institutionalization of science and its history, an important fact is when the National Commission for the Academy of Sciences of Cuba created in 1962, the Carlos J. Finlay Historical Museum of Medical Sciences, occasion of 130th Anniversary of the birth of the outstanding Cuban scholar. The museum was located at the facilities and library of the former Academy of Medical, Physical and Natural Sciences Havana, at 460 Cuba Street.

In 1977, the Center for Studies of History and Organization of Sciences (CEHOC) was incorporated into the museum, this institution has extraordinary relevance in terms of the study of the history of science and technology, by systematically performing research on the development of science in Cuba. The researchers linked to this study center have enriched the bibliography on the historiography of science, mainly through publications that under the Editorial Academia seal appeared after *Anales de la Academia* journal ceased to be published from 1957 to 2011 (Valero & Jiménez, 2022). The CEHOC, in addition to bringing together prestigious historians of science, through talks, courses, and lectures by national and international guests, became a promoting center for the history of science and technology and for the preparation and improvement of science historians.

The Museum of Medical Sciences became the Carlos J. Finlay National Museum of the History of Sciences in 1999, thus attaching more significance to this institution subordinated to the Ministry of Science, Technology and Environment (CITMA) and holds important documentary and bibliographical funds as well as collections of pieces from the history of science in Cuba.

20.6 The Cuban Society for the History of Science and Technology (SCHCT)

The fundamental characteristic of the development of science and technology in the period that began in 1959 has been the necessary formation of human potential and the creation of research entities. Of the few professionals existing in Cuba, before 1959, many emigrated and, as stated before, there was no system of science and technology institutions that could support the development of the country. There were only three universities that participated in scientific research with personal initiatives and some institutions such as: the Santiago de las Vegas Agronomic Experimental Station, established in 1904, which initially operated with foreign personnel and the Institute of Tropical Medicine founded in 1937 by Pedro Kourí (1900–1964), an institution that still functions with relevant scientific results that contributed to confronting Covid-19 in Cuba.

Starting from a few scientists and a dozen research centers, Cuba has managed to form a network of more than two hundred research centers, which extend throughout the country, with tens of thousands of researchers, directly linked to the research to which they serve. Professors from more than 40 universities are too joined. From this scientific will and training of professionals, with the cooperation of the Union of Soviet Socialist Republics (USSR) and other countries of the extinct socialist camp, important scientific and technological results have emerged, from which the guild of historians of science and technology arises and is nourished.

The Cuban Society for the History of Science and Technology (SCHCT) was officially established in 1985. It was not the first society of this type to exist in Cuba, there were precedents with the history of medicine before. The process followed for the constitution of the SCHCT, as expressed by Pruna (2015), begins in September 1981, during the celebration of the fifteenth and first International Congress of the History of Science, which took place in Bucharest, Romania, when, at the initiative of the prominent Colombian mathematician and historian of science, Luis Carlos Arboleda, brought together seven Latin American researchers who were participating in that congress, from Argentina, Brazil, Colombia, Cuba, Chile and Mexico. Among these were Dr. José López Sánchez and Dr. Pedro Marino Pruna for Cuba.

At the meeting promoted by Professor Arboleda, the agreement was adopted that advocated attention to the regrettable degree of isolation in which the majority of Latin American historians of science and technology found themselves, without the possibility of exchanging among themselves, nor with colleagues from the same country. The absence of an adequate institutional framework for research in this field of knowledge was also identified, even though at the international level the history of science and technology was advancing promisingly as it had not been before and in some teaching entities there were chairs or seminars of History of science.

Several proposals emerged from the meeting of science historians from the five Latin American countries: publishing a newsletter, forming a directory of Latin American science historians, and holding a First Meeting of Latin American Science Historians. The three goals were met in the next 3 years. The bulletin was transformed into a magazine, the directory of historians was made and in 1982 a meeting was held in the city of Puebla, in Mexico and there the Latin American Society for the History of Science and Technology (SLHCT) was founded, chaired by Juan José Saldaña González, Full Professor of the Faculty of Philosophy and Letters of the National Autonomous University of Mexico (UNAM), passionate about the history of science, also founder of the magazine Quipu in 1984, organ of the SLHCT. By passing away in 2022, Professor Saldaña has left a void in the leadership of the history of science in Latin America and Mexico that will be impossible to fill.

At the proposal of José López Sánchez that Havana hosted the First Latin American Congress of the History of Science and Technology, the Organizing Committee of the Cuban Society of History of Science and Technology was created in Cuba in 1984, including government support and institutions linked to science and their history. The commission was chaired by Professor Salvador Vilaseca Forné (1909–2003), who would become the first president of the SCHCT when this association was founded within the framework of the First Latin American Congress of the History of Science and Technology.

Professor Vilaseca, who held important academic responsibilities such as rector of the Institute of International and Diplomatic Relations, serving as Cuban ambassador to Italy, Malta and Greece, dedicated his entire life to clarifying various aspects of the history of science in Cuba, and contributed substantially to the enrichment of the national bibliographic fund related to the history of science and technology by gathering important annotations and a library of the history of science, which he donated to the Center for Studies of History and Organization of Science, of the Cuban Academy of Sciences (CEHOC).

The First National Congress of the History of Science and Technology, convened by the SCHCT, was held in 1994 and was held in the halls of the historic building where, since 1867, the Royal Academy of Medical, Physical and Natural Sciences of Havana was located. It is home to the National Science Museum and other institutions related to the history of science such as the CEHOC. The responsibility for organizing the congress fell to the recently elected new board of directors of the SCHCT, chaired by Dr. José Altshuler Gutwert, a distinguished professor at the University of Havana who participated in the University Reform of the 60s and who with great dedication in the history of science, he has published important books and articles that have to do with the history of science and technology in Cuba and the world. The celebration of the SCHCT congresses has had the participation of senior State officials, including the Minister of Science and Technology and representatives of UNESCO in Cuba.

Under the expert leadership of Professor Altshuler, which extended until 2017 when he retired, the SCHCT has held five conferences (Altshuler, 2015), has maintained international relations with several institutions, in particular the SMHCT, the SLHCT, the Smithsonian Institution, the Center of Historical Studies of the Higher Council of Scientific Research of Spain, the Society for the History of Science of the United States. These international relations have facilitated the exchange of specialists, although on occasions the United States Departments of State and Treasury have denied permission to North American historians to attend exchanges in Cuba and offer conferences and courses.

The SCHCT has spread to the country through the creation of branches in provinces and sections in institutions, research centers and universities. Between congresses, the Society has sponsored multiple events and has maintained its internal functioning by periodically holding meetings of the board of directors and general assemblies, which have had on their agenda the renewal through election of members of the board of directors. Since 1994, the SHCT has published a total of 44 newsletters about its activities, with scientific anniversaries from Cuba and the world, performance of its members in participating in international events, courses and internships, as well as references to published books and awards granted to honorary members and benefactors, through these bulletins the history of the SCHCT can be followed in particular since 1994.

The SCHCT, which emerged within the framework of the First Congress of the SLHCT, has maintained close links to this society, such that one of the first issues of the magazine Quipu were devoted to the history of science in Cuba and Cuban historians have integrated as members of the board of directors of the SLHCT and publications by Cuban authors have been present in the Quipu journal. The Mexican Society for the History of Science and Technology has also been close to the SCHCT, which as an older sister and with the encouragement of Professor Saldaña has facilitated the systematic exchange of researchers in Congresses and Seminars. Prestigious researchers from the history of Mexican science and technology have been present at the SCHCT congresses, sharing with Cuban colleagues.

20.7 Conclusions

The moments that authors considered most significant have been described to meet one of the purpose of the essay to showing the process of constitution of a genuinely Cuban science until the first half of the twentieth century. This exhibition was necessary to make known how and when historians of science and technology showed up in the Cuban context, first individually and self-taught, until they came to establish institutions, publications and associations in which they have found support for their research and possibilities for national and international exchange. Attention was paid to the constitutive process the current Academy of Sciences of Cuba that run in parallel with the Ministry of Science, Technology and Environment (CITMA). Outstanding interest was devoted in showing the emergence and main activities of the Cuban Society for the History of Science and Technology, which following a vocation of full inclusion brings together researchers and those interested in the history of science and technology in the country and provides spaces to foreign guests. It should be taken into account that this essay, necessary synthetic, will leave some gaps for interested readers can fill or complement by consulting the references.

Acknowledgement This work would not have been possible without the help received from the Members of Merit of the Cuban Society of History of Science and Technology, Doctors José Altshuler Gutwer and Roberto Díaz Martín. Thanks to Professors María de la Paz Ramos-Lara and Luis Carlos Arboleda for taking into account us in order to participate in this Proyect, and Mr. Horacio Mederos Acosta for the careful reading of the manuscript. Thanks to my daughter Lisandra Alamino her support and assistance in the final review of this work.

References

Alamino, Diego de Jesús. (2001). Félix Varela: religión, ciencia e independencia. *XXI international congress of history of science*. Mexico City 8–14 July 2001.

Altshuler, J. (2015). Palabras inaugurales Acto por el xxx aniversario de la fundación de la sociedad cubana de historia de la ciencia y la tecnología. *Revista Anales de la Academia de Ciencias de Cuba., 5*(2), 1–2.

Altshuler, J., & Baracca, A. (2004). La enseñanza de la física en Cuba, desde la Colonia hasta 1959. *Llull (Zaragoza), 27*, 557–608.

Batard, L. F. (2010). Historia del desarrollo de las ciencias exactas y naturales en Cuba: La eclosión científica en la etapa colonial. *ISLAS, 52*(163), 8–27.

Batista Salvador, A., Mena Campo, A., & García González, I. (2018). La Sociedad Económica de Amigos del País de La Habana: contribución al progreso sociocultural. *VARONA, Revista Científico-Metodológica* (pp. 1–8). Edición especial.

Capote, E. (2015). La historia de la ciencia y la organización de la ciencia. *Revista Anales de la Academia de Ciencias de Cuba., 5*(2), 1–5.

Clark, I. (2018). *La Ciencia en Cuba*. Editorial Científico Técnica.

Editorial Rev Cubana Med Trop. (1996). La Academia de Ciencias en Cuba: 135 años de reflexión y promoción en la ciencia y la sociedad. *Revista Cubana Medicina Tropical, 48*(3), 147–148. Ciudad de la Habana.

Escrigas, Juan. El real Arsenal de La Habana. (2020). *Revista general de marina,* Vol. 279 (Ejemplar dedicado a: 250 Aniversario del Cuerpo de Ingenieros de la Armada) 315–333.

Espinosa, J. A. (2007). Cronología Enero 11 de 1723.Apariciónn del primer incunable cubano. *ACIMED* vol.15 n.4 Ciudad de la Habana.

Funes, R. (2002). Azúcar, deforestación y medio ambiente. Los bosques de Cuba entre 1792 y 1926, Tesis doctoral, Universidad Jaume I España.

García Galló Gaspar Jorge. (1978). *Bosquejo Histórico de la Educación en Cuba.* Editorial de Libros para la Educación. La Habana.

González, L., & Arduengo, D. A. (2021). Calle Cuba 460, monumento de la ciencia cubana. *Anales de la Academia de Ciencias de Cuba, 11*(2).

Gouzévitch, D., & Gouzevitch, I. (2022). La machine à vapeur de Betancourt pour l'industrie sucrière cubaine: un épisode inédit de l'histoire du machinisme (fin XVIIIe siècle). *Cuadernos Dieciochistas, 23*, 59–86.

LeRoy, L. F. (1971). Casaseca, maestro y precursor de Reynoso. *Revista de la Biblioteca Nacional José Martí 3ra. época-vol XIII Número 1 Enero-Abril* (pp. pp. 5–58). La Habana, Cuba.

LeRoy, L. F. (1979). *Profesores de Física de la Universidad de La Habana desde su secularización en 1842 hasta Manuel F. Gran Gilledo*. Editorial Academia. La Habana.

López, José A. (1998) La Primera Revista Médica Cubana. *Rev Cubana Salud Pública, 24*(1), 53–63. Ciudad de La Habana ene.

López Sánchez, J. (1998). Génesis histórica de la cultura científica cubana. Conferencia magistral impartida en el V Congreso Nacional de Historia de la Ciencia y la Tecnología. https://web.archive.org/web/20160116172632/http://revistas.bnjm.cu/index.php/revista-bncjm/article/download/909/861

Mesa Trujillo, D., Espinosa Ferro, Y., & García Mesa, I. (2019). Dr. Nicolás José Gutiérrez Hernández. Pionero de la medicina y la ciencia en Cuba. *Revista Médica Electrónica, 41*(4), 1076–1081.

Ortiz, F. (1940). *Contrapunteo cubano del tabaco y el azúcar*. Jesús Montero Editor.

Piñera, H. (1960). *La Polémica Filosófica*. Panorama de la Filosofía cubana. Proyecto Filosofía en español (pp. 51–76). Washington DC.

Pogollotti, G. (2015). Ciencia y Cultura. Juventud Rebelde, 26 de abril.

Pruna, P. M. (2015). Palabras en el Acto por el xxx aniversario de la fundación de la sociedad cubana de historia de la ciencia y la tecnología. *Revista Anales de la Academia de Ciencias de Cuba., 5*(2), 1–4.

Rosell, W., Gómez, M., & Paneque, E. R. (2006). Real Tribunal del Protomedicato de La Habana. *Educ Med Super, 20*(2), 1–5.

Torres, C., et al. (1997). *Félix Varela, Obras t 1*. Editorial Cultura Popular.

Valero, M., & Jiménez, A. (2022). Anales de la Academia: mirada en la tradición. *Anales de la Academia de Ciencias de Cuba, 12*, 3.

Varela, F. (1941). *Lecciones de Filosofía*. Quinta edición corregida y aumentada por el autor. Editorial de la Universidad de La Habana. Tomada de la edición Imprenta de Don Juan de la Granja.

Diego de Jesús Alamino Ortega: Graduate in Physics at the University of Havana (UH). Doctor in Physical Sciences (UH). Titular Professor and Consulting Professor at the University of Matanzas. Attend national events and international scientific meetings. He has exhibited works and taught courses in Russia, Argentina, Mexico, Brazil, England and Venezuela. He has published articles and chapters in books such as The History of Physics in Cuba, Springer Dordrech Editor. Master's and Doctorate thesis advisor.

Yenile Aguilar Rodríguez: Graduate in Physics and Astronomy at the Higher Pedagogical Institute of Matanzas. Master in Physics Didactics and Doctor in Pedagogical Sciences from the University of Pedagogical Sciences of Havana. Titular Professor at the University of Matanzas. Director of the "Enrique Rodríguez-Loeches Fernández" University Center of the University of Matanzas. Award by the International Astronomical Union for naming an exo-planet and a star after the Cuban scientists Félix Varela and Carlos Juan Finlay.

Index

M. de la P. Ramos-Lara, L. C. Arboleda (eds.), *History of Science in Latin America: The Construction of an Intellectual Field (20th century)*, https://doi.org/10.1007/978-3-031-75536-1

www.ingramcontent.com/pod-product-compliance
Lightning Source LLC
Chambersburg PA
CBHW051647110225
21773CB00002B/11
9783031755354